LONDON MATHEMATICAL SOCIETY LECTURE NOTE SERIES

Managing Editor: Professor Endre Süli, Mathematical Institute, University of Oxford,
Woodstock Road, Oxford OX2 6GG, United Kingdom

The titles below are available from booksellers, or from Cambridge University Press at
www.cambridge.org/mathematics

London Mathematical Society Lecture Note Series: 464

Lectures on Orthogonal Polynomials and Special Functions

Sixth Summer School, Maryland, 2016

Edited by

HOWARD S. COHL
National Institute of Standards and Technology

MOURAD E. H. ISMAIL
University of Central Florida

CAMBRIDGE
UNIVERSITY PRESS

CAMBRIDGE
UNIVERSITY PRESS

University Printing House, Cambridge CB2 8BS, United Kingdom

One Liberty Plaza, 20th Floor, New York, NY 10006, USA

477 Williamstown Road, Port Melbourne, VIC 3207, Australia

314-321, 3rd Floor, Plot 3, Splendor Forum, Jasola District Centre, New Delhi - 110025, India

79 Anson Road, #06-04/06, Singapore 079906

Cambridge University Press is part of the University of Cambridge.

It furthers the University's mission by disseminating knowledge in the pursuit of education, learning and research at the highest international levels of excellence.

www.cambridge.org
Information on this title: www.cambridge.org/9781108821599
DOI: 10.1017/9781108908993

© Cambridge University Press 2021

First published 2021

A catalogue record for this publication is available from the British Library

ISBN 978-1-108-82159-9 Paperback

Contents

Contributors

Antonio J. Durán   *Departamento de Análisis Mathemático, Universidad de Sevilla, Sevilla, Spain.*

Mourad E.H. Ismail   *Department of Mathematics, University of Central Florida, Orlando, Florida 32816, USA.*

Erik Koelink   *Department of Mathematics, Radboud Universiteit Nijmegen, Nijmegen, The Netherlands.*

Hjalmar Rosengren   *Chalmers University of Technology and University of Gothenburg, Gothenburg, Sweden.*

Jiang Zeng   *Institute Camille Jordan, Université Claude Bernard Lyon 1, Villeurbanne, Lyon, France.*

Preface

On July 11–15, 2016, we organized a summer school, *Orthogonal Polynomials and Special Functions*, Summer School 6 (OPSF-S6) which was hosted at the University of Maryland, College Park, Maryland. This summer school was co-organized with Kasso Okoudjou, Professor and Associate Chair, Department of Mathematics, and Norbert Wiener Center for Harmonic Analysis and Applications. OPSF-S6 was a National Science Foundation (NSF) supported summer school on orthogonal polynomials and special functions which received partial support from the Institute for Mathematics and its Applications (IMA), Minneapolis, Minnesota. Twenty-two undergraduates, graduate students, and young researchers attended the summer school from the USA, China, Europe, Morocco and Tunisia, hoping to learn a new state of the art in these subject areas.

Since 1970, the subjects of *special functions* and special families of *orthogonal polynomials*, have gone through major developments. The Wilson and Askey–Wilson polynomials paved the way for a better understanding of the theory of hypergeometric and basic hypergeometric series and shed new light on the pioneering work of Rogers and Ramanujan. This was combined with advances in the applications of q-series in number theory through the theory of partitions and allied subjects. When quantum groups arrived, the functions which appeared in their representation theory turned out to be the same q-functions which were recently developed at that time. This motivated researchers to revisit the old Bochner problem of studying polynomial solutions to second order differential, difference, or q-difference equations, which are of Sturm–Liouville type and have polynomial coefficients. This led to a generalization where having a solution of degree n for all $n = 0, 1, 2, \ldots$ is now weakened to having a solution of degree n for all $n = N, N + 1, \ldots$, for some finite N. The polynomial solutions are required to be orthogonal. This gave birth to the subject of exceptional orthogonal polynomials. What is amazing is that these orthogonal polynomials continued to be complete in the L_2 space weighted by their orthogonality measure. At the same time, harmonic analysis of the newly developed functions attracted the attention of many mathematicians with different backgrounds. Alongside the analytic theory of the newly discovered functions and polynomials, a powerful combinatorial theory of orthogonal polynomials evolved. This also included a combinatorial theory of continued fractions. In the late 1990s, the elliptic gamma function was introduced and the theory of elliptic special functions followed. This area overlaps with mathematical physics because there are physical models whose solutions use elliptic special functions. The theory of multivariate orthogonal polynomials and special functions, as well as the harmonic analysis of root systems was also developed. The Selberg integral from 1944 became a prime example of integrals over root systems.

In this summer school, we felt that we needed to cover topics of current interest, as well as newly evolving topics which are expected to evolve and blossom into very active areas of research. This is hard to do in a classical subject like special functions. We also tried to cover areas within the broad scope of this classical and well established subject which have not been covered in recent summer schools. The fruits of our labor are in the lecture notes written by the lecturers of the summer school contained in this volume. We believe the these notes are detailed enough to learn or teach graduate level classes from, or cover in advanced seminars. They are full of insights into the subjects covered, and are written by experts whose research is at the leading edge of the subject.

The lecturers were Antonio Durán (Departamento de Análisis Matemático, Universidad de Sevilla, Sevilla, Spain), Mourad Ismail (Department of Mathematics, University of Central Florida, Orlando, Florida, USA), Erik Koelink (Department of Mathematics, Radboud Universiteit Nijmegen, Nijmegen, The Netherlands), Hjalmar Rosengren (Chalmers University of Technology and University of Gothenburg, Gothenburg, Sweden), and Jiang Zeng (Institut Camille Jordan, Université Claude Bernard Lyon 1, Villeurbanne, Lyon, France).

The first set of these lecture notes is by Antonio Durán who has made significant contributions in the subject of matrix orthogonal polynomials and exceptional orthogonal polynomials. Both subjects are relatively new, but their roots go back to the 1940s and have evolved very rapidly since the 1990s. These notes give a very nice introduction to the subject of classical orthogonal polynomials, and covers some very recent results on two expansions of the Askey tableau: Krall and exceptional polynomials; both in their continuous and discrete versions. Critical to Durán's discussion is the theory of $\mathscr{D}$-operators, how they can be used to construct Krall polynomials, and the use of duality to construct exceptional discrete polynomials. Durán clearly explains the relations between all of these families via limiting processes and using Christoffel transformations of the measure (Darboux transformations). Examples are treated in detail including the orthogonal polynomial families of Krall–Laguerre, Krall–Meixner, Krall–Charlier, exceptional Hermite, exceptional Laguerre, exceptional Meixner, and exceptional Charlier polynomials. These notes discuss the above developments and describes as well the current state of the art.

Mourad Ismail's lectures represent a nice tutorial which gives a brief review of many useful and recent results in the theory of q-series. He uses the Askey–Wilson operators, polynomial expansions, and analytic continuation to derive many of the fundamental identities and transformations of q-series. The Askey–Wilson calculus plays a major role in his approach, and even though this approach was also used in his 2005 book, our understanding of these analytic and more conceptional techniques has clearly advanced in the last few years. Some of these techniques were already used for the development of the theory of hypergeometric functions on root systems. This approach is more conceptual than the classical approach presented in most books on the subject. We expect this tutorial to be an advantageous text for learners of the subject.

The lectures by Erik Koelink cover the spectral theory of self-adjoint operators related to, or arising from, problems involving special functions and orthogonal polynomials. Of paramount importance is the spectral decomposition (the spectral theorem) of compact bounded and unbounded self-adjoint operators on infinite-dimensional Hilbert spaces. In particular, the emphasis in these notes is on the spectral theory of Jacobi operators, and especially on Jacobi operators corresponding to matrix-valued orthogonal polynomials. The theory of such Jacobi operators is built up, and several explicit examples are presented. In addition, he surveys the rigorous mathematical treatment of the J-matrix method and the more general concept of tridiagonalization. The J-matrix method was originally introduced by physicists and has been applied very successfully to many quantum mechanical systems.

Hjalmar Rosengren's lectures provides an excellent introduction to elliptic hypergeometric functions, a relatively recent class of special functions. These lecture notes fill a hole in the demand for a clear introduction to the field. He dedicates the first part of his lecture to a review of important topics connected to the subject, such as additive and multiplicative elliptic functions, theta functions, factorization, Weierstrass's three-term identity, even elliptic functions, partial fraction decomposition, modularity and elliptic curves. Throughout his lectures, through examples and a host of carefully constructed exercises, he connects his discussion with important and illustrative facts in complex function theory and to generalized and basic hypergeometric functions. He closes the lectures with a highly relevant historical application from statistical mechanics which is how elliptic hypergeometric series first appeared in the literature. This application concerns fused Boltzmann weights for Baxter's solvable elliptic solid-on-solid lattice model.

Note that in the abstract of his lectures, Rosengren says "By contrast, there is no hint about elliptic hypergeometric series in the literature until 1988." This statement is definitely true, but Mourad Ismail would like to shed some light on this issue. In 1982, he visited the Chudnovsky brothers and Gregory Chudnovsky asked him to explore orthogonal polynomials whose three-term recurrence coefficients are elliptic functions. Gregory also said that an elliptic generalization of the continuous q-ultraspherical polynomials should exist. Mourad was unable to solve this problem, but Gregory Chudnovsky anticipated the *elliptic revolution* in hypergeometric functions.

It is our opinion that a mathematical theory is best understood when formulated at the correct level of generality. This is indeed the case with the one variable theory of special functions. It is conceptually simpler at the most general level, the elliptic level.

The lecture notes by Jiang Zeng report on some recent topics developed in the cross-cutting field of combinatorics, special functions and orthogonal polynomials. This subject is relatively recent, although it is rooted in earlier works from the 1940s by Touchard, Kaplansky, Riordan, and Erdös. Zeng presents side-by-side, the two dual combinatorial approaches to orthogonal polynomials, namely that of Flajolet and Viennot, and that of Foata.

The lectures begin by reviewing essential results in the theory of general orthogonal polynomials, namely Favard's theorem, three-term recurrence relations and their connection to continued fraction expansions through generating functions for the moments. Since, orthogonal polynomials can be considered as enumerators of finite structures with respect to some statistics, Zeng illustrates how to use combinatorial models to explore orthogonal polynomial identities. As examples, Zeng develops combinatorial models for orthogonal Sheffer polynomials (Hermite, Laguerre, Charlier, Meixner, and Meixner–Pollaczek) through exponential generating functions, and for Al-Salam–Chihara polynomials (and their rescaled special sub-families including continuous q-Hermite, q-Charlier, and q-Laguerre polynomials) to combinatorially compute their moments and linearization coefficients. Zeng gives some details about Ismail, Stanton and Viennot's groundbreaking combinatorial evaluation of the Askey–Wilson integral and inquires whether there may exist combinatorial proofs of the Nassrallah–Rahman integral and Askey–Wilson moments. There are even recent applications of this approach to the asymmetric simple exclusion process (ASEP), where the moments of the Askey–Wilson polynomials play a central role.

Organizing such a summer school was a pleasure, although we encountered many challenges along the way. Kasso Okoudjou led the local organizational efforts for OPSF-S6. It was a pleasure working with Kasso throughout the entire process, both before and after the summer school. He skillfully maneuvered all aspects of the summer school, including the successful procurement of NSF and IMA funding, without which OPSF-S6 would not have been nearly as successful as it was. Kasso smoothly interacted with graduate students, early career researchers, and lecturers, for travel planning, booking, and ultimately for reimbursement. It was both an honor and a pleasure working with Kasso whose influence and organization with the Norbert Wiener Center for Harmonic Analysis and Applications is greatly appreciated. The lecture rooms we used, that Kasso obtained, were both useful and comfortable; snacks and coffee were always available; and the local staff was extremely pleasant and efficient. We would also like to acknowledge the support of Dan Lozier and Stephen Casey for their help and advice in planning the summer school.

Ladera Ranch, CA   Howard S. Cohl
Orlando, FL   Mourad E. H. Ismail
                                                       November 2019

1

Exceptional Orthogonal Polynomials via Krall Discrete Polynomials

Antonio J. Durán

Abstract: We consider two important extensions of the classical and classical discrete orthogonal polynomials: namely, Krall and exceptional polynomials. We also explore the relationship between both extensions and how they can be used to expand the Askey tableau.

Introduction

In these lectures, we will consider two important extensions of the classical and classical discrete orthogonal polynomials: namely, Krall and exceptional polynomials. On the one hand, Krall or bispectral polynomials are orthogonal polynomials that are also common eigenfunctions of higher-order differential or difference operators. On the other hand, exceptional polynomials have recently appeared in connection with quantum mechanical models associated to certain rational perturbations of classical potentials. We also explore the relationship between both extensions and how they can be used to expand the Askey tableau.

Section 1.1. *Background on classical and classical discrete polynomials.* The explicit solution of certain mathematical models of physical interest often depends on the use of special functions. In many cases, these special functions turn out to be certain families of orthogonal polynomials which, in addition, are also eigenfunctions of second-order operators of some specific kind. We can consider these families as the workhorse of all classical mathematical physics, ranging from potential theory, electromagnetism, etc. through the successes of quantum mechanics in the 1920s in the hands of Schrödinger. These families are called "classical orthogonal polynomials". E.J. Routh proved in 1884 (see [77]) that the only families of orthogonal polynomials (with respect to a positive weight) that can be simultaneous eigenfunctions of a second-order differential operator are those going with the names of Hermite, Laguerre and Jacobi. This result is also a consequence of the Bochner classification theorem of 1929 [7]. The similar question for second-order difference operators gave rise to the classical discrete families of orthogonal polynomials (Charlier, Meixner, Krawtchouk, Hahn), classified by Lancaster in 1941 [61]. Finally, second-order q-difference operators gave rise to the q-classical families of orthogonal polynomials (Askey–Wilson, q-Racah, etc.), although the q-families will not be considered in these lectures.

Sections 1.2, 1.3 and 1.4. *The Askey tableau. Constructing Krall and Krall discrete orthogonal polynomials using $\mathscr{D}$-operators.*

Since all these families of orthogonal polynomials can be represented by hypergeometric functions, they are also known as hypergeometric orthogonal polynomials, and they are organized as a hierarchy in the so-called Askey tableau.

As an extension of the classical families, more than 75 years ago the first families of orthogonal polynomials which are also eigenfunctions of higher-order differential operators were discovered by H.L. Krall, who classified orthogonal polynomials which are also eigenfunctions of differential operators of order 4. Because of that, orthogonal polynomials which are eigenfunctions of differential or difference operators of higher order are usually called Krall polynomials and Krall discrete polynomials, respectively. Following the terminology of Duistermaat and Grünbaum (see [11]), they are also called bispectral, and this is because of the following reason. In the continuous parameter, they are eigenfunctions of the above-mentioned operators, and in the discrete parameter, they are eigenfunctions of a second-order difference operator: the three-term recurrence relation (which is equivalent to orthogonality with respect to a measure supported in the real line).

Since the 1980s, a lot of effort has been devoted to studying Krall polynomials, with contributions by L.L. Littlejohn, A.M. Krall, J. Koekoek and R. Koekoek. A. Grünbaum and L. Haine (and collaborators), K. H. Kwon (and collaborators), A. Zhedanov, P. Iliev, and many others. The orthogonality of all these families is with respect to particular cases of Laguerre and Jacobi weights together with one or several Dirac deltas (and its derivatives) at the end points of their interval of orthogonality.

Surprisingly enough, until very recently no example was known in the case of difference operators, despite the problem being explicitly posed by Richard Askey in 1990, more than twenty five years ago. The results known for the continuous case do not provide enough help: indeed, adding Dirac deltas to the classical discrete weights does not seem to work. The first examples of Krall discrete polynomials appeared three years ago in a paper of mine, where I proposed some conjectures on how to construct Krall discrete polynomials by multiplying the classical discrete weights by the "annihilator polynomial" of certain finite sets of numbers. These conjectures have been already proved by using a new concept: $\mathscr{D}$-operators. $\mathscr{D}$-operators provide a unified approach to construct Krall, Krall discrete or q-Krall orthogonal polynomials.

This approach has also led to the discovery of new and deep symmetries for determinants whose entries are classical and classical discrete orthogonal polynomials, and has led to an unexpected connection of these symmetries with Selberg type formulas and constant term identities.

Sections 1.5 and 1.6. *Exceptional orthogonal polynomials. The dual connection with Krall polynomials at the discrete level. Expanding the Askey tableau.*

Exceptional orthogonal polynomials are complete orthogonal polynomial systems with respect to a positive measure which in addition are eigenfunctions of a second-order differential operator. They extend the classical families of Hermite, Laguerre and Jacobi.

The most apparent difference between classical orthogonal polynomials and exceptional orthogonal polynomials is that the exceptional families have a finite number of gaps in their degrees. That is, not all degrees are present in the sequence of polynomials (as it happens with the classical families). Besides that, they form a complete orthonormal set in the underlying Hilbert space defined by the orthogonalizing positive measure.

This means in particular that they are not covered by the hypotheses of Bochner's classification theorem. Each family of exceptional polynomials is associated to a quantum-mechanical potential whose spectrum and eigenfunctions can be calculated using the exceptional family. These potentials turn out to be, in each case, a rational perturbation of the classical potentials associated to the classical polynomials. Exceptional orthogonal polynomials have been applied to shape-invariant potentials, supersymmetric transformations, discrete quantum mechanics, mass-dependent potentials, and quasi-exact solvability. Exceptional polynomials appeared some nine years ago, but there has been a remarkable activity around them mainly by theoretical physicists (with contributions by D. Gómez-Ullate,

N. Kamran and R. Milson, Y. Grandati, C. Quesne, S. Odake and R. Sasaki, and many others).

In the same way, exceptional discrete orthogonal polynomials are an extension of discrete classical families such as Charlier, Meixner, Krawtchouk and Hahn. They are complete orthogonal polynomial systems with respect to a positive measure (but having gaps in their degrees) which in addition are eigenfunctions of a second-order difference operator.

Taking into account these definitions, it is scarcely surprising that no connection has been found between Krall and exceptional polynomials. However, if one considers Krall discrete polynomials, something very exciting happens: duality (roughly speaking, swapping the variable with the index) interchanges Krall discrete and exceptional discrete polynomials. This unexpected connection of Krall discrete and exceptional polynomials allows a natural and important extension of the Askey tableau. Also, this worthy fact can be used to solve some of the most interesting questions concerning exceptional polynomials; for instance, to find necessary and sufficient conditions such that the associated second-order differential operators do not have any singularity in their domain. This important issue is very much related to the existence of an orthogonalizing measure for the corresponding family of exceptional polynomials.

1.1 Background on classical and classical discrete polynomials

1.1.1 Weights on the real line

These lectures deal with some relevant examples of orthogonal polynomials. Orthogonality here is with respect to the inner product defined by a weight in the linear space of polynomials with real coefficients (which we will denote by $\mathbb{R}[x]$).

Definition 1.1   A weight μ is a positive Borel measure with support in the real line satisfying

(1) μ has finite moments of every order: i.e., the integrals $\int x^n d\mu(x)$ are finite for $n \in \mathbb{N} := \{0, 1, 2, \ldots\}$;

(2) μ has infinitely many points in its support (x is in the support of μ if $\mu(x - \varepsilon, x + \varepsilon) > 0$ for all $\varepsilon > 0$).

Condition (2) above is equivalent to saying that

$$\int p^2(x) d\mu(x) > 0, \quad \text{for } p \in \mathbb{R}[x], \ p \neq 0. \tag{1.1}$$

We associate to the weight μ the inner product defined in $\mathbb{R}[x]$ by

$$\langle p,q \rangle_\mu = \int p(x)q(x)d\mu(x). \tag{1.2}$$

By applying the Gram–Schmidt orthogonalization process to the sequence $1,x,x^2,x^3,\ldots$ (condition (1.1) is required), one can generate a sequence $(p_n)_n$, $n = 0,1,2,3,\ldots$, of polynomials with $\deg p_n = n$ that satisfy the orthogonality condition

$$\int p_n(x)p_m(x)d\mu(x) = c_n \delta_{n,m}, \quad c_n > 0.$$

We then say that the polynomial sequence $(p_n)_n$ is orthogonal with respect to the weight μ. When $c_n = 1$, we say that the polynomial p_n is orthonormal. It is not difficult to see that the orthogonal polynomial p_n is unique up to multiplicative constants.

Except for multiplicative constants, the nth orthogonal polynomial p_n is characterized because $\int p_n(x)q(x)d\mu(x) = 0$ for $q \in \mathbb{R}[x]$ with $\deg q \leq n-1$. This is trivially equivalent to saying that $\int p_n(x)x^k d\mu(x) = 0$ for $0 \leq k \leq n-1$.

Condition (2) above can be weakened; if we assume that the positive measure μ has N points in its support (say, x_i, $1 \leq i \leq N$), with $N < \infty$, we can only guarantee the existence of N orthogonal polynomials p_n, $n = 0,1,\ldots,N-1$, with respect to μ. Indeed, up to multiplicative constants, the polynomial $p(x) = \prod_{i=1}^{N}(x-x_i)$ is the only one with degree N orthogonal to p_n, $n = 0,1,\ldots,N-1$. But the norm of p is zero.

1.1.2 The three-term recurrence relation

Orthogonality with respect to a weight can be characterized in terms of an algebraic equation. Indeed, since xp_n is a polynomial of degree $n+1$, we can expand it in terms of the polynomials $p_0, p_1, \ldots, p_{n+1}$:

$$xp_n(x) = a_{n+1}p_{n+1}(x) + b_n p_n(x) + c_n p_{n-1}(x) + d_n p_{n-2}(x) + \cdots + e_n p_0(x). \tag{1.3}$$

Using the orthogonality of the polynomial p_n to the polynomials of lower degree, we have

$$d_n \langle p_{n-2}(x), p_{n-2}(x) \rangle_\mu = \langle xp_n(x), p_{n-2}(x) \rangle_\mu = \int_\mathbb{R} xp_n(x)p_{n-2}(x)d\mu$$
$$= \langle p_n(x), xp_{n-2}(x) \rangle_\mu = 0.$$

That is $d_n = 0$. Proceeding in a similar way, we deduce that in the expansion (1.3), only the polynomials p_{n+1}, p_n and p_{n-1} appear:

$$xp_n(x) = a_{n+1}p_{n+1}(x) + b_np_n(x) + c_np_{n-1}(x). \qquad (1.4)$$

It is not difficult to see that the positivity of μ implies that $a_nc_n > 0$. For the orthonormal polynomials we have the symmetry condition $c_n = a_n$.

The converse of (1.4) is also true and it is the spectral theorem for orthogonal polynomials (it is also known as Favard's theorem [32], although the result seems to be known already to Stieltjes, Chebyshev, and others, and is contained in the book by Stone [81] which appeared a couple of years before Favard's paper).

Theorem 1.2 *If the sequence of polynomials* $(p_n)_n$ *with* $\deg p_n = n$ *satisfies the three-term recurrence relation (1.4) with* $a_nc_n > 0$, *then they are orthogonal with respect to a weight.*

For a proof see [3], [10] or [81].

1.1.3 The classical orthogonal polynomial families

The most important examples of orthogonal polynomials are the so-called classical families. There are three such families (see, for instance, [10, 31, 50, 53, 67, 82], and for a good historical account see [4, 10]).

(1) The Jacobi polynomials $(P_n^{(\alpha,\beta)})_n$, $\alpha, \beta > -1$. They are a double parametric family of orthogonal polynomials with respect to the weight $d\mu = (1-x)^\alpha(1+x)^\beta dx$ on the interval $(-1, 1)$. There are some relevant particular values of the parameters which deserve special interest. The simplest case is when $\alpha = \beta = 0$ which results in Legendre polynomials. Legendre introduced them at the end of the 18th century and they are the first example of orthogonal polynomials in history (more about that later). The cases $\alpha = \beta = -1/2$ and $\alpha = \beta = 1/2$ are called Chebyshev polynomials of the first and second kind, respectively, and they were introduced and studied by this Russian mathematician in the second half of the 19th century. When $\alpha = \beta$ we have Gegenbauer or ultraspherical polynomials.

(2) The Laguerre polynomials $(L_n^\alpha)_n$, $\alpha > -1$. They are a single parametric family of orthogonal polynomials with respect to the weight $d\mu(x) = x^\alpha e^{-x} dx$ on the half line $x > 0$.

(3) The Hermite polynomials $(H_n)_n$. They are a single family of orthogonal polynomials with respect to the weight $d\mu(x) = e^{-x^2} dx$ on the real line.

I would like to note that, as usual in mathematics, many of the orthogonal families which appear in these lectures are misnamed, in the sense that someone else introduced them earlier than the person after whom the family is named. The classical families are related to the three most important continuous distribution of probabilities: Beta, Gamma and Normal, respectively. The classical families enjoy a set of important characterization properties that we will discuss in later sections.

There are many other examples of orthogonal polynomials. For instance, the so-called Heine polynomials, a single parametric family of orthogonal polynomials with respect to the weight $d\mu(x) = [x(1-x)(a-x)]^{-1/2} dx$ on the interval $(0,1)$ where $a > 1$ [10]. Of course there are sequences of orthogonal polynomials with no name: as far as I know the orthogonal polynomials with respect to the weight $d\mu(x) = |\cos(x^2 + 1)| dx$ on the interval $(0,2)$ have not yet been baptized. However, any careful reader will be concerned by the previous examples because actually they are not examples of orthogonal polynomials: they are just examples of weights!

Orthogonal polynomials can be explicitly computed only for a few weights. *Explicitly* means here a closed-form expression for each of the polynomials p_n in the family. For an arbitrary weight (as the one with no name above) one can hardly compute an approximation of the first few orthogonal polynomials. The classical families are among those happy cases in which we can compute *everything* explicitly (most of the identities we will consider next can be found in many books, see, for instance, [10, 31, 50, 53, 67, 82]).

To start with, let us consider the Legendre polynomials $(P_n)_n$, i.e., the Jacobi case with $\alpha = \beta = 0$. They are orthogonal with respect to the Lebesgue measure on the interval $[-1,1]$. Here is a hint for finding explicitly the Legendre polynomials: try with the differentiation formula

$$P_n(x) = ((1-x)^n(1+x)^n)^{(n)}. \tag{1.5}$$

Trivially we get $\deg P_n = n$. We then have to prove that

$$\int_{-1}^{1} P_n(x) x^k dx = 0, \quad 0 \le k \le n-1.$$

This can be easily proved using the following integration by parts:

$$\int_{-1}^{1} ((1-x)^n (1+x)^n)^{(n)} x^k dx$$

$$= ((1-x)^n (1+x)^n)^{(n-1)} x^k \Big|_{-1}^{1} - k \int_{-1}^{1} ((1-x)^n (1+x)^n)^{(n-1)} x^{k-1} dx$$

$$= -k \int_{-1}^{1} ((1-x)^n (1+x)^n)^{(n-1)} x^{k-1} dx$$

$$= \cdots = (-1)^k k! \int_{-1}^{1} ((1-x)^n (1+x)^n)^{(n-k)} dx$$

$$= (-1)^k k! ((1-x)^n (1+x)^n)^{(n-1-k)} \Big|_{x=-1}^{x=1} = 0.$$

Formula (1.5) is called the Rodrigues' formula for the Legendre polynomials, honoring the French mathematician Olinde Rodrigues who discovered it at the beginning of the 19th century.

Each classical family of orthogonal polynomials has a corresponding Rodrigues' formula of the form

$$p_n(x) = (a_2^n(x)\mu(x))^{(n)} / \mu(x), \qquad (1.6)$$

where a_2 is a polynomial of degree at most 2 and μ is the corresponding weight function for the family. More explicitly

(1) Jacobi polynomials

$$P_n^{(\alpha,\beta)}(x) = \frac{(-1)^n}{2^n n!} ((1-x^2)^n (1-x)^\alpha (1+x)^\beta)^{(n)} (1-x)^{-\alpha} (1+x)^{-\beta}.$$

(2) Laguerre polynomials

$$L_n^\alpha(x) = \frac{1}{n!} (x^n x^\alpha e^{-x})^{(n)} e^x x^{-\alpha}.$$

(3) Hermite polynomials

$$H_n(x) = (-1)^n (e^{-x^2})^{(n)} e^{x^2}.$$

One of the characteristic properties of the classical families is precisely this kind of Rodrigues' formula. These Rodrigues' formulas allow an explicit computation for each family. For instance, one get for the Legendre and Laguerre polynomials, respectively

$$P_n(x) = \frac{1}{2^n} \sum_{k=0}^{[n/2]} (-1)^k \binom{n}{k} \binom{2n-2k}{n} x^{n-2k},$$

$$L_n^\alpha(x) = \sum_{k=0}^{n} \binom{n+\alpha}{n-k} \frac{(-x)^k}{k!}. \qquad (1.7)$$

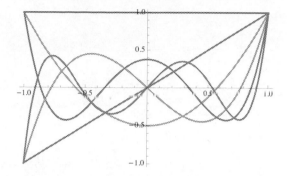

Figure 1.1 The first few Legendre polynomials

Exercise 1.3 Prove the explicit expression for the Laguerre polynomials.

Using the explicit formulas for the Legendre polynomials one can draw the first few of them, as shown in Figure 1.1.

One can then check that each one of these first few Legendre polynomials has real and simple zeros which live in the interval of orthogonality $(-1,1)$. Actually this is a general property of the zeros of orthogonal polynomials.

Theorem 1.4 *The zeros of an orthogonal polynomial p_n with respect to a weight μ are real, simple and live in the convex hull of the support of μ.*

Exercise 1.5 Prove this theorem.
Hint: let $x_1,\ldots,x_m$ be the real zeros of p_n with odd multiplicity. Consider then the polynomial $q(x) = (x - x_1) \cdots (x - x_m)$. Show that $p_n(x)q(x)$ has constant sign in $\mathbb{R}$. Use that fact and the orthogonality of p_n to conclude that $m \geq n$.

Rodrigues' formulas can also be used to explicitly compute the three-term recurrence relation (1.4). That is the case for the Hermite polynomials:

$$H_{n+1}(x) = (-1)^{n+1}(e^{-x^2})^{(n+1)}e^{x^2} = (-1)^{n+1}(-2xe^{-x^2})^{(n)}e^{x^2}$$
$$= (-1)^{n+1}\left(-2x(e^{-x^2})^{(n)} - 2n(e^{-x^2})^{(n-1)}\right)e^{x^2}$$
$$= 2xH_n(x) - 2nH_{n-1}(x).$$

The three-term recurrence relation for Laguerre and Jacobi polynomials are, respectively

$$L_{n+1}^{\alpha}(x) = \frac{1}{n+1}\left((2n+1+\alpha-x)L_n^{\alpha}(x) - (n+\alpha)L_{n-1}^{\alpha}(x)\right),$$

$$2(n+1)(n+\alpha+\beta+1)(2n+\alpha+\beta)P_{n+1}^{(\alpha,\beta)}(x)$$
$$= -2(n+\alpha)(n+\beta)(2n+\alpha+\beta+2)P_{n-1}^{(\alpha,\beta)}(x)$$
$$+ (2n+\alpha+\beta+1)\big[(\alpha^2-\beta^2)$$
$$+ (2n+\alpha+\beta)(2n+\alpha+\beta+2)x\big]P_n^{(\alpha,\beta)}(x).$$

1.1.4 Second-order differential operator

Legendre polynomials appeared in the context of planetary motion. Indeed, Legendre studied the following potential function related to planetary motion

$$V(t,y,z) = \int\int\int \frac{\rho(u,v,w)}{r}\,du\,dv\,dw,$$

where $r = \sqrt{(t-u)^2+(y-v)^2+(z-w)^2}$, and $\rho(u,v,w)$ stands for the density at the point (u,v,w) (Cartesian coordinates). This integral is easier after performing the change of variable

$$r(\rho,x) = (1-2\rho x+\rho^2)^{1/2}, \quad x = \cos\gamma.$$

Legendre then expanded $1/r(\rho,x)$ in power series of ρ

$$\frac{1}{(1-2\rho x+\rho^2)^{1/2}} = \sum_{n=0}^{\infty}\rho^n P_n(x), \tag{1.8}$$

and he realized that the functions $P_n(x)$ are actually polynomials in x of degree n: they are the polynomials we now call Legendre polynomials.

The function in the left-hand side of identity (1.8) is called the generating function for the Legendre polynomials. The generating function was an invention of the Swiss mathematician Leonard Euler: his genius idea was to pack a sequence $(a_n)_n$ in an analytic function f,

$$f(z) = \sum_{n=0}^{\infty} a_n z^n,$$

so that one can find interesting properties of the sequence $(a_n)_n$ from the function f, as long as one can explicitly find the function f. As the identity (1.8) shows, that is the case of the Legendre polynomials. A very interesting property one can deduce for the Legendre polynomials from its generating function is that they satisfy a second-order differential equation.

Indeed, by a direct computation one can see that the function $f(\rho,x)$ in (1.8) satisfies the partial differential equation

$$(1-x^2)\frac{\partial^2}{\partial x^2}f(\rho,x) - 2x\frac{\partial}{\partial x}f(\rho,x) = -\rho\frac{\partial^2}{\partial\rho^2}(\rho f(\rho,x)).$$

Substituting the function $f(\rho,x)$ by its power expansion (1.8), one can straightforwardly find the following second-order differential equation for the Legendre polynomials

$$(1-x^2)P_n''(x) - 2xP_n'(x) = -n(n+1)P_n(x). \tag{1.9}$$

Since in this differential equation the coefficients of the second and first derivatives do not depend on n, the differential equation is actually the eigenfunction equation for the second-order differential operator

$$D = (1-x^2)\frac{\partial^2}{\partial x^2} - 2x\frac{\partial}{\partial x}.$$

The differential equation (1.9) can also be proved using the orthogonality of the Legendre polynomials. Indeed, write $q_n(x) = (1-x^2)p_n''(x) - 2xp_n'(x)$. Obviously $\deg q_n = n$. For $0 \le k \le n-1$, an integration by parts gives

$$\int_{-1}^{1}\left((1-x^2)P_n''(x) - 2xP_n'(x)\right)x^k dx$$

$$= (1-x^2)P_n'(x)x^k\Big|_{-1}^{1} - \int_{-1}^{1}P_n'(x)\left(-2xx^k + kx^{k-1}(1-x^2)\right)dx$$

$$- \int_{-1}^{1}P_n'(x)2xx^k dx$$

$$= -k\int_{-1}^{1}P_n'(x)\left(x^{k-1}(1-x^2)\right)dx$$

$$= -k(1-x^2)P_n(x)x^{k-1}\Big|_{-1}^{1}$$

$$+ k\int_{-1}^{1}P_n(x)\left((k-1)x^{k-2}(1-x^2) - 2xx^{k-1}\right)dx$$

$$= 0.$$

That is, $(1-x^2)P_n''(x) - 2xP_n'(x)$ is orthogonal to the polynomials of degree at most $n-1$. Since orthogonal polynomials are unique up to multiplicative constants we get $(1-x^2)P_n''(x) - 2xP_n'(x) = \lambda_n P_n(x)$. Comparing leading coefficients in both sides of this equality, we recover the differential equation (1.9).

Each classical family of orthogonal polynomials is also the eigenfunctions of a second-order differential operator:

(1) Jacobi polynomials

$$(1-x^2)\left(P_n^{(\alpha,\beta)}\right)''(x) + (\beta - \alpha - (\alpha+\beta+2)x)\left(P_n^{(\alpha,\beta)}\right)'(x)$$
$$= \lambda_n P_n^{(\alpha,\beta)}(x),$$
$$\lambda_n = -n(n+\alpha+\beta+1).$$

(2) Laguerre polynomials

$$x\left(L_n^\alpha\right)''(x) + ((\alpha+1)-x)\left(L_n^\alpha\right)'(x) = -nL_n^\alpha(x). \tag{1.10}$$

(i) Hermite polynomials

$$H_n''(x) - 2xH_n'(x) = -2nH_n(x). \tag{1.11}$$

Being eigenfunctions of a second-order differential operator is another characteristic property of the classical families. Probably this is the most relevant of these properties if we take into account the important physical applications of these second-order differential equations (more about this later).

1.1.5 Characterizations of the classical families of orthogonal polynomials

The last characteristic property of the classical families considered in these lectures is the so-called Pearson's equation. The Rodrigues' formula (1.6) for $n = 1$ gives the following first-order differential equation for the classical weights

$$(a_2(x)\mu(x))' = a_1(x)\mu(x),$$

where a_i is a polynomial of degree at most i. This is the so-called Pearson equation for the classical weights, where the coefficients a_2 and a_1 are precisely the differential coefficients of the second-order differential operator for the corresponding classical family.

We summarize some of the characteristic properties of the classical families in the following theorem.

Theorem 1.6   *Let $(p_n)_n$ be a sequence of orthogonal polynomials with respect to a weight $d\mu$. Let a_2 and a_1 be polynomials of degree at most 2 and 1 respectively. Then the following conditions are equivalent:*

(1) $a_2(x)p_n''(x) + a_1(x)p_n'(x) = \lambda_n p_n(x), \quad n \geq 0.$

(2) *The weight is absolutely continuous with respect to the Lebesgue measure $d\mu(x) = \mu(x)dx$ with μ an smooth function satisfying $(a_2\mu)' = a_1\mu$.*

(3) *Up to a linear change of variable, $(p_n)_n$ are the classical families: Jacobi, Laguerre or Hermite,*

(4) *The polynomials can be defined by means of the Rodrigues' formula $p_n(x) = (a_2^n\mu)^{(n)}/\mu$.*

The implication $(1) \Rightarrow (3)$ in this theorem was proved by E.J. Routh in 1884 (see [77]), but it is known as the Bochner classification theorem (see Bochner's paper [7] of 1929). Actually Bochner classified polynomial solutions to Sturm–Liouville problems (which, beside the classical orthogonal polynomials, also include the solution $(x^n)_n$). For a complete treatment of this theorem we refer to [53] (see also [67]).

Exercise 1.7 Under the additional assumptions

(1) $d\mu(x) = \mu(x)dx$, with μ a twice differentiable positive function supported on an interval (a, b) of the real line (not necessarily bounded),

(2) $x^n a_2\mu, x^n(a_2\mu)'$ and $x^n a_1\mu$ have vanishing limits at a and b,

prove the implications $(2) \Rightarrow (1)$, $(2) \Rightarrow (4)$ and $(2) \Leftrightarrow (3)$.
Hint: The two first implications are similar to what we did for Legendre polynomials. The last one is a straightforward integration once one has reduced to the cases $a_2(x) = 1, x, 1 - x^2, x^2$, after a linear change of variable.

(The classical families satisfy other important characteristic properties that we do not consider in these lectures; for instance the sequence of derivatives $(p'_{n+1})_n$ is again orthogonal with respect to a positive measure (with respect to $a_2\mu$)).

1.1.6 The classical families and the basic quantum models

The classical families of orthogonal polynomials appear when the Schrödinger equation is integrated in basic models for non-relativistic quantum mechanics (see [67]). Let $\psi(t,x) = \psi_t(x)$ be a wave function for a quantum system; then the Schrödinger equation is

$$i\hbar\frac{\partial}{\partial t}\psi(x,t) = H(\psi),$$

where H is the Hamiltonian of the system and $\hbar \equiv$ Planck constant. When the Hamiltonian H is independent of time, this equation can be reduced to

$$H(\phi)(x) = E\phi(x).$$

This equation is usually called the time-independent Schrödinger equation and coincides with the eigenvalue equation for the energy of the system, because the Hamiltonian H is the operator associated to the energy (so that the corresponding eigenvalue E gives the energy of the system).

Consider now the one-dimensional harmonic oscillator, i.e., the system consisting of a particle attracted to the origin with force proportional to the displaced distance: $F = -kx$ where k is a constant of the system (for instance, the system formed by a vibrating diatomic molecule). The potential is then $V(x) = -\int F(t)dt = \frac{1}{2}kx^2$ and the Hamiltonian is given by

$$H = -\frac{\hbar^2}{2m}\frac{d^2}{dx^2} + \frac{1}{2}kx^2,$$

which is obviously independent of time, and where m is the particle mass.

The Schrödinger equation is then

$$-\frac{\hbar^2}{2m}\psi''(x) + \frac{1}{2}kx^2\psi(x) = E\psi(x). \tag{1.12}$$

This equation has a square integrable solution only when

$$E_n = (2n+1)\frac{\hbar}{2}\sqrt{\frac{k}{m}},$$

which provides the spectrum of energies for the system.

The second-order differential equation (1.12) can then be reduced to the corresponding one for the Hermite polynomials (1.11), so that the wave functions for the different states of the system can be constructed by means of the Hermite polynomials

$$\psi_n(x) = \left(\frac{\sqrt{b}}{n!2^n}\right)^{\frac{1}{2}} H_n(\sqrt{b}x)e^{-\frac{b}{2}x^2}, \quad b = \frac{\sqrt{km}}{\hbar}.$$

Consider now the model for the hydrogen atom according to the Schrödinger equation, [79]. This is a three-dimensional model describing the interaction between the proton (at the nucleus) and the electron given by the Coulombian potential

$$V(x,y,z) = -\frac{Ze^2}{4\pi\varepsilon_0}\frac{1}{\sqrt{x^2+y^2+z^2}},$$

with $Z \equiv$ atomic number and $e \equiv$ electron charge.

The equation was proposed and solved by Erwin Schrödinger. It gives the values of the energy spectrum found experimentally. Because of that, he won in 1933 the Nobel Physics prize. The Schrödinger equation for the hydrogen atom can be solved by separation of variables. One then gets a second-order differential equation for the radial part and another second-order differential equation for the angular part. The radial equation can be reduced to the second-order differential equation for the Laguerre polynomials. In the angular equation the so-called spherical harmonics appear. They are given by

$$Y_n^\lambda(\theta,\rho) = Ne^{i\lambda\rho}C_n^\lambda(\cos\theta),$$

where C_n^λ, $\lambda > -1/2$, $\lambda \neq 0$, stands for the nth ultraspherical polynomial (Jacobi family with $\alpha = \beta = \lambda - 1/2$).

For a trio of quantum numbers $n, l, m \in \mathbb{N}$, the wave function corresponding to this quantum state can then be constructed by using Laguerre and ultraspherical (Gegenbauer) polynomials; it has the form (spherical coordinates):

$$\psi_{n,l,m}(r,\theta,\rho) = A_{n,l,m}e^{-r/2}r^l L_{n+l}^{2l+1}(r)Y_l^m(\theta,\rho).$$

Most of the images for the hydrogen atom are drawn using this formula. For instance in the *Encyclopaedia Britannica* one can find the images, reproduced in Figure 1.2 for the electron density functions $((x,z)$-plane) for the quantum numbers $n = 8, m = 0$ and $l = 0, 2, 6$ and 7.

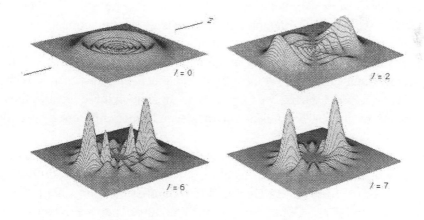

Figure 1.2

1.1.7 The classical discrete families

The classical discrete families follow the classical families in the ranking of the most important examples of orthogonal polynomials. Roughly speaking they are obtained by substituting derivatives with differences in the operational calculus of the classical families (most of the identities we will consider next can be found in [10, 50, 53, 67]). In order to be more precise, we need some notation.

For $a \in \mathbb{R}$, we denote by $\mathfrak{s}_a$ the shift operator

$$\mathfrak{s}_a f(x) = f(x+a).$$

As usual Δ and ∇ denote the basic delta and nabla first-order difference operators

$$\Delta f = f(x+1) - f(x) = (\mathfrak{s}_1 - \mathrm{Id})f, \quad \nabla f = f(x) - f(x-1) = (\mathrm{Id} - \mathfrak{s}_{-1})f.$$

They are also known in the literature of finite differences as forward and backward difference operators.

A classical discrete family is formed by orthogonal polynomials that in addition are also eigenfunctions of a second-order difference operator with polynomial coefficients. We have several equivalent possibilities for writing such second-order difference operators:

$$\left.\begin{aligned} D &= a_2\Delta^2 + a_1\Delta + a_0, \\ D &= a_2\Delta\nabla + a_1\Delta + a_0, \\ D &= a_1\mathfrak{s}_1 + a_0\mathfrak{s}_0 + a_{-1}\mathfrak{s}_{-1}. \end{aligned}\right\} \tag{1.13}$$

We prefer to work with the last one. Since we want the operator D (1.13) to have a sequence of orthogonal polynomials as eigenfunctions, D has to satisfy the necessary condition $\deg(D(p)) \le \deg(p)$ for $p \in \mathbb{R}[x]$. It is easy to see that this is equivalent to saying that the coefficients a_1, a_0 and a_{-1} are polynomials of degree at most 2 satisfying $\deg(a_1 + a_0 + a_{-1}) = 0$ and $\deg(a_1 - a_{-1}) = 1$. Hence we implicitly assume along these lectures those assumptions for the coefficients of the second-order difference operator (1.13).

Lancaster classified in 1941 (see [61]) all families of orthogonal polynomials which, in addition, are eigenfunctions of a second-order difference operator of the form (1.13). He found four families of orthogonal polynomials which had already appeared in the literature. They are associated to the names of Charlier, Meixner, Krawtchouk and Hahn. They are known as the classical discrete families and are related with some of the most important discrete probability distributions: Poisson, Pascal, binomial and hypergeometric, respectively. More precisely:

(1) Charlier polynomials $(c_n^a)_n, a > 0$, orthogonal with respect to (Poisson distribution)

$$\sum_{n=0}^{\infty} \frac{a^n}{n!} \delta_n. \tag{1.14}$$

(2) Meixner polynomials $(m_n^{a,c})_n$, $0 < a < 1$, $c > 0$, orthogonal with respect to (Pascal distribution)

$$\sum_{n=0}^{\infty} \frac{a^n \Gamma(n+c)}{n!} \delta_n.$$

(3) Krawtchouk polynomials $(k_n^{a,N})_{0 \leq n \leq N}$, $a > 0$, $N \in \mathbb{N}$, orthogonal with respect to (binomial distribution)

$$\sum_{n=0}^{N} a^n \binom{N}{n} \delta_n.$$

(4) Hahn polynomials $(h_n^{a,b,N})_{0 \leq n \leq N}$, $a > 0$, $N \in \mathbb{N}$, orthogonal with respect to (hypergeometric distribution)

$$\sum_{n=0}^{N} \binom{a+n}{n} \binom{b+N-n}{N-n} \delta_n.$$

Notice that both Krawtchouk and Hahn polynomials are finite families.

All the classical discrete weights are discrete measures supported either in the non-negative integers (Charlier and Meixner) or the first $N+1$ non-negative integers (Krawtchouk and Hahn). The classical discrete families have relevant applications in many areas of mathematics and other sciences. For instance, one of the most important ball and urn models due to P. and T. Ehrenfest (1907) can be modeled by means of Krawtchouk polynomials (see [51], also [42]).

Let us recall that for a real number u the Dirac delta δ_u is the Borel measure defined by

$$\delta_u(A) = \begin{cases} 1, & u \in A, \\ 0, & u \notin A. \end{cases}$$

Hence, an integral with respect to δ_u just picks up the value of the corresponding function at u: $\int f d\delta_u = f(u)$. Hence, integrals with respect to the classical discrete weights reduce to sums. For instance if we write μ for the Charlier weight (1.14), we have

$$\int f d\mu = \sum_{n=0}^{\infty} \frac{a^n}{n!} f(n).$$

The classical discrete families enjoy a number of important characteristic properties, some of which are summarized in the following Theorem.

Theorem 1.8 *Let $(p_n)_n$ be a sequence of orthogonal polynomials with* *respect to a weight* $d\mu = \sum_{n=0}^{N}\mu(n)\delta_n$, *where N is infinite or a positive integer. Let a_1, a_0 and a_{-1} be polynomials of degree at most 2 satisfying that* $\deg(a_1 + a_0 + a_{-1}) = 0$, $\deg(a_1 - a_{-1}) = 1$. *Then the following conditions are equivalent:*

(1) $a_1 p_n(x+1) + a_0 p_n(x) + a_{-1} p_n(x-1) = \lambda_n p_n(x)$.

(2) $a_{-1}(x+1)\mu(x+1) = a_1(x)\mu(x)$, $x \in \{0,\dots,N-1\}$, *with initial condition $a_{-1}(0) = 0$ and $a_1(N) = 0$ (if N is finite).*

(3) *Up to a linear change of variable $(p_n)_n$ are some of the classical discrete families: Charlier, Meixner, Krawtchouk or Hahn.*

(4) $p_n(x) = \nabla^n(\prod_{k=1}^{n} a_{-1}(x+k)\mu(x+n))/\mu(x)$.

Implication $(1) \Rightarrow (2)$ is the Lancaster classification theorem already mentioned. For a complete treatment of this theorem we refer to [53] (see also [68]).

Exercise 1.9   Prove the implication $(2) \Rightarrow (1)$.
Hint: Show that the polynomial

$$q_n(x) = a_1(x)p_n(x+1) + a_0(x)p_n(x) + a_{-1}(x)p_n(x-1),$$

which has degree at most n, satisfies

$$\sum_{x=0}^{N} q_n(x)x^k\mu(x) = 0, \quad k = 0,\dots,n-1.$$

As with the classical families, *everything* related to the classical discrete polynomials can be computed explicitly, including the polynomials themselves, three-term recurrence relations, etc. For instance, the three-term recurrence relation for the Charlier polynomials $(c_n^a)_n$, normalized with leading coefficient $1/n!$, is given by

$$xc_n^a(x) = (n+1)c_{n+1}^a(x) + (n+a)c_n^a(x) + ac_{n-1}^a(x). \tag{1.15}$$

1.2 The Askey tableau. Krall and exceptional polynomials. Darboux Transforms

1.2.1 The Askey tableau

The classical and classical discrete families can be represented using (generalized) hypergeometric functions. This representation gives rise to the Askey scheme or Askey tableau: an organized hierarchy of these families of orthogonal polynomials (for a historical account see the Foreword by T. Koornwinder in [53]).

The hypergeometric function $_rF_s$ is defined by

$$_rF_s\left(\begin{matrix} a_1 & \cdots & a_r \\ b_1 & \cdots & b_s \end{matrix};z\right) = \sum_{k=0}^{\infty} \frac{(a_1,\ldots,a_r)_k}{(b_1,\ldots,b_s)_k} \frac{z^k}{k!}$$

$(b_i \notin -\mathbb{N})$, where $(a_1,\ldots,a_r)_k = (a_1)_k \cdots (a_r)_k$, and $(a)_j$ is the Pochhammer symbol

$$(a)_0 = 1, \quad (a)_j = a(a+1)\cdots(a+j-1), \quad j \geq 1.$$

The Gauss hypergeometric function $_2F_1\left(\begin{matrix} a & b \\ c \end{matrix};z\right)$ satisfies the second-order differential equation

$$z(1-z)w'' + (c-(a+b+1)z)w' - abw = 0.$$

When $a_i = -n$, for some i, $1 \leq i \leq r, n \in \mathbb{N}$, then $_rF_s$ is a polynomial.

Some hypergeometric representations for the classical and classical discrete families are the following (see [53]):

(1) Jacobi polynomials

$$P_n^{(\alpha,\beta)}(x) = \frac{(\alpha+1)_n}{n!} {}_2F_1\left(\begin{matrix} -n & n+\alpha+\beta+1 \\ \alpha+1 \end{matrix};\frac{1-x}{2}\right).$$

(2) Laguerre polynomials

$$L_n^\alpha(x) = \frac{(\alpha+1)_n}{n!} {}_1F_1\left(\begin{matrix} -n \\ \alpha+1 \end{matrix};x\right).$$

(3) Hermite polynomials

$$H_n(x) = (2x)^n {}_2F_0\left(\begin{matrix} -n/2 & -(n-1)/2 \\ - \end{matrix};-\frac{1}{x^2}\right).$$

(4) Hahn polynomials

$$h_n^{\alpha,\beta,N}(x) = {}_3F_2\left(\begin{matrix} -n & n+\alpha+\beta+1 & -x \\ \alpha+1 & -N \end{matrix};1\right).$$

(5) Meixner polynomials

$$m_n^{a,c}(x) = \frac{a^n(c)_n}{(a-1)^n n!} {}_2F_1\left(\begin{matrix} -n & -x \\ c \end{matrix};1-\frac{1}{a}\right). \tag{1.16}$$

(6) Krawtchouk polynomials

$$k_n^{a,N}(x) = {}_2F_1\left(\begin{matrix} -n & -x \\ -N \end{matrix};\frac{a}{1+a}\right).$$

(7) Charlier polynomials

$$c_n^a(x) = \frac{(-a)^n}{n!}\,{}_2F_0\left(\begin{matrix} -n & -x \\ - & \end{matrix};-\frac{1}{a}\right).$$

We have normalized Charlier and Meixner polynomials (which will appear in Sections 1.3, 1.4, 1.5 and 1.6) so that the leading coefficient is $1/n!$. This is a much more convenient normalization for the purpose of these lectures than the one in [53] (Krawtchouk and Hahn, which will not appear again in these lectures, are normalized as in [53]).

In each case these hypergeometric representations are an easy consequence of the explicit expression for the classical and classical discrete families one can get from the corresponding Rodrigues' formula. For instance, check that the hypergeometric expression for the Laguerre polynomials can be easily derived from the explicit expression (1.7).

Using their hypergeometric representations, the classical and classical discrete families can be arranged in a tableau: the bigger the number of parameters and the indices of the hypergeometric function corresponding to a classical or classical discrete family, the higher this family is located in the tableau. This gives rise to the Askey tableau. A very important fact concerning the Askey tableau is that one can navigate through it from north to south by taking the limit in some of the parameters. That is, the southern families are limits of the northern families. Some examples of these limit relationships are the following:

$$\lim_{\beta\to\infty} P_n^{(\alpha,\beta)}(1-2x/\beta) = L_n^\alpha(x),$$

$$\lim_{\alpha\to\infty} \alpha^{-n/2} P_n^{(\alpha,\alpha)}(x/\sqrt{\alpha}) = \frac{H_n(x)}{2^n n!},$$

$$\lim_{\alpha\to\infty} \left(\frac{2}{\alpha}\right)^{n/2} L_n^\alpha(\sqrt{(2\alpha)}x+\alpha) = \frac{(-1)^n}{n!}H_n(x),$$

$$\lim_{a\to 1}(a-1)^n m_n^{a,\alpha+1}\left(\frac{x}{1-a}\right) = L_n^\alpha(x), \tag{1.17}$$

$$\lim_{N\to\infty} h_n^{\alpha,\beta,N}(Nx) = \frac{P_n^{(\alpha,\beta)}(1-2x)}{P_n^{(\alpha,\beta)}(1)},$$

$$\lim_{a\to\infty} \left(\frac{2}{a}\right)^{n/2} c_n^a(\sqrt{(2a)}x+a) = \frac{H_n(x)}{n!}. \tag{1.18}$$

These limit relationships can be proved easily by induction on n from the corresponding three-term recurrence relations.

Exercise 1.10 Prove the limit relationships between Charlier and Hermite polynomials.

Besides the classical and classical discrete families, the Askey tableau includes some more families of orthogonal polynomials (Wilson, Racah, etc.) which we do not consider in these lectures. There is also a q-Askey tableau for the so-called q-hypergeometric orthogonal polynomials (which can be represented by using basic hypergeometric functions), see [53].

Figures 1.3 and 1.4 show the shortened Askey tableau containing the orthogonal polynomials we consider in these lectures and also the full Askey tableau.

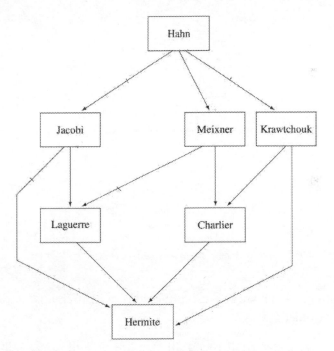

Figure 1.3 Shortened Askey tableau (classical and classical discrete families)

1.2.2 Krall and exceptional polynomials

The rest of these lectures will be devoted to the expansion of the Askey tableau by including two of the most important extensions of the classical and classical discrete orthogonal polynomials: namely, Krall and exceptional polynomials.

Definition 1.11 Krall (or bispectral) polynomials are orthogonal polyno-

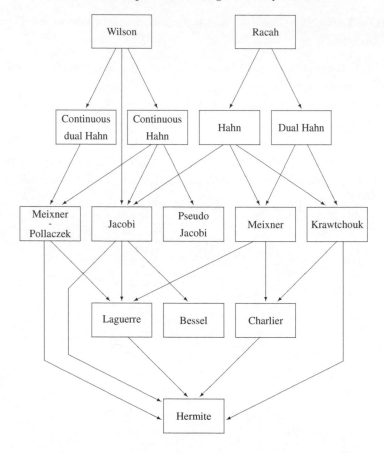

Figure 1.4 Askey tableau (full)

mials that in addition are also eigenfunctions of a differential or difference operator of order $k, k > 2$.

In his 1941 paper Krall found all families of orthogonal polynomials which are common eigenfunctions of a fourth-order differential operator. Since the 1980s a lot of effort has been devoted to this issue (with contributions by L.L. Littlejohn, A.M. Krall, J. and R. Koekoek, F.A. Grünbaum and L. Haine (and collaborators), K.H. Kwon (and collaborators), A. Zhedanov, P. Iliev, myself and many others).

Definition 1.12 Exceptional orthogonal polynomials share with classical and classical discrete families the property of being eigenfunctions of a second-order differential or difference operator, respectively. But they differ from them in that their degree sequence does not contain every possible degree, i.e., there are a finite number of degrees for which no polynomial eigenfunction exists. The gaps in the degrees allow the differential operators to have rational coefficients.

Exceptional polynomials came from mathematical physics. These polynomials allow one to write exact solutions of the Schrödinger equation corresponding to certain rational extensions of the classical quantum potentials. Exceptional polynomials appeared some nine years ago, but there has been a remarkable activity around them mainly by theoretical physicists (with contributions by D. Gómez-Ullate, N. Kamran and R. Milson, Y. Grandati, C. Quesne, S. Odake and R. Sasaki, the author and many others). We will construct exceptional polynomials in Sections 1.5 and 1.6.

1.2.3 Krall polynomials

In this section we introduce the different recipes used in the construction of the Krall polynomials. Since there are important differences between the cases of differential or difference operators, we consider each case separately. Most of the results we will display below will be proved in Sections 1.3 and 1.4 by using my approach based on the concept of $\mathscr{D}$-operators (see Section 1.3).

Krall polynomials: differential operators

When differential operators are considered, the procedure to construct Krall polynomials is the following (we will only consider positive measures): take the Laguerre or Jacobi weights, assume one or two of the parameters are nonnegative integers and add a Dirac delta at one or two of the end points of the interval of orthogonality. No example of Krall polynomials associated with the Hermite polynomials is known.

We display the Laguerre case in full detail.

Theorem 1.13 *If $\alpha \in \mathbb{N}$, $\alpha \geq 1$, the orthogonal polynomials with respect to the weight*

$$\mu_{M,\alpha} = M\delta_0 + x^{\alpha-1}e^{-x}, \quad x > 0, \ M \geq 0, \tag{1.19}$$

are eigenfunctions of a differential operator of order $2\alpha + 2$.

Krall (see [57, 58]) proved the case $\alpha = 1$. In the 1980s the cases $\alpha = 2, 3$ were proved by L.L. Littlejohn [63, 64]. J. Koekoek and R. Koekoek proved the general case $\alpha \in \mathbb{N}$, $\alpha \geq 1$ in 1991 [54] (see also [14, 44, 49]). For the Jacobi case see [14, 45, 56, 55, 60, 48, 86].

F.A. Grünbaum and L. Haine (et al.) proved that polynomials satisfying fourth- or higher-order differential equations can be generated by applying Darboux transforms to certain instances of the classical polynomials (see [43, 44, 45, 46]). P. Iliev refined this approach by proving that the order $2\alpha + 2$ in the previous theorem is minimal (see [49] and also [48]). More about this in the next section.

Krall discrete polynomials

The problem of finding Krall discrete polynomials was open for many decades. What we know for the differential case has seemed to be of little help because adding Dirac deltas to the classical discrete measures seems not to work. Indeed, Richard Askey in 1991 explicitly posed the problem of finding the first examples of Krall-discrete polynomials (see page 418 of [8]). He suggested to study measures which consist of some classical discrete weights together with a Dirac delta at the end point(s) of the interval of orthogonality. Three years later, Bavinck, van Haeringen and Koekoek gave a negative answer to Askey's question: they added a Dirac delta at zero to the Charlier and Meixner weights and constructed difference operators with respect to which the corresponding orthogonal polynomials are eigenfunctions, but these difference operators always have infinite order [5, 6].

In 2012, I found the first examples of measures whose orthogonal polynomials are also eigenfunctions of higher-order difference operator (see [13]). My recipe to construct Krall discrete polynomials is the following: apply a suitable Christoffel transform to the classical discrete measures; that is, multiply the classical discrete measures by suitable polynomials.

Depending on the classical discrete measure, we have found several classes of suitable polynomials for which this procedure works (one class for Charlier, two classes for Meixner and Krawtchouk and four classes for Hahn; the cross product of these classes also works). We next display the Charlier case in full detail.

Let F be a finite set of positive integers, and write n_F for the number of elements of F. Assume that the annihilator polynomial of F,

$$\prod_{f \in F}(x - f),$$

is non-negative in $\mathbb{N}$ (actually, this is needed only if one wants to work with positive measures). This set F is the new parameter of the Krall–Charlier family.

Theorem 1.14 *The orthogonal polynomials with respect to the weight*

$$\sum_{x=0}^{\infty} \left(\prod_{f \in F} (x - f) \right) \frac{a^x}{x!} \delta_x,$$

are common eigenfunctions of a difference operator D of order

$$r = 2 \left(\sum_{x \in F} x - \frac{n_F (n_F - 1)}{2} + 1 \right).$$

This theorem was conjectured by the author in [13]. I proved it for the particular case $F = \{1, 2, \ldots, k\}$ in [14]; in that paper I introduced the concept of $\mathscr{D}$-operators. Together with my colleague M. D. de la Iglesia, the general case was proved in [25]. Whether the order in the previous theorem is minimal or not is still an open problem.

The same procedure as in the previous theorem works for the other classical discrete families. For the Charlier family this is the only known process which produces Krall polynomials, although for the other classical families there are other Christoffel transform for constructing Krall polynomials (see [13, 14, 25, 27]).

1.2.4 Darboux transforms

The Darboux transform is a useful tool for constructing Krall and exceptional polynomials. Although we will not use Darboux transforms in these lectures (our main tool is $\mathscr{D}$-operators), we introduce here some basic definitions and facts because it will be useful for comparing the differences between the cases of differential and difference operators in Krall polynomials (see Section 1.5). The Darboux transform (actually credited to Moutard by Darboux) has been reinvented several times in the context of differential operators. We should also note that the same term *Darboux transformation* has been used in different fields with different meanings. Sometimes this causes a certain amount of confusion (it is also known as the factorization method, and in the theory of integrable systems, yet other transformations receive the same name). In our context, we mean the factorization of a second-order differential or difference operator and permutation of the two factors. The term Darboux transformation is also applied to factorizations of the Jacobi matrix in upper-triangular and lower-triangular factors.

Definition 1.15 Given a system $(T, (\phi_n)_n)$ formed by a second-order differential or difference operator T and a sequence $(\phi_n)_n$ of eigenfunctions for T, $T(\phi_n) = \pi_n \phi_n$, by a Darboux transform of the system $(T, (\phi_n)_n)$ we mean the following. For a real number λ, we factorize $T - \lambda \,\mathrm{Id}$ as the product of two first-order differential or difference operators

$$T = AB + \lambda \,\mathrm{Id}$$

(Id denotes the identity operator). We then produce a new system consisting in the operator $\hat{T}$, obtained by reversing the order of the factors,

$$\hat{T} = BA + \lambda \,\mathrm{Id},$$

and the sequence of eigenfunctions $\hat{\phi}_n = B(\phi_n)$: $\hat{T}(\hat{\phi}_n) = \pi_n \hat{\phi}_n$. We say that the system $(\hat{T}, (\hat{\phi}_n)_n)$ has been obtained by applying a Darboux transformation with parameter λ to the system $(T, (\phi_n)_n)$.

Darboux transform (of Jacobi matrices) and Krall polynomials

In the context of orthogonal polynomials, we always have a second-order difference operator at hand: the three-term recurrence relation. Indeed, a sequence $(p_n)_n$ of monic orthogonal polynomials with respect to a weight μ always satisfies a three-term recurrence relation of the form

$$xp_n = p_{n+1} + b_n p_n + c_n p_{n-1}, \quad n \geq 0, \tag{1.20}$$

where $c_n > 0, n \geq 1$, and $p_{-1} = 0$.

The polynomials $p_n(x)$ can be considered as functions of two variables: the continuous variable x and the discrete variable n. Hence, if we write

$$D = \mathfrak{s}_1 + b_n \mathfrak{s}_0 + c_n \mathfrak{s}_{-1}, \tag{1.21}$$

where the shift operators $\mathfrak{s}_a$ act on the discrete variable n, the three-term recurrence relation (1.20) gives a system $(D, (p_n(x))_x)$ formed by the second-order difference operator D (acting on the discrete variable n), and the eigenfunctions $p_n(x)$ (as functions of n) with eigenvalues x. In other words, $D(p_n(x)) = xp_n(x)$.

By using the so-called Jacobi matrix (associated to the polynomials $(p_n)_n$):

$$J = \begin{pmatrix} b_0 & 1 & 0 & 0 & \cdots \\ c_1 & b_1 & 1 & 0 & \cdots \\ 0 & c_2 & b_2 & 1 & \cdots \\ \vdots & \vdots & \vdots & \vdots & \ddots \end{pmatrix}, \tag{1.22}$$

the recurrence relation (1.20) can be expressed in matrix form as

$$JP = xP,$$

where $P = (p_0, p_1, p_2, \ldots)^t$.

The Darboux transform of the operator D (1.21) can then be managed as follows: for any $\lambda \in \mathbb{R}$, decompose J into

$$J - \lambda I = AB, \tag{1.23}$$

$$A = \begin{pmatrix} \alpha_0 & 1 & 0 & 0 & \cdots \\ 0 & \alpha_1 & 1 & 0 & \cdots \\ 0 & 0 & \alpha_2 & 1 & \cdots \\ \vdots & \vdots & \vdots & \vdots & \ddots \end{pmatrix} \quad \text{and} \quad B = \begin{pmatrix} 1 & 0 & 0 & 0 & \cdots \\ \beta_1 & 1 & 0 & 0 & \cdots \\ 0 & \beta_2 & 1 & 0 & \cdots \\ \vdots & \vdots & \vdots & \vdots & \ddots \end{pmatrix}.$$

We then call $\tilde{J} = BA + \lambda I$ a Darboux transform of J at λ. Notice that $\tilde{J}$ is again a tridiagonal matrix. Of course, this can be written in terms of difference operators

$$A = \mathfrak{s}_1 + \alpha_n \mathfrak{s}_0, \quad B = \mathfrak{s}_0 + \beta_n \mathfrak{s}_{-1},$$

so that one can see that the matrix $\tilde{J}$ represents a Darboux transform of the operator D (1.21).

Equation (1.23) amounts to

$$b_{n-1} = \alpha_{n-1} + \beta_n - \lambda, \quad c_n = \alpha_n \beta_n, \quad n \geq 1.$$

This allows one to solve recursively for $\alpha_0, \alpha_1, \beta_2, \alpha_2, \ldots$ in terms of β_1, λ and the elements of J. β_1 is then called a free parameter of the Darboux process. We can associate a sequence of monic polynomials $(q_n)_n$ to $\tilde{J}$ as before: $Q = BP$, where $Q = (q_0, q_1, q_2, \ldots)^t$. This gives $\tilde{J}Q = xQ$. Since the polynomials $(q_n)_n$ satisfy the three-term recurrence relation defined by the matrix $\tilde{J}$, Favard's theorem says that, as far as $\beta_n \neq 0$, $n \geq 1$, the polynomials $(q_n)_n$ are orthogonal with respect to a measure $\tilde{\mu}$ (take into account that this measure may not be positive).

It was proved by G.J. Yoon that the measures μ and $\tilde{\mu}$ are related by the formula $(x - \lambda)\tilde{\mu} = \mu$ (see [85], Th. 2.4) and then

$$\tilde{\mu} = \frac{1}{x - \lambda}\mu + M_{\beta_1}\delta_\lambda, \tag{1.24}$$

where M_{β_1} is certain number which depends on the free parameter β_1. The measure $\tilde{\mu}$ in (1.24) is sometimes called a Geronimus transform of the measure μ (Geronimus transform is reciprocal to the Christoffel transform; see [34] or for a modern systematic treatment [9, 65]).

According to this definition, one can see that the Krall–Laguerre weight (1.19) in Theorem 1.13 is a Geronimus transform of the Laguerre measure and therefore the Krall–Laguerre polynomials can be obtained by applying a Darboux transform to the Jacobi matrix associated to the Laguerre polynomials.

A form of the Darboux process was used in the context of orthogonal polynomials and their q versions by V. Spiridonov and A. Zhedanov [80]. Grünbaum and Haine (and some collaborators: Hozorov, Yakimov) have used the Darboux transform as explained above to construct Krall polynomials: they have applied it to the Jacobi matrix of the classical families of Laguerre and Jacobi with parameter at the endpoints of the support of the corresponding classical measures (see [43, 44, 45, 46]). The method was improved later on by P. Iliev (see [48, 49]). Notice that this Darboux process does not always seem to work. For instance, in the case of the Laguerre family, the parameter α has to be a non-negative integer and the parameter λ of the Darboux process has to be 0 (the end point of the support of the weight), because otherwise the corresponding orthogonal polynomials do not seem to be the eigenfunctions of any finite-order differential operator. On the other hand, the Darboux process always works with respect to the free parameter introduced by the Darboux process (the mass M at the Dirac delta in the weight (1.19)) because, no matter the value of this free parameter, the corresponding orthogonal polynomials are always eigenfunctions of a differential operator of order $2\alpha + 2$.

Duistermaat and Grünbaum introduced the terminology bispectral in [11] (see also [43]) which is now sometimes used for Krall polynomials. Indeed, as we have already explained, each sequence of orthogonal polynomials can be understood as solutions of an eigenfunction equation for a second-order difference operator (acting in the discrete variable n) associated to the three-term recurrence relation. In addition, the classical polynomials are also eigenfunctions of a second-order differential operator, acting now in the continuous variable x, while the Krall polynomials are also eigenfunctions of a differential operator of higher order. The word *bispectral* is then stressing the existence of two eigenfunctions problems associated to two different operators in relation to Krall polynomials.

The Darboux process seems to work in a very different way when it is applied to the classical discrete families (more about this later in Section 1.5).

Darboux transforms (of differential operators) and exceptional
polynomials

The Darboux transform is also a useful tool for constructing exceptional polynomials, but now it has to be applied to the second-order differential operator associated to the classical families.

The relationship between exceptional polynomials and the Darboux transform (differential operator) was shown by C. Quesne (2008) [76], shortly after the first examples of exceptional polynomials were introduced by Gómez-Ullate, Kamran and Milson in 2007. Here is an example (we omit the computations).

Consider the Laguerre second-order differential operator and its factorization:

$$D_L = -x\frac{d^2}{dx^2} - (\alpha + 1 - x)\frac{d}{dx} = AB + (\alpha + 2)\,\mathrm{Id}, \qquad (1.25)$$

where

$$A = -\frac{x}{1+\alpha+x}\frac{d}{dx} - \frac{\alpha+1}{1+\alpha+x},$$
$$B = -(1+\alpha+x)\frac{d}{dx} + (2+\alpha+x).$$

Applying a Darboux transform with parameter $\lambda = \alpha + 2$ as explained in Definition 1.15, we get the polynomials

$$q_{n+1}(x) = (1+\alpha+x)L_{n-1}^{\alpha+1}(x) + (2+\alpha+x)L_n^\alpha(x), \quad n \geq 0.$$

Notice that the degree 0 corresponding to q_0 is missing.

They are automatically eigenfunctions of the second-order differential operator

$$\tilde{D}_L = x\frac{d^2}{dx^2} + \left(\alpha+2-x-\frac{2x}{\alpha+1+x}\right)\frac{d}{dx} + \frac{2x-\alpha-1}{\alpha+1+x}\,\mathrm{Id},$$

obtained by reversing the order of the factors in the factorization (1.25), so that $\tilde{D}_L = BA + (\alpha + 2)\,\mathrm{Id}$.

For $\alpha > -1$, it turns out that they are orthogonal with respect to the weight

$$\omega(x) = \frac{x^{\alpha+1}e^{-x}}{(1+\alpha+x)^2}, \quad x > 0.$$

(This last can be proved by checking that this weight satisfies the Pearson equation $(a_2\omega)' = a_1\omega$, where a_2 and a_1 are the coefficient of the second and first derivative of the second-order differential operator $\tilde{D}_L$). However,

I will not use the Darboux transform in my approach to exceptional polynomials in Sections 1.5 and 1.6.

1.3 $\mathscr{D}$-operators

In this section, we introduce our main tool for constructing Krall polynomials: $\mathscr{D}$-operators. This is a new concept introduced by the author in
[14] that provides a unified approach for constructing orthogonal polynomial which are eigenfunctions of higher-order differential, difference or
q-difference operators, respectively (see [2, 14, 18, 25, 26, 27, 28]).

1.3.1 $\mathscr{D}$-operators

We introduce the concept of $\mathscr{D}$-operators in an abstract setting. The starting
point is a sequence of polynomials $(p_n)_n$, p_n of degree n, and an algebra
of operators $\mathscr{A}$ acting in the linear space of polynomials. In addition, we
assume that the polynomials p_n, $n \geq 0$, are eigenfunctions of certain operator $D_p \in \mathscr{A}$. We write $(\theta_n)_n$ for the corresponding eigenvalues, so that
$D_p(p_n) = \theta_n p_n, n \geq 0$.

For a fixed positive integer K, consider $K + 1$ sequences of numbers
$(\beta_{n,j})_n$, $j = 0, \ldots, K$, and define the polynomials:

$$q_n(x) = \beta_{n,0} p_n(x) + \beta_{n,1} p_{n-1}(x) + \cdots + \beta_{n,K} p_{n-K}(x). \tag{1.26}$$

We are interested in solving the following question:

Question 1   *Is it possible to determine the structure of the sequences
$(\beta_{n,j})_n$ such that it guarantees the existence of an operator $D_q \in \mathscr{A}$ for
which $(q_n)_n$ are eigenfunctions: $D_q(q_n) = \lambda_n q_n$?*

(We point out that orthogonality is not going to play any role at this
stage.) The answer is yes, it is! The main tool to find a complete answer
to the Question 1 is using $\mathscr{D}$-operators. We have two different types of
$\mathscr{D}$-operators, depending on whether the sequence of eigenvalues $(\theta_n)_n$ is
linear in n or not. Let us start with the simplest one, the $\mathscr{D}$-operator of
type 1. This is useful when the eigenvalue sequence $(\theta_n)_n$ is linear in n.

Given a sequence of numbers $(\varepsilon_n)_n$, a $\mathscr{D}$-operator of type 1 associated to
the algebra $\mathscr{A}$ and the sequence of polynomials $(p_n)_n$ is defined as follows.
Consider the operator $\mathscr{D} : \mathbb{P} \to \mathbb{P}$ defined by linearity

$$\mathscr{D}(p_n) = \sum_{j=1}^{n} (-1)^{j+1} \varepsilon_n \cdots \varepsilon_{n-j+1} p_{n-j}. \tag{1.27}$$

We then say that $\mathscr{D}$ is a $\mathscr{D}$-operator if $\mathscr{D} \in \mathscr{A}$.

This is a somewhat mysterious definition, but we will show in the next section that it appears in a rather natural way when one tries to solve Question 1 for $K = 1$. The definition of $\mathscr{D}$-operators of type 2 (the general case) is slightly different and will be considered later.

We now give a couple of examples of $\mathscr{D}$-operators (see [14]).

Exercise 1.16 Prove that the sequence defined by $\varepsilon_n = -1$ defines a $\mathscr{D}$-operator for the Laguerre polynomials $p_n = L_n^\alpha$ and the algebra $\mathscr{A}$ of differential operators with polynomial coefficients. Moreover $\mathscr{D} = d/dx$.
Hint: Use the following two well-known identities for the Laguerre polynomials:

$$\frac{d}{dx}(L_n^\alpha) = -L_{n-1}^{\alpha+1},$$

$$L_n^{\alpha+1} = \sum_{j=0}^{n} L_{n-j}^\alpha.$$

Exercise 1.17 Prove that the sequence defined by $\varepsilon_n = 1$ defines a $\mathscr{D}$-operator for the Charlier polynomials $p_n = c_n^a$ and the algebra $\mathscr{A}$ of difference operators with polynomial coefficients. Moreover $\mathscr{D} = \nabla$ ($\nabla f = f(x) - f(x-1)$).
Hint: Use the following identities:

$$\nabla = \sum_{j=1}^{\infty} (-1)^{j+1} \Delta^j,$$

$$\Delta(c_n^a) = c_{n-1}^a.$$

We next show that $\mathscr{D}$-operators are *rara avis*. Indeed, consider the Hermite polynomials $p_n = H_n$ and the algebra $\mathscr{A}$ of differential operators with polynomial coefficients. We are interested in finding a $\mathscr{D}$-operator for the Hermite polynomials and this algebra $\mathscr{A}$. According to Definition 1.27, we have to find a sequence of numbers $(\varepsilon_n)_n$ and a differential operator $D \in \mathscr{A}$ such that

$$D(H_n) = \sum_{j=1}^{n} (-1)^{j+1} \varepsilon_n \varepsilon_{n-1} \cdots \varepsilon_{n-j+1} H_{n-j}. \tag{1.28}$$

Notice that ε_n is the coefficient of H_{n-1} in the expansion of $D(H_n)$.

Since D reduces the degree of the polynomials (see (1.28)), we first try with the simplest operator in $\mathscr{A}$ with this property: $D = d/dx$. Using the well-known formula for the derivative of Hermite polynomials, $H_n' =$

$2nH_{n-1}$, we get

$$D(H_n) = H'_n(x) = 2nH_{n-1}(x). \tag{1.29}$$

Hence, necessarily $\varepsilon_n = 2n$. But this is impossible because the coefficients of H_{n-2} in (1.28) and (1.29) are $\varepsilon_n\varepsilon_{n-1} = 4n(n-1)$ and 0, respectively, and $4n(n-1) \neq 0$ for $n \neq 0,1$. So $D = d/dx$ can not be a $\mathscr{D}$-operator for the Hermite polynomials.

The simplest example of a second-order differential operator in $\mathscr{A}$ reducing the degree is $D = (ax+b)d^2/dx^2 + cd/dx$. We now see that D cannot be a $\mathscr{D}$ operator for the Hermite polynomials. Indeed, we have $D(H_n) = (ax+b)H''_n + cH'_n = 4n(n-1)(ax+b)H_{n-2}(x) + 2ncH_{n-1}$. Using the three-term recurrence relation for the Hermite polynomials we get

$$D(H_n) = (2nc + 2n(n-1)a)H_{n-1} + \beta_n H_{n-2} + \gamma_n H_{n-3}, \tag{1.30}$$

for certain sequences $(\beta_n)_n$, $(\gamma_n)_n$. As before, we have $\varepsilon_n = 2nc + 2na(n-1)$ and therefore $a \neq 0$ or $c \neq 0$. But again, this is impossible because the coefficients of H_{n-4} in (1.28) and (1.30) are $\varepsilon_n\varepsilon_{n-1}\varepsilon_{n-2}\varepsilon_{n-4}$ (which is a non-null polynomial in n) and 0, respectively. In the same way, one can see that the Hermite polynomials do not have any $\mathscr{D}$-operator.

1.3.2 $\mathscr{D}$-operators on the stage

In this section we show how to use $\mathscr{D}$-operators to answer Question 1 above. Assume we have a sequence of polynomials $(p_n)_n$, p_n of degree n, an algebra of operators $\mathscr{A}$ acting in the linear space of polynomials and an operator $D_p \in \mathscr{A}$ for which the polynomials p_n, $n \geq 0$, are eigenfunctions with linear eigenvalues. To simplify, we assume $D_p(p_n) = np_n$, $n \geq 0$. We also have a $\mathscr{D}$-operator defined by the sequence of numbers $(\varepsilon_n)_n$ as in (1.27). To the sequence $(\varepsilon_n)_n$ we associate new numbers defined by

$$\xi_{n,j} = (-1)^{j+1}\prod_{i=0}^{j-1}\varepsilon_{n-i}. \tag{1.31}$$

We fix a non-negative integer K and define a new sequence of polynomials $(q_n)_n$ by using a linear combination of $K+1$ consecutive polynomials p_n as in (1.26). First of all, we rewrite Question 1 in a much more convenient way:

Question 2 *Find numbers $a_n^{i,j}$, $i, j = 0, \ldots, K$, such that for the sequence* *of polynomials defined by*

$$
q_n(x) =
\begin{vmatrix}
p_n(x) & a_n^{0,1}p_{n-1}(x) & \cdots & a_n^{0,K}p_{n-K}(x) \\
a_n^{1,0} & a_n^{1,1} & \cdots & a_n^{1,K} \\
\vdots & \vdots & \ddots & \vdots \\
a_n^{K,0} & a_n^{K,1} & \cdots & a_n^{K,K}
\end{vmatrix},
\tag{1.32}
$$

there exists $D_q \in \mathscr{A}$ with $D_q(q_n) = \lambda_n q_n$.

Notice that only the first row in the determinant on the right-hand side of (1.32) depends on x. Hence, by expanding this determinant by its first row we see that the polynomials q_n are actually a linear combination of the polynomials $p_n, \ldots, p_{n-K}$ as in (1.26).

The concept of 𝒟-operators provides the following complete and constructive answer to Question 2. First of all, we carefully have to choose the numbers $a_n^{0,j}$, $j = 1, \ldots, K$, in the first row of (1.32). More precisely, we choose $a_n^{0,j} = \xi_{n,j}$, where the numbers $\xi_{n,j}$ are defined by (1.31).

We now have a lot of freedom to choose the other sequences given by $a_n^{i,j}$, $i \geq 1$, in (1.32). Indeed, take any polynomials $R_1, \ldots, R_K$, and define $a_n^{i,j} = R_i(n - j)$, $1 \leq i \leq K$, $0 \leq j \leq K$, $0 \leq n$. With this choice, the polynomials q_n in (1.32) can be written as

$$
q_n(x) =
\begin{vmatrix}
p_n(x) & \xi_{n,1}p_{n-1}(x) & \cdots & \xi_{n,K}p_{n-K}(x) \\
R_1(n) & R_1(n-1) & \cdots & R_1(n-K) \\
\vdots & \vdots & \ddots & \vdots \\
R_K(n) & R_K(n-1) & \cdots & R_K(n-K)
\end{vmatrix}.
\tag{1.33}
$$

One can then prove that for any choice of the polynomials $R_1, \cdots, R_K$, there exists an operator $D_q \in \mathscr{A}$ such that $D_q(q_n) = \lambda_n q_n$, where q_n is defined by (1.33).

Moreover, we can do some more magic with 𝒟-operators: we can explicitly construct the operator D_q. Indeed, consider the Casorati determinant

$$
\Omega(x) = \det\left(R_l(x-j)\right)_{l,j=1}^{K},
\tag{1.34}
$$

and define the polynomial P as a solution of the first-order difference equation $P(x) - P(x-1) = \Omega(x)$. Then

$$
D_q = P(D_p) + \sum_{h=1}^{K} M_h(D_p) \circ \mathscr{D} \circ R_h(D_p),
\tag{1.35}
$$

where M_h are the polynomials defined by

$$M_h(x) = \sum_{j=1}^{K} (-1)^{h+j} \det \left(R_l(x+j-r) \right)_{l \in \mathbb{I}_h; r \in \mathbb{I}_j}, \qquad \mathbb{I}_j := \{1, \ldots, K\} \setminus \{j\}.$$

(1.36)

As usual, given a polynomial $p(x) = \sum_{j=0}^{n} a_j x^j$, and an operator D, we define the operator $p(D) = \sum_{j=0}^{n} a_j D^j$.

Actually, the operator D_q can be considered as a perturbation of the operator $P(D_p)$ by adding the sum of the operators on the right-hand side of (1.35). Indeed, the operator $P(D_p)$ can be considered the main part of D_q because of the following two reasons:

(1) It provides the eigenvalues of D_q because $D_q(q_n) = P(n)q_n$.
(2) It provides the order of the operator D_q when we apply this technique of $\mathscr{D}$-operators to construct Krall polynomials from the classical and classical discrete families.

This technique of $\mathscr{D}$-operators was introduced in [14] for $K = 1$. The case for arbitrary K was worked out with my colleague M.D. de la Iglesia in [25]. In fact, in this last paper, we improved the method of $\mathscr{D}$-operators. On the one hand the determinant in (1.33) can be slightly modified so that one can use several $\mathscr{D}$-operators at once to construct more sequences of polynomials $(q_n)_n$ that are eigenfunctions of an operator in the algebra $\mathscr{A}$. On the other hand one can construct not only one operator in $\mathscr{A}$ having the polynomials $(q_n)_n$ as eigenfunctions but a whole algebra of them (in fact, I guess that this technique allows one to construct the whole subalgebra of operators in $\mathscr{A}$ for which the polynomials $(q_n)_n$ are eigenfunctions). For details see [25].

We will also want the polynomials $(q_n)_n$ to be orthogonal with respect to a weight. Hence $\deg q_n$ has to be equal to n. This implies a technical condition on the polynomials $R_1, \ldots, R_K$. Indeed, from (1.33) it follows easily that $\deg q_n = n$ if and only if $\Omega(n) \neq 0, n \geq 0$, where Ω is the polynomial defined by (1.34). This technical condition will be implicitly assumed in what follows.

Here, we will only prove the case $K = 1$ (see [14]; for a proof for arbitrary size K see [25]). The proof of this case is surprisingly easy, but more important it plainly shows from where the mysterious definition (1.27) for $\mathscr{D}$-operators is coming.

The starting point is a sequence of polynomials $(p_n)_n$, p_n of degree n, an algebra of operators $\mathscr{A}$ acting in the linear space of polynomials and an operator $D_p \in \mathscr{A}$ for which the polynomials $p_n, n \geq 0$, are eigenfunctions with linear eigenvalues (to simplify, we assume $D_p(p_n) = np_n, n \geq 0$). We stress that for any polynomial $P \in \mathbb{R}[x]$ the operator $P(D_p)$ also belongs to the algebra $\mathscr{A}$, and

$$P(D_p)(p_n) = P(n)p_n. \tag{1.37}$$

For a sequence of numbers $(\beta_n)_n$, we consider the polynomials ((1.26) for $K = 1$)

$$q_n = p_n + \beta_n p_{n-1}. \tag{1.38}$$

We want to find the structure of the sequences $(\beta_n)_n$ for which there exists an operator $D_q \in \mathscr{A}$ such that $D_q(q_n) = \lambda_n q_n$. Since the polynomials $(q_n)_n$ form a basis of $\mathbb{R}[x]$ we have that the degree of $D_q(p)$ is at most the degree of p, for $p \in \mathbb{R}[x]$. Hence for such an operator D_q we can write

$$D_q(p_n) = \sum_{j=0}^{n} \lambda_{n,j} p_{n-j}. \tag{1.39}$$

Using the definition of q_n (1.38), we have on the one hand

$$D_q(q_n) = D_q(p_n) + \beta_n D_q(p_{n-1}) = \sum_{j=0}^{n} \lambda_{n,j} p_{n-j} + \beta_n \sum_{j=0}^{n-1} \lambda_{n-1,j} p_{n-1-j}, \tag{1.40}$$

and on the other hand

$$D_q(q_n) = \lambda_n q_n = \lambda_n p_n + \lambda_n \beta_n p_{n-1}. \tag{1.41}$$

Comparing (1.40) and (1.41), we get

$$\lambda_{n,0} = \lambda_n,$$
$$\lambda_{n,1} + \beta_n \lambda_{n-1,0} = \lambda_n \beta_n,$$
$$\lambda_{n,2} + \beta_n \lambda_{n-1,1} = 0,$$
$$\vdots$$

This gives

$$\lambda_{n,0} = \lambda_n, \tag{1.42}$$
$$\lambda_{n,1} = \beta_n(\lambda_n - \lambda_{n-1}), \tag{1.43}$$
$$\lambda_{n,j} = (-1)^{j+1} \beta_n \beta_{n-1} \cdots \beta_{n-j+1}(\lambda_{n-j+1} - \lambda_{n-j}), \quad j \geq 1. \tag{1.44}$$

Equations (1.43) and (1.44) suggests that one should define the function R by

$$R(n-1) = \lambda_n - \lambda_{n-1}, \tag{1.45}$$

and decompose the numbers β_n in the form $\beta_n = \varepsilon_n \frac{R(n)}{R(n-1)}$, where the sequence ε_n does depend on $(p_n)_n$ but not on R. This gives from (1.44)

$$\lambda_{n,j} = (-1)^{j+1} \varepsilon_n \varepsilon_{n-1} \cdots \varepsilon_{n-j+1} R(n).$$

We can then rewrite (1.39) in the form

$$D_q(p_n) = \lambda_n p_n + R(n) \sum_{j=1}^{n} (-1)^{j+1} \varepsilon_n \varepsilon_{n-1} \cdots \varepsilon_{n-j+1} p_{n-j}. \qquad (1.46)$$

Consider now the operator $\mathscr{D}$ defined by linearity from

$$\mathscr{D}(p_n) = \sum_{j=1}^{n} (-1)^{j+1} \varepsilon_n \varepsilon_{n-1} \cdots \varepsilon_{n-j+1} p_{n-j}. \qquad (1.47)$$

Since the numbers ε_n do not depend on R, we have from (1.46)

$$D_q(p_n) = \lambda_n p_n + R(n) \mathscr{D}(p_n) = \lambda_n p_n + \mathscr{D}(R(n) p_n). \qquad (1.48)$$

Hence if R is a polynomial in n, (1.45) implies that λ_n is a polynomial in n as well, say $\lambda_n = P(n)$. Hence, using (1.37), we can rewrite (1.48) in the form

$$D_q(p_n) = P(D_p)(p_n) + \mathscr{D}(R(D_p)(p_n)).$$

This gives that the polynomials $(q_n)_n$ are eigenfunctions of the operator

$$D_q = P(D_p) + \mathscr{D} \circ R(D_p).$$

Since $P(D_p), R(D_p) \in \mathscr{A}$, we can conclude that also $D_q \in \mathscr{A}$ assuming that $\mathscr{D} \in \mathscr{A}$ as well.

If we combine the definition (1.47) for the operator $\mathscr{D}$ together with the condition $\mathscr{D} \in \mathscr{A}$, what we get is the mysterious definition of the $\mathscr{D}$-operators (1.27)! Moreover, this provides a proof for our result above for $K = 1$. Indeed, in this case we have only one polynomial R and then $\Omega(x) = R(x-1)$ (1.34). The polynomials q_n (1.33) take then the form

$$q_n(x) = p_n(x) + \varepsilon_n \frac{R(n)}{R(n-1)} p_{n-1}(x). \qquad (1.49)$$

The polynomial P then satisfies $P(x) - P(x-1) = R(x-1)$ and there is only one polynomial M_h with $M_h = 1$. Finally, for the differential operator D_q (1.35) we have

$$D_q = P(D_p) + \mathscr{D} \circ R(D_p). \qquad (1.50)$$

1.3.3 𝒟-operators of type 2

When the eigenvalues θ_n of p_n with respect to D_p are not linear in n, we have to modify the definition of the $\mathcal{D}$-operators (1.27) slightly. The reason is the following. Assume that $D_p(p_n) = \theta_n p_n$ and θ_n are not linear in n. For a polynomial P, we have now $P(D_p)(p_n) = P(\theta_n)p_n$. Hence, proceeding as in the proof of the case $K = 1$ in the previous section, the difference equation (1.45) takes the form $\lambda_n - \lambda_{n-1} = R(\theta_{n-1})$, with R a polynomial. This means that if, for instance, θ_n is quadratic in n, since $R(\theta_{n-1})$ is a polynomial of even degree, λ_n has to be a polynomial of odd degree. But we also want to have $\lambda_n = P(\theta_n)$, for a certain polynomial P, which implies that λ_n has to be a polynomial of even degree. This is obviously impossible.

This problem can be fixed with a slight change in Definition 1.27. We now have two sequences of numbers $(\varepsilon_n)_n$ and $(\sigma_n)_n$ which define by linearity the operator $\mathcal{D}$

$$\mathcal{D}(p_n) = -\frac{1}{2}\sigma_{n+1}p_n + \sum_{j=1}^{n}(-1)^{j+1}\varepsilon_n\varepsilon_{n-1}\cdots\varepsilon_{n-j+1}\sigma_{n+1-j}p_{n-j}.$$

We then say that the operator $\mathcal{D}$ is a $\mathcal{D}$-operator (of type 2) if $\mathcal{D} \in \mathscr{A}$. It is easy to see that $\mathcal{D}$ operators of type 1 (Definition 1.27) correspond with the case when the numbers σ_n are constant (do not depend on n). Using this type of $\mathcal{D}$-operators we have developed a technique similar to the one in the previous section to solve Question 1 above, but it is technically more involved and it will not be considered in these lectures (see [27] for details).

We have found $\mathcal{D}$-operators for most of the orthogonal polynomials in the Askey and q-Askey tableau. A summary is included in the following tables.

Classical family	# $\mathcal{D}$-operators
Hermite	0
Laguerre	1
Jacobi	2

q-Classical family	# $\mathcal{D}$-operators
Discrete q-Hermite	2
q-Charlier	2
q-Laguerre	2
Al-Salam–Carlitz	2
Little q-Jacobi	3
q-Meixner	3
Big q-Jacobi	4
Askey–Wilson	4

Classical discrete and Wilson family	# $\mathscr{D}$-operators
Charlier	1
Meixner	2
Krawtchouk	2
Hahn	4
Dual Hahn	3
Racah	4
Wilson	4

Exercise 1.18 Prove that the sequences defined by $\varepsilon_n = \frac{n+\alpha}{n+\alpha+\beta}$ and $\sigma_n = 2n+\alpha+\beta-1$ define a $\mathscr{D}$-operator for the Jacobi polynomials $p_n = P_n^{(\alpha,\beta)}$ and the algebra $\mathscr{A}$ of differential operators with polynomial coefficients. Moreover $\mathscr{D} = -\frac{\alpha+\beta+1}{2}I + (1-x)\frac{d}{dx}$.
(Hint: Use the following identities:

$$\frac{d}{dx}\left(P_n^{(\alpha,\beta)}(x)\right) = \frac{n+\alpha+\beta+1}{2}P_{n-1}^{(\alpha+1,\beta+1)}(x),$$

$$(n+\alpha/2+\beta/2+1)(1-x)P_n^{(\alpha+1,\beta)}(x) = (n+\alpha+1)P_n^{(\alpha,\beta)}(x)$$
$$- (n+1)P_{n+1}^{(\alpha,\beta)}(x),$$

$$(2n+\alpha+\beta+1)P_n^{(\alpha,\beta)}(x) = (n+\alpha+\beta+1)P_n^{(\alpha,\beta+1)}(x)$$
$$+ (n+\alpha)P_{n-1}^{(\alpha,\beta+1)}(x).$$

Actually we have found another $\mathscr{D}$-operator for Jacobi polynomials: the sequences $\varepsilon_n = \frac{n+\beta}{n+\alpha+\beta}$, $\sigma_n = 2n+\alpha+\beta-1$ define a $\mathscr{D}$-operator for the Jacobi polynomials $p_n = P_n^{(\alpha,\beta)}$ and the algebra $\mathscr{A}$ of differential operators with polynomial coefficients. Moreover (see [14])

$$\mathscr{D} = -\frac{\alpha+\beta+1}{2}I - (1+x)\frac{d}{dx}.$$

1.4 Constructing Krall polynomials by using $\mathscr{D}$-operators

In this section, we construct Krall polynomials using the concept of $\mathscr{D}$-operators, introduced in the previous section. The method is the same regardless of whether one is considering differential, difference (or q-difference) operators. In all these cases, we start with a family of polynomials $(p_n)_n$ in the Askey tableau and some of its $\mathscr{D}$-operators. The polynomials in this family are eigenfunctions of a second-order differential or difference operator, respectively. Fixing a positive integer K and taking arbitrary polynomials $R_1, \ldots, R_K$, the technique of $\mathscr{D}$-operators explained in Section 1.3.2 shows that the polynomials $(q_n)_n$ defined by (1.33) are eigenfunctions of higher-order differential or difference operators, respectively. These polynomials $(q_n)_n$ are not going to be orthogonal in general, but there are some good choices of the R_i that will guarantee the orthogonality of the polynomials $(q_n)_n$ with respect to a weight. For these good choices, the polynomials $(q_n)_n$ are then Krall polynomials. We will present a method for finding these good choices. In particular, using this technique we will prove Theorems 1.13 (Krall–Laguerre polynomials) and 1.14 (Krall–Charlier polynomials) in Section 1.2.

1.4.1 Back to the orthogonality

Let us recall the basic steps in the method of $\mathscr{D}$-operators described in the previous section. The starting point is a sequence of polynomials $(p_n)_n$, p_n of degree n, an algebra of operators $\mathscr{A}$ acting in the linear space of polynomials and an operator $D_p \in \mathscr{A}$ for which the polynomials $p_n, n \geq 0$, are eigenfunctions. We will consider in what follows only the simplest case when the eigenvalues of p_n with respect to D_p are linear in n. Hence to simplify, we assume $D_p(p_n) = np_n, n \geq 0$. We also have a $\mathscr{D}$-operator defined by the sequence of numbers $(\varepsilon_n)_n$ as in (1.27). To the sequence $(\varepsilon_n)_n$ we associate a new sequence of numbers given by $\xi_{n,j}$ defined in (1.31). For a fixed non-negative integer K, we take arbitrary polynomials $R_1, \ldots, R_K$, and define the polynomials q_n as

$$
q_n(x) = \begin{vmatrix} p_n(x) & \xi_{n,1}p_{n-1}(x) & \cdots & \xi_{n,K}p_{n-K}(x) \\ R_1(n) & R_1(n-1) & \cdots & R_1(n-K) \\ \vdots & \vdots & \ddots & \vdots \\ R_K(n) & R_K(n-1) & \cdots & R_K(n-K) \end{vmatrix}. \tag{1.51}
$$

One then has that there always exists an operator $D_q \in \mathscr{A}$ (which can be explicitly constructed) such that $D_q(q_n) = \lambda_n q_n$.

To stress that the polynomials q_n depend on $R_1, \ldots, R_K$ and $\mathscr{D}$, we sometimes write

$$
q_n(x) = \mathrm{Cas}_n^{\mathscr{D}, R_1, \ldots, R_K}(x).
$$

We also assume the technical condition $\Omega(n) \neq 0, n \geq 0$, where Ω is the polynomial defined by (1.34). As explained above, this condition is necessary and sufficient to guarantee that $\deg q_n = n, n \geq 0$. As we will later see, when the orthogonalizing measure for the polynomials $(q_n)_n$ is positive, this technical condition always holds.

We will apply this technique of $\mathscr{D}$-operators starting with a family $(p_n)_n$ in the Askey tableau. But it turns out that even if the polynomials $(p_n)_n$ are orthogonal, the polynomials $(\mathrm{Cas}_n^{\mathscr{D},R_1,\ldots,R_K})_n$ are in general not orthogonal (for an arbitrary choice of $R_1,\ldots,R_K$). The examples we have worked out show that they are close to being, but are not necessarily, orthogonal. More precisely: Favard's Theorem 1.2 establishes that a sequence of polynomials is orthogonal with respect to a weight if and only if they satisfy a three-term recurrence relation. It turns out that the polynomials $(\mathrm{Cas}_n^{\mathscr{D},R_1,\ldots,R_K})_n$ do not always satisfy a three-term recurrence relation but they do always satisfy a higher-order recurrence relation of the form

$$h(x)\mathrm{Cas}_n^{\mathscr{D},R_1,\ldots,R_K}(x) = \sum_{j=-H}^{H} a_{n,j}\mathrm{Cas}_{n+j}^{\mathscr{D},R_1,\ldots,R_K}(x),$$

where h is certain polynomial of degree H.

The good news is that, starting from a family in the Askey tableau, there are good choices of the polynomials $R_1,\ldots,R_K$ for which the polynomials $(\mathrm{Cas}_n^{\mathscr{D},R_1,\ldots,R_K})_n$ are orthogonal as well. In the next sections we will demonstrate how one might find these good choices.

1.4.2 Krall–Laguerre polynomials

We start with the case of the Laguerre polynomials. We have already constructed a $\mathscr{D}$-operator for the Laguerre polynomials (see Exercise 1.16 in the previous section). This $\mathscr{D}$-operator is given by the derivative $\mathscr{D} = d/dx$ and is defined from the sequence of numbers defined by $\varepsilon_n = -1$.

Using the technique of $\mathscr{D}$-operators for the case $K = 1$, we find that for an arbitrary polynomial R with $R(n-1) \neq 0, n \geq 0$, the polynomials

$$q_n(x) = L_n^\alpha(x) - \frac{R(n)}{R(n-1)}L_{n-1}^\alpha(x), \qquad (1.52)$$

are eigenfunctions of a higher-order differential operator; see (1.49). Moreover, this differential operator can be explicitly constructed: if we define the polynomial P by $P(x) - P(x-1) = R(x-1)$ then $D_q = P(D_p) + \mathscr{D} \circ R(D_p)$ (1.50). It is not difficult to see now that the order of this differential operator is $2 \deg R + 2$.

The above discussion provides a very short proof for the following theorem (Theorem 1.13 in Section 1.2) where Krall–Laguerre polynomials are constructed.

Theorem 1.19 *If $\alpha \in \mathbb{N}$, $\alpha \geq 1$, the orthogonal polynomials with respect to the measure*

$$\mu_{M,\alpha} = M\delta_0 + x^{\alpha-1}e^{-x}, \quad x > 0, \tag{1.53}$$

are eigenfunctions of a differential operator of order $2\alpha + 2$.

To prove this theorem, we will explicitly construct orthogonal polynomials with respect to the weight (1.53) and will check that they are precisely of the form (1.52) for a certain polynomial R.

Lemma 1.20 *Let $M \geq 0$ be and consider the positive measure $\mu_{M,\alpha}$ (1.53). Then the polynomials*

$$q_n(x) = L_n^\alpha(x) - \frac{\Gamma(\alpha) + M\binom{n+\alpha}{n}}{\Gamma(\alpha) + M\binom{n-1+\alpha}{n-1}} L_{n-1}^\alpha(x) \tag{1.54}$$

are orthogonal with respect to $\mu_{M,\alpha}$.

Proof To simplify the writing, set

$$\beta_n = \frac{\Gamma(\alpha) + M\binom{n+\alpha}{n}}{\Gamma(\alpha) + M\binom{n-1+\alpha}{n-1}}.$$

The proof is just a matter of computation. Indeed, for $1 \leq k \leq n-1$ with $n \geq 1$, the orthogonality of $(L_n^\alpha)_n$ with respect to $x^\alpha e^{-x}$ gives

$$\langle L_n^\alpha + \beta_n L_{n-1}^\alpha, x^k \rangle_\mu = \int_0^\infty (L_n^\alpha(x) + \beta_n L_{n-1}^\alpha(x))x^k x^{\alpha-1}e^{-x}dx$$

$$= \int_0^\infty (L_n^\alpha(x) + \beta_n L_{n-1}^\alpha(x))x^{k-1}x^\alpha e^{-x}dx = 0.$$

To compute the case $k = 0$, we use the well-known formulas $L_n^\alpha(0) = \binom{n+\alpha}{n}$ and $L_n^\alpha = \sum_{j=0}^n L_{n-j}^{\alpha-1}$. Then

$$\langle L_n^\alpha + \beta_n L_{n-1}^\alpha, 1 \rangle_\mu$$

$$= \int_0^\infty (L_n^\alpha(x) + \beta_n L_{n-1}^\alpha(x))x^{\alpha-1}e^{-x}dx + M(L_n^\alpha(0) + \beta_n L_{n-1}^\alpha(0))$$

$$= \int_0^\infty (1 + \beta_n)x^{\alpha-1}e^{-x}dx + M\left(\binom{n+\alpha}{n} + \beta_n\binom{n-1+\alpha}{n-1}\right)$$

$$= \Gamma(\alpha)(1 + \beta_n) + M\left(\binom{n+\alpha}{n} + \beta_n\binom{n-1+\alpha}{n-1}\right) = 0. \qquad \square$$

Notice that in the previous lemma, α is a real number with the only restriction $\alpha > -1$. Notice also that the orthogonal polynomials $(q_n)_n$ with respect to $\mu_{M,\alpha}$ in the Lemma are of the form (1.52), with $R(n) =$

$\Gamma(\alpha) + M\binom{n+\alpha}{n}$. Hence, for those α for which this function $R(n)$ is a polynomial, the technique of $\mathscr{D}$-operators guarantees then that the orthogonal polynomials (1.52) are also eigenfunctions of a higher-order differential operator of order $2 \deg R + 2$. That is precisely the case when α is a positive integer:

$$\Gamma(\alpha) + M\binom{n+\alpha}{n} = \Gamma(\alpha) + M\frac{(n+\alpha)!}{\alpha!(n)!} = \Gamma(\alpha) + M\frac{(n+1)_\alpha}{\alpha!} \in \mathbb{R}[n].$$

Since $\deg R = \alpha$, this provides a proof for Theorem 1.19 (the proof is also constructive because, as explained above, the differential operator can be explicitly constructed).

1.4.3 Krall discrete polynomials

The method of $\mathscr{D}$-operators can also be used to construct Krall discrete polynomials. In fact, we will use it to prove the next theorem (Theorem 1.14 in Section 1.2) which provides Krall–Charlier polynomials.

Theorem 1.21   *Let F be a finite set of positive integers satisfying, for $x \in \mathbb{N}$, $\prod_{f \in F}(x - f) \geq 0$. Then, the orthogonal polynomials with respect to the weight*

$$\sum_{x=0}^{\infty} \left(\prod_{f \in F}(x - f) \right) \frac{a^x}{x!} \delta_x, \tag{1.55}$$

are common eigenfunctions of a difference operator D of order

$$2\left(\sum_{x \in F} x - \frac{n_F(n_F - 1)}{2} + 1 \right).$$

We will use three ingredients to prove this theorem.

The first one is the method of $\mathscr{D}$-operators developed in the previous section. But we need two more ingredients because the proof of the previous theorem is not going to be as straightforward as in the Krall–Laguerre case. The reason is the following. First of all, we need to explicitly compute the orthogonal polynomials with respect to the weight (1.55). This weight is a Christoffel transform of the Charlier weight. This is good news because we have an expression at hand for the orthogonal polynomials with respect to the Christoffel transform of a weight in terms of the orthogonal polynomials with respect to that weight! This formula is due to Christoffel (see also [82]): let μ and $(p_n)_n$ be a weight and its sequence of orthogonal polynomials, respectively. Write $F = \{f_1, \ldots, f_k\}$ for a finite set of real numbers (written with increasing size), and consider the Christoffel transform μ_F of μ associated to F

$$\mu_F = (x - f_1) \cdots (x - f_k)\mu.$$

Exercise 1.22 Prove that if we assume $\Phi_n = |p_{n\,|\,j-1}(f_i)|_{i,j=1,\ldots,k} \neq 0$, then the polynomials

$$q_n(x) = \frac{1}{(x - f_1) \cdots (x - f_k)} \begin{vmatrix} p_n(x) & p_{n+1}(x) & \cdots & p_{n+k}(x) \\ p_n(f_1) & p_{n+1}(f_1) & \cdots & p_{n+k}(f_1) \\ \vdots & \vdots & \ddots & \vdots \\ p_n(f_k) & p_{n+1}(f_k) & \cdots & p_{n+k}(f_k) \end{vmatrix}, \quad (1.56)$$

are orthogonal with respect to μ_F. Moreover if we write λ_n^P and λ_n^Q for the leading coefficient of the polynomials p_n and q_n, respectively, then

$$\lambda_n^Q = (-1)^k \lambda_{n+k}^P \Phi_n.$$

The problem is that the expression (1.56) is far from being of the form (1.51), which is the standard expression coming from the method of $\mathscr{D}$-operators. Hence, the expression (1.56) is useless to apply in the method of $\mathscr{D}$-operators to the classical discrete families.

Here the second ingredient mentioned above enters into the picture. This establishes how to chose the polynomials R_j such that the polynomials $(q_n)_n$ (1.51) are also orthogonal with respect to a measure.

When the polynomials $(p_n)_n$ are the classical discrete families, this second ingredient turns into a surprising and very nice symmetry between the family $(p_n)_n$ and the polynomials R_j's. Once we have made a good choice for the R's which guarantees the orthogonality of the polynomials $(q_n)_n$ (1.51), the third and last ingredient of our method will show how to identify the weight with respect to which the polynomials $(q_n)_n$ are orthogonal.

We now explain, in some detail, the second ingredient. The main tool we use to guarantee a good choice of the polynomials R_j is a certain second-order difference equation. Indeed, on the one hand, we start with a family $(p_n)_n$ of orthogonal polynomials, and so they satisfy a three-term recurrence relation of the form

$$x p_n(x) = a_{n+1} p_{n+1}(x) + b_n p_n(x) + c_n p_{n-1}(x). \quad (1.57)$$

On the other hand, we have a $\mathscr{D}$-operator for the polynomials $(p_n)_n$ defined from the sequence $(\varepsilon_n)_n$. With the recurrence coefficients in (1.57) and the numbers ε_n, we construct the following second-order difference equation

$$-a_{n+1}\varepsilon_{n+1}Y(n+1) + b_n Y(n) - \frac{c_n}{\varepsilon_n}Y(n-1) = \lambda Y(n), \quad (1.58)$$

where λ is certain number which does not depend on n.

We now show that if we want to find polynomials R such that the polynomials $(q_n)_n$ defined by (1.51) are also orthogonal with respect to a weight, then we have to find the polynomial solutions of the difference equation (1.58). Indeed, assume that each of the polynomials R_i in (1.51) satisfies the difference equation

$$-a_{n+1}\varepsilon_{n+1}R_i(n+1) + b_nR_i(n) - \frac{c_n}{\varepsilon_n}R_i(n-1) = \lambda_iR_i(n), \qquad (1.59)$$

where λ_i depends on the polynomial R_i but not on n.

The polynomial sequence $(q_n)_n$ (1.51) is orthogonal if there exists a weight $\tilde{\rho}$ such that $\int q_n(x)x^kd\tilde{\rho}(x) = 0$, for $k = 0,\ldots,n-1$. Taking into account that only the first row in the determinant in the right-hand side of (1.51) depends on x, we have

$$
\int x^kq_n(x)d\tilde{\rho}(x)
$$

$$
= \begin{vmatrix} \int x^kp_n(x)d\tilde{\rho} & \xi_{n,1}\int x^kp_{n-1}(x)d\tilde{\rho} & \cdots & \xi_{n,K}\int x^kp_{n-K}(x)d\tilde{\rho} \\ R_1(n) & R_1(n-1) & \cdots & R_1(n-K) \\ \vdots & \vdots & \ddots & \vdots \\ R_K(n) & R_K(n-1) & \cdots & R_K(n-K) \end{vmatrix} = 0.
$$

This follows if the first row is a linear combination of the other rows, i.e., for $n \geq 0$ and $k = 0,\ldots,n-1$, there exist numbers $w_{n,k,i}$, $i = 1,\ldots,K$,

$$\xi_{n,j}\int x^kp_{n-j}(x)d\tilde{\rho} = \sum_{i=1}^K w_{n,k,i}R_i(n-j), \quad j = 0,\ldots,K. \qquad (1.60)$$

We now prove the identities (1.60) by induction on k, assuming that the case $k = 0$ holds. That is, we assume the following claim.

Claim 1.23 *There exist a weight $\tilde{\rho}$ and numbers $w_{n,0,i}$, $n \geq 0$ and $i = 1,\ldots,K$, for which*

$$\xi_{n,j}\int p_{n-j}(x)d\tilde{\rho} = \sum_{i=1}^K w_{n,0,i}R_i(n-j), \quad j = 0,\ldots,K. \qquad (1.61)$$

Then using the three-term recurrence relation (1.57), the definition $\xi_{n,j} =$

$(-1)^{j+1}\prod_{i=0}^{j-1}\varepsilon_{n-i}$, the difference equation (1.59) and (1.61) we have

$$\xi_{n,j}\int xp_{n-j}(x)d\tilde{\rho}$$

$$=\xi_{n,j}\int\left(a_{n-j+1}p_{n-j+1}+b_{n-j}p_{n-j}+c_{n-i}p_{n-j-1}\right)d\tilde{\rho}$$

$$=\int\left(-a_{n-j+1}\varepsilon_{n-j+1}\xi_{n,j-1}p_{n-j+1}+b_{n-j}\xi_{n,j}p_{n-j}-\right.$$

$$\left.c_{n-j}\frac{\xi_{n,j+1}}{\varepsilon_{n-j}}p_{n-j-1}\right)d\tilde{\rho}$$

$$=\sum_{i=1}^{K}w_{n,0,i}\left(-a_{n-j+1}\varepsilon_{n-j+1}R_i(n-j+1)+b_{n-j}R_i(n-j)-\right.$$

$$\left.\frac{c_{n-j}}{\varepsilon_{n-j}}R_i(n-j-1)\right)$$

$$=\sum_{i=1}^{K}w_{n,0,i}\lambda_iR_i(n-j)=\sum_{i=1}^{K}w_{n,1,i}R_i(n-j),$$

where $w_{n,1,i}=w_{n,0,i}\lambda_i$. Proceeding in the same way, the identity (1.60) can be proved for $k\geq1$.

It turns out that for the classical discrete families, when looking for polynomial solutions of the second-order difference equation (1.58) one finds a very nice symmetry between the family $(p_n)_n$ and the polynomials R_j. We will demonstrate this symmetry for the Charlier polynomials. In Exercise 1.17, we show that the difference operator ∇ is a $\mathcal{D}$-operator for the Charlier polynomials $(c_n^a)_n$ associated to the sequence of numbers $\varepsilon_n=1$. Using the three-term recurrence relation (1.15) for the Charlier polynomials, we find that the associated difference equation (1.58) is the following:

$$-(n+1)R_i(n+1)+(n+a)R_i(n)-aR_i(n-1))=\lambda_iR_i(n).$$

By writing $x=-n-1$ we get

$$xR_i(-x)+(-x-1+a)R_i(-x-1)-aR_i(-x-2)=\lambda_iR_i(-x-1),$$

which can be rewritten in the form

$$xR_i(-x)-(x-a)R_i(-x-1)-aR_i(-x-2)=(\lambda_i+1)R_i(-x-1). \quad (1.62)$$

If we compare now with the second-order difference equation for the Charlier polynomials

$$xc_i^a(x-1)-(x+a)c_i^a(x)+ac_i^a(x+1)=-ic_i^a(x), \quad i\geq0,$$

we can conclude that by setting $R_i(-x-1) = c_i^{-a}(x)$, the identity (1.62) holds for $\lambda_i = -i-1$. This allows us to conclude that the polynomial $R_i(x) = c_i^{-a}(-x-1)$, $i \geq 0$, is a solution of the difference equation (1.62) for $\lambda_i = -i-1$.

This is the nice symmetry mentioned above for the Charlier polynomials. When we start with the Charlier polynomials $p_n = c_n^a$, $R_i(x) = c_i^{-a}(-x-1)$ are promising candidates for the polynomials R_i in (1.51), such that the polynomial sequence $(q_n)_n$ is orthogonal as well. That is: we have to choose the polynomials R_i to be Charlier polynomials again but with a change of sign in the parameter and a linear change in the variable!

But we still have some work to do, because we have not yet proved Claim 1.23. It is here that the third and last ingredient enters into the picture. With it, we identify the measure $\tilde{\rho}$ with respect to which the polynomial sequence $(q_n)_n$ defined by (1.51) is orthogonal, and prove Claim 1.23. First, we chose a finite set of non-negative integers $G = \{g_1, \ldots, g_m\}$ and define the polynomials $R_i(x) = c_{g_i}^{-a}(-x-1)$. With these polynomials R_i we construct the polynomials q_n defined by (1.51). This polynomial sequence $(q_n)_n$ depends on the set of indices G. According to the Theorem 1.21, the orthogonalizing measure for $(q_n)_n$ has the form

$$\prod_{f \in F}(x-f)\rho_a,$$

for a certain finite set F of positive integers. The third ingredient establishes the relationship between the sets F and G. This relationship is given by the involution defined by

$$I(F) = \{1, 2, \ldots, f_k\} \setminus \{f_k - f, f \in F\}, \tag{1.63}$$

where $f_k = \max F$ and k is the number of elements of F.

For the involution I, the bigger the holes in F (with respect to the set $\{1, 2, \ldots, f_k\}$), the bigger the involuted set $I(F)$. Here are a couple of examples

$$I(\{1, 2, 3, \ldots, k\}) = \{k\}, \qquad I(\{1, k\}) = \{1, 2, \ldots, k-2, k\}.$$

We now put together the three ingredients to prove Theorem 1.21.

The first ingredient is the $\mathscr{D}$-operator for the Charlier polynomials (Exercise 1.17): $\mathscr{D} = \nabla$ and is defined from the sequence of numbers defined by $\varepsilon_n = 1$, and the associated sequence is then defined by $\xi_{n,j} = (-1)^{j+1}$ (1.31). According to the method of $\mathscr{D}$-operators for arbitrary polynomials

$R_1, \ldots, R_m$, the polynomials q_n defined by

$$
q_n(x) = \begin{vmatrix} c_n^a(x) & -c_{n-1}^a(x) & \cdots & (-1)^m c_{n-m}^a(x) \\ R_1(-n-1) & R_1(-n) & \cdots & R_1(-n+m-1) \\ \vdots & \vdots & \ddots & \vdots \\ R_m(-n-1) & R_m(-n) & \cdots & R_m(-n+m-1) \end{vmatrix}
$$

are eigenfunctions of a higher-order difference operator. This difference operator can be explicitly constructed using (1.35), from where one can prove that its order is equal to

$$
2 \left(\sum_{x \in F} x - \frac{n_F(n_F-1)}{2} + 1 \right)
$$

(see [25] for the details).

The second ingredient provides the polynomial sequence R_i for which the polynomials $(q_n)_n$ is also orthogonal with respect to a weight. After solving the associated difference equation (1.58), we found that

$$
R_i(x) \in \{c_0^{-a}(-x-1), c_1^{-a}(-x-1), c_2^{-a}(-x-1), \ldots\}.
$$

Given the set $F = \{f_1, \ldots, f_k\}$, consider the involuted set $G = I(F) = \{g_1, \ldots, g_m\}$, where I is the involution (1.63) (third step). We then consider the polynomials

$$
q_n(x) = \begin{vmatrix} c_n^a(x) & -c_{n-1}^a(x) & \cdots & (-1)^m c_{n-m}^a(x) \\ c_{g_1}^{-a}(-n-1) & c_{g_1}^{-a}(-n) & \cdots & c_{g_1}^{-a}(-n+m-1) \\ \vdots & \vdots & \ddots & \vdots \\ c_{g_m}^{-a}(-n-1) & c_{g_m}^{-a}(-n) & \cdots & c_{g_m}^{-a}(-n+m-1) \end{vmatrix}, \tag{1.64}
$$

and the measure

$$
\rho_a^F = \sum_{x=-f_k-1}^{\infty} \left(\prod_{f \in F} (x+f_k+1-f) \right) \frac{a^{x+f_k+1}}{(x+f_k+1)!} \delta_x. \tag{1.65}
$$

One can then prove the Claim 1.23. More precisely

$$
(-1)^{j+1} \int c_{n-j}^a(x) d\rho_a^F = \sum_{i=1}^m w_{n,0,i} c_{g_i}^{-a}(-n+j-1), \quad j=0, \ldots, m,
$$

where

$$
w_{n,0,i} = \frac{(-1)^{n+m} e^a a^{g_m}}{p'(g_i+1) c_{g_i}^{-a}(0)}, \quad n \geq 0,
$$

and p is the polynomial $p(x) = \prod_{i=1}^m (x - g_i - 1)$.

Hence, since $R_i(x) = c_{g_i}^{-a}(-x-1)$ satisfies the associated second-order difference equation for the R_i and the Claim 1.23 for the measure ρ_a^F, the polynomials $(q_n)_n$ are orthogonal with respect to the measure ρ_a^F. Theorem 1.21 follows after taking into account that the measure (1.55) is equal to $\rho_a^F(x - f_k - 1)$.

1.5 First expansion of the Askey tableau. Exceptional polynomials: discrete case

1.5.1 Comparing the Krall continuous and discrete cases (roughly speaking): Darboux transform

In this section we analyze the recipes we have shown in Section 1.2 for constructing Krall continuous (differential operators) and discrete (difference operators) polynomials. To simplify, we consider the Krall–Laguerre and Krall–Charlier cases.

On the one hand (see below), the Krall–Laguerre measure is obtained by assuming that the parameter α is a positive integer, dividing the Laguerre weight by x and adding a Dirac delta at 0 (Geronimus transform):

$$M\delta_0 + \frac{x^\alpha}{x}e^{-x}, \quad \alpha = 1, 2, \ldots, \text{ and } M \geq 0. \tag{1.66}$$

Notice that 0 is the endpoint of the Laguerre weight. Taking into account what we explained in Section 1.2.4 on the Darboux transform for Jacobi matrices, Krall–Laguerre weights are obtained by applying a single Darboux transform to the Laguerre Jacobi matrix at the endpoint of the Laguerre weight and assuming that the parameter α is a positive integer. The Krall–Laguerre family has two parameters: the Laguerre parameter α which is now assumed to be a positive integer and the new parameter M which corresponds to the free parameter in the Darboux transform (the parameter denoted by M_{β_1}) in (1.24).

On the other hand, Krall–Charlier weights are obtained by applying the Christoffel transform associated to a finite set F of positive integers to the Charlier measure. For instance, for $F = \{1, 2\}$, we get the Charlier weight

$$\sum_{x=0}^{\infty} \frac{(x-1)(x-2)a^x}{x!}\delta_x, \quad a > 0. \tag{1.67}$$

The Krall–Charlier family also has two parameters: the Charlier parameter a, with only the assumption $a > 0$ as in the Charlier weight, and the finite set F of positive integers (which is equal to $\{1, 2\}$ for this particular example).

At first glance, both procedures seem to be rather different, but one should take into account that appearances can be very deceptive! Indeed, we can rewrite the Krall–Charlier weight (1.67) as follows

$$\sum_{x=0}^{\infty} \frac{(x-1)(x-2)a^x}{x!}\delta_x = 2\delta_0 + \sum_{x=3}^{\infty} \frac{a^x}{x(x-3)!}\delta_x = 2\delta_0 + \frac{a^3}{x}\rho_a(x-3),$$

where $\rho_a = \sum_{x=0}^{\infty} \frac{a^x}{x!}\delta_x$ is the Charlier measure.

Shifting three units to the left, we find the weight

$$2\delta_{-3} + \frac{a^3}{(x+3)}\rho_a, \quad a > 0. \tag{1.68}$$

According to Section 1.2.4, this is a particular case of the family obtained by applying a Darboux transform at -3 to the Jacobi matrix of the Charlier polynomials (or, equivalently, applying a Geronimus transform at -3 to the Charlier weight).

Comparing (1.66) and (1.68), both procedures for generating Krall–Laguerre and Krall–Charlier weights seem now to be rather similar, but one should again take into account that appearances can be very deceiving! Indeed, if we compare the structure of the Krall–Laguerre weight (1.66) with that of the Krall–Charlier weight (1.68), we can see three important differences.

(1) The parameter of the family. In the Krall–Laguerre case the parameter α changes from being a real number with $\alpha > -1$ for the Laguerre polynomials, to being a positive integer for the Krall–Laguerre polynomials. In the Krall–Charlier case, the assumption on the parameter a of being a positive real number is the same for both the Charlier and Krall–Charlier polynomials.

(2) The location of the Dirac delta. At which point λ do we apply the Darboux transform? In the Krall–Laguerre case, 0 is the endpoint of the (convex hull) of the support of the Laguerre weight. But in the Krall–Charlier case, -3 is by no means the endpoint of the (convex hull) of the support of the Charlier weight.

(3) The free parameter of the Darboux transform has disappeared in the Krall–Charlier case. Indeed, as explained above, in the Krall–Laguerre case, the free parameter, which the Darboux transform introduces, corresponds to the mass M of the Dirac delta δ_0. In the same way, if we apply a Darboux transform to the Charlier weight at -3 we get the measures (see (1.24)):

$$M_{\beta_1}\delta_{-3} + \frac{a^3}{x+3}\sum_{x=0}^{\infty}\frac{a^x}{x!}\delta_x, \quad a > 0. \tag{1.69}$$

Notice that the Krall–Charlier weight (1.68) is a particular case of (1.69) for $M_{\beta_1} = 2$. We have computational evidence which shows that when $M_{\beta_1} \neq 2$. The orthogonal polynomials with respect to (1.69) seem not to be eigenfunctions of any finite-order difference operators.

1.5.2 First expansion of the Askey tableau

We can produce a first expansion of the Askey tableau by including in it the Krall polynomials (see Figure 1.5). The two bold arrows on the left of Figure 1.5 have the following meanings. As explained in Section 1.2.1, one can navigate through the Askey scheme by passing to the limits in the parameters. Some examples of these limits are included in Section 1.2.1. In particular, one can get the classical families by taking limits of the classical discrete families. A natural question concerning our expansion of the Askey tableau is: can one get in the same form the Krall polynomials (at the north-west part of the expanded tableau) from the Krall discrete polynomials (at the south-west part of the expanded tableau)? The answer is no. As we will show below when we take limits in the Krall discrete polynomials, proceeding as in the Askey tableau, we get classical polynomials instead of Krall polynomials. This is the meaning of the struck and bend arrows on the left in the tableau above, respectively.

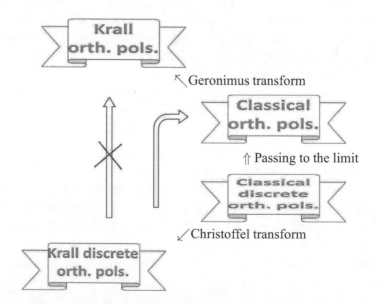

Figure 1.5

The purpose of this section is to show that one can actually get the Krall polynomials from the Krall discrete polynomials by taking limits in a different way from the procedure used in the Askey scheme. For simplicity, we display an example of those limits (see [22]). Consider the Krall–Laguerre polynomials $(L_n^{\kappa;u})_n$ orthogonal with respect to the weight

$$x^{\kappa-1}e^{-x} + \kappa!u\delta_0, \quad \kappa \in \mathbb{N}, \ \kappa \geq 1, \ u \geq 0.$$

They are eigenfunctions of a $2\kappa + 2$-order differential operator. They can be expanded in terms of Laguerre polynomials (see Lemma 1.20)

$$L_n^{\kappa;u}(x) = L_n^{\kappa}(x) - \frac{(\kappa-1)! + u(n+1)_\kappa}{(\kappa-1)! + u(n)_\kappa} L_{n-1}^{\kappa}(x). \tag{1.70}$$

We also consider the Krall–Meixner polynomials $(m_n^{a,c;\kappa})_n$ orthogonal with respect to the weight

$$\sum_{x=0}^{\infty} \prod_{j=1}^{\kappa}(x+c-j)\frac{a^x\Gamma(x+c-\kappa-1)}{x!}\delta_x,$$

$$\kappa \in \mathbb{N}, \ \kappa \geq 1, \ 0 < a < 1, \ c > \kappa+1.$$

They were introduced in [14] where, using the technique of $\mathscr{D}$-operators, we proved that they are eigenfunctions of a $2\kappa + 2$-order difference operator and can be expanded in terms of two consecutive Meixner polynomials

$$m_n^{a,c;\kappa}(x) = m_n^{a,c}(x) - \frac{am_\kappa^{1/a,2-c}(-n-1)}{(a-1)m_\kappa^{1/a,2-c}(-n)}m_{n-1}^{a,c}(x). \tag{1.71}$$

Let us remind the reader of the limit relationship between Meixner and Laguerre polynomials

$$\lim_{a\to1}(a-1)^n m_n^{a,c}\left(\frac{x}{1-a}\right) = L_n^{c-1}(x). \tag{1.72}$$

If we proceed in the same way with the Krall–Meixner polynomials (1.71), we get the following.

Exercise 1.24

$$\lim_{a\to1}(a-1)^n m_n^{a,c;\kappa}\left(\frac{x}{1-a}\right) = \begin{cases} L_n^{c-2}(x), & c \neq 2,\ldots,\kappa+1, \\ L_n^{c-1}(x) - \frac{n+\kappa}{n}L_{n-1}^{c-1}(x), & c = 2,\ldots,\kappa+1. \end{cases}$$

Hint: use the hypergeometric representation (1.16) for the Meixner polynomials to show that

$$\lim_{a \to 1} (a-1)^\kappa m_\kappa^{1/a,2-c}(z) = (-1)^\kappa \binom{\kappa+1-c}{\kappa}, \quad c \neq 2, \ldots, \kappa+1,$$

$$\lim_{a \to 1} (a-1)^{\kappa+1-c} m_\kappa^{1/a,2-c}(z) = (-1)^{\kappa+1-c} \binom{z}{c-1}, \quad c = 2, \ldots, \kappa+1.$$

Notice that for $c = 2, \ldots, \kappa + 1$ the limit above is the degenerate case of the polynomials $L_n^{\kappa;u}$ (1.70) for $u = \infty$.

We next show how the Krall–Laguerre polynomials $L_n^{\kappa;u}$ (1.70) can be reached from the Krall–Meixner polynomials $m_n^{a,c;\kappa}$ (1.71) by taking a limit but in a different way as in the Askey tableau. For the benefit of the reader we first consider the case $\kappa = 1$. The numerator in the coefficient of $m_{n-1}^{a,c}$ in (1.71) is $m_1^{1/a,2-c}(-n-1)$. This can explicitly be computed:

$$m_1^{1/a,2-c}(x) = x + \frac{c-2}{a-1}.$$

We see then that if we write $c = \phi(a)$ with $\phi(1) = 2$, we have

$$\lim_{a \to 1} \frac{c-2}{a-1} = \phi'(1),$$

and $\phi'(1)$ can be used as a free parameter. Hence, by setting $c = 2 + (1 - a)/u$, we straightforwardly get using (1.71), (1.72) and (1.70):

$$\lim_{a \to 1} (a-1)^n m_n^{a,2+(1-a)/u;1}\left(\frac{x}{1-a}\right) = L_n^{1;u}(x).$$

In order to manage the case for an arbitrary positive integer κ, we need the following limit.

Exercise 1.25

$$\lim_{a \to 1} m_\kappa^{1/a,-k+1-(1-a)^\kappa/u}(z) = \frac{(-1)^\kappa}{\kappa u} + \frac{(z-\kappa+1)_\kappa}{\kappa!}.$$

Hint: use again the hypergeometric representation (1.16) for the Meixner polynomials.

Using this limit and (1.71), (1.72) and (1.70) we get

$$\lim_{a \to 1} (a-1)^n m_n^{a,\kappa+1+(1-a)^\kappa/u;\kappa}\left(\frac{x}{1-a}\right) = L_n^{\kappa;u}(x).$$

In summary, to navigate from classical discrete to classical polynomials in the Askey tableau, one takes a limit in only one parameter of the classical discrete family while the other parameters are fixed and independent of the parameter in which one takes the limit. In the expanded Askey tableau we can also navigate from Krall discrete polynomials to Krall polynomials but one has to carefully choose suitable functions and equate all those parameters to such functions by the parameter in which one takes the limit. This is indicated in the scheme in Figure 1.6 by inserting an arrow from south-east to north-east and writing in italics the phrase *Passing to the limit*: the italics stress that by taking a limit we mean something slightly different to the way we take the limit in the Askey tableau.

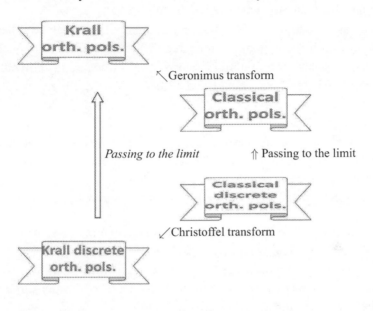

Figure 1.6

1.5.3 Exceptional polynomials

In what follows, we expand the Askey tableau by including exceptional polynomials. Although some examples of exceptional polynomials were investigated back in the early 1990s, [30], their systematic study started a few years ago with the full classification for codimension one, [36, 37]. Quesne identified the role of Darboux transformations in the construction process and discovered new examples, [76], and, soon after, Odake and Sasaki improved the method, [69]. The role of Darboux transformations was further clarified in a number of works, [38, 35, 40, 78]. Other equivalent approaches for building exceptional polynomial systems have been developed in the physics literature using the prepotential approach, [47], or the symmetry group preserving the form of the Rayleigh–Schrödinger equation, [41]. This led to rational extensions of the well-known solvable potentials. The mathematical physics community has played an important role in the conception and development of these ideas, and exceptional polynomial systems have found physical applications mostly as solutions to exactly solvable quantum mechanical problems, describing both bound states and scattering amplitudes. Exceptional orthogonal polynomials have been applied to shape-invariant potentials, [76]; to supersymmetric transformations, [38]; to discrete quantum mechanics, [70]; to mass-dependent potentials, [66]; and to quasi-exact solvability, [83].

As far as the author is aware, the first example of what can be called exceptional Charlier polynomials appeared in 1999, [84]. If orthogonal discrete polynomials on non-uniform lattices and orthogonal q-polynomials are considered, then one should add [70, 71, 72, 73] where exceptional Wilson, Racah, Askey–Wilson and q-Racah polynomials are considered (the list of references is not exhaustive).

However, exceptional polynomials are going to be constructed here by using the unexpected connection between exceptional discrete polynomials and Krall discrete polynomials I discovered in [15] (see also [16, 19, 21, 29]). Let us recall the basic definition of exceptional polynomials.

Definition 1.26 Let X be a subset of non-negative integers such that the set $\mathbb{N} \setminus X$ is finite. The polynomials p_n, $n \in X$, with $\deg p_n = n$, are called exceptional polynomials if they are orthogonal with respect to a weight and, in addition, they are also eigenfunctions of a second-order differential or difference operator.

Since we now have gaps in the degrees, the differential or difference operators can have rational coefficients. Here is an example (they are called Type II exceptional Laguerre polynomials, see [39]). Fix α and $m \geq 1$ with $\alpha > m - 1$ and consider the polynomial of degree n

$$q_n(x) = -xL_{m-1}^{-\alpha}(x)L_{n-m}^{\alpha+1}(x) - (\alpha+1+n-m)L_m^{-\alpha-1}(x)L_{n-m}^{\alpha}(x), \quad n \geq m.$$

Notice that the degrees $n = 0, \ldots, m-1$ are missing. They are orthogonal with respect to the weight

$$\frac{x^\alpha e^{-x}}{(L_m^{-\alpha-1}(x))^2}, \quad x > 0,$$

and also eigenfunctions of the second-order differential operator

$$D_\alpha(y) + 2x(\log L_m^{-\alpha-1}(x))'(y - y') - my.$$

Let us remind the reader of the Krall–Laguerre polynomials

$$q_n(x) = L_n^\alpha(x) - \frac{\Gamma(\alpha) + M\binom{n+\alpha}{n}}{\Gamma(\alpha) + M\binom{n-1+\alpha}{n-1}} L_{n-1}^\alpha(x),$$

where α is now a positive integer. They are orthogonal with respect to the weight $\mu = x^{\alpha-1}e^{-x} + M\delta_0$ and are eigenfunctions of a differential operator of order $2\alpha + 2$.

If we compare these exceptional Laguerre polynomials with the Krall–Laguerre polynomials, we observe very important differences: in the range of the parameter α, in the range of the degree n, in the orthogonalizing weights, in the way they are constructed from the Laguerre polynomials, and in the order of the associated differential operators. In fact, there is not any known connection between both families. This is, of course, rather understandable because exceptional and Krall polynomials correspond to rather different extensions of the classical families.

If we include exceptional and exceptional discrete polynomials in the expanded Askey tableau above one gets the scheme in Figure 1.7. Why have we not included at the foot of the figure, between Krall discrete and exceptional discrete polynomials, a similar indication as at the top between Krall and exceptional polynomials? Because, in the former, a very exciting situation happens: Krall discrete and exceptional discrete polynomials are actually the same objects!

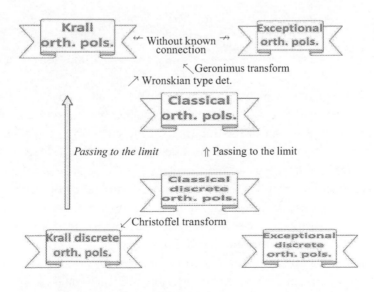

Figure 1.7

1.5.4 Constructing exceptional discrete polynomials by using duality

To show this, we need to introduce a well-known and very useful tool for the classical discrete polynomials: duality.

Definition 1.27   Given two sets of non-negative integers $U, V \subset \mathbb{N}$, we say that the two sequences of polynomials $(p_u)_{u \in U}$, $(q_v)_{v \in V}$ are dual if there exist a couple of sequences of numbers $(\xi_u)_{u \in U}$, $(\zeta_v)_{v \in V}$ such that

$$\xi_u p_u(v) = \zeta_v q_v(u), \quad u \in U, \ v \in V. \tag{1.73}$$

The notion of duality (see [62]) is actually a little bit more complicated, but since we are going to consider here only the cases of Charlier and Meixner polynomials, this simplified version is enough for our purpose. It turns out that Charlier and Meixner polynomials are self-dual for this notion of duality. This can be easily checked from the hypergeometric representation for these families of polynomials:

$$c_n^a(x) = \frac{(-a)^n}{n!} {}_2F_0\left(\begin{matrix} -n, -x \\ - \end{matrix}; -\frac{1}{a}\right),$$

$$m_n^{a,c}(x) = \frac{a^n(c)_n}{(a-1)^n n!} {}_2F_1\left(\begin{matrix} -n, -x \\ c \end{matrix}; 1 - \frac{1}{a}\right).$$

Indeed, for $n, x \in \mathbb{N}$ and except for the normalization constants, one can interchange the variables n, x in the hypergeometric functions. Hence

$$\frac{n!}{(-a)^n} c_n^a(m) = \frac{m!}{(-a)^m} c_m^a(n), \qquad n, m \in \mathbb{N}, \tag{1.74}$$

$$\frac{n!(a-1)^n}{a^n(c)_n} m_n^{a,c}(m) = \frac{m!(a-1)^m}{a^m(c)_m} m_m^{a,c}(n), \qquad n, m \in \mathbb{N}. \tag{1.75}$$

But, what happens if we apply duality to the Krall discrete polynomials? The answer is very exciting: we get exceptional discrete polynomials! A short explanation of this fact is the following. As any sequence of orthogonal polynomials, a Krall discrete family satisfies a three-term recurrence relation which, as we have explained above, can be considered as the eigenfunction equations for a second-order difference operator acting in the discrete variable n. Roughly speaking duality consists in swapping the continuous variable x with the discrete variable n; therefore the three-term recurrence relation for the Krall discrete polynomials is transformed by duality in the eigenfunction equation for a second-order difference operator acting now in the continuous variable x. Hence, duality provides a very efficient approach to constructing exceptional discrete polynomials.

To show the technique in detail we need to revisit the background of the Christoffel transform.

Let μ and $(p_n)_n$ be a weight and its sequence of orthogonal polynomials. They satisfy a three-term recurrence relation

$$xp_n(x) = a_{n+1}^P p_{n+1}(x) + b_n^P p_n(x) + c_n^P p_{n-1}(x). \qquad (1.76)$$

Given a finite set $F = \{f_1, \ldots, f_k\}$ of real numbers (written in increasing size), consider the Christoffel transform μ_F of μ associated to F

$$\mu_F = (x - f_1) \cdots (x - f_k)\mu.$$

In Section 1.4.3, Exercise 1.22, we construct orthogonal polynomials $(q_n)_n$ with respect to the measure μ_F by means of the determinantal formula

$$q_n(x) = \frac{1}{(x - f_1) \cdots (x - f_k)} \begin{vmatrix} p_n(x) & p_{n+1}(x) & \cdots & p_{n+k}(x) \\ p_n(f_1) & p_{n+1}(f_1) & \cdots & p_{n+k}(f_1) \\ \vdots & \vdots & \ddots & \vdots \\ p_n(f_k) & p_{n+1}(f_k) & \cdots & p_{n+k}(f_k) \end{vmatrix}, \qquad (1.77)$$

assuming that $\Phi_n = |p_{n+j-1}(f_i)|_{i,j=1,\ldots,k} \neq 0$. Moreover, if we write λ_n^P and λ_n^Q for the leading coefficient of the polynomials p_n and q_n, respectively, then

$$\lambda_n^Q = (-1)^k \lambda_{n+k}^P \Phi_n.$$

One can also construct the three-term recurrence relation of the polynomials $(q_n)_n$ in terms of the three-term recurrence relation for the polynomials $(p_n)_n$.

Exercise 1.28 Define the sequence of numbers $(\Psi_n)_n$ by

$$\Psi_n = \begin{vmatrix} p_n(f_1) & p_{n+1}(f_1) & \cdots & p_{n+k-2}(f_1) & p_{n+k}(f_1) \\ \vdots & \vdots & \ddots & \vdots & \vdots \\ p_n(f_k) & p_{n+1}(f_k) & \cdots & p_{n+k-2}(f_k) & p_{n+k}(f_k) \end{vmatrix}.$$

Prove that the polynomials q_n satisfy the three-term recurrence relation

$$xq_n(x) = a_n^Q q_{n+1}(x) + b_n^Q q_n(x) + c_n^Q q_{n-1}(x),$$

where

$$a_n^Q = \frac{\lambda_n^Q}{\lambda_{n+1}^Q} = \frac{\lambda_{n+k}^P}{\lambda_{n+k+1}^P} \frac{\Phi_n}{\Phi_{n+1}},$$

$$b_n^Q = b_{n+k}^P + \frac{\lambda_{n+k}^P}{\lambda_{n+k+1}^P} \frac{\Psi_{n+1}}{\Phi_{n+1}} - \frac{\lambda_{n+k-1}^P}{\lambda_{n+k}^P} \frac{\Psi_n}{\Phi_n},$$

$$c_n^Q = c_n^P \frac{\Phi_{n+1}}{\Phi_n}.$$

Moreover,

$$\langle q_n, q_n \rangle_{\mu_F} = (-1)^k \frac{\lambda_{n+k}^P}{\lambda_n^P} \Phi_n \Phi_{n+1} \langle p_n, p_n \rangle_\mu. \tag{1.78}$$

Hint: Prove the identities in this order: a_n^Q, $\langle q_n, q_n \rangle_{\mu_F}$, c_n^Q and b_n^Q.

We now rely on the above analysis to construct exceptional discrete polynomials from Krall discrete polynomials. To simplify, we first consider the case of Krall–Charlier polynomials (see [15] for a detailed explanation of what follows). In this case we have

$$\mu = \sum_{x=0}^{\infty} \frac{a^x}{x!} \delta_x, \quad p_n = c_n^a,$$

and a finite set $F = \{f_1, \ldots, f_k\}$ of positive integers (increasing size). The starting point is the Krall–Charlier weight

$$\mu_F = (x - f_1) \cdots (x - f_k)\mu.$$

Assuming $\Phi_n = |c_{n+j-1}^a(f_i)|_{i,j=1,\ldots,k} \neq 0$, the polynomials

$$q_n(x) = \frac{\begin{vmatrix} c_n^a(x - u_F) & c_{n+1}^a(x - u_F) & \cdots & c_{n+k}^a(x - u_F) \\ c_n^a(f_1) & c_{n+1}^a(f_1) & \cdots & c_{n+k}^a(f_1) \\ \vdots & \vdots & \ddots & \vdots \\ c_n^a(f_k) & c_{n+1}^a(f_k) & \cdots & c_{n+k}^a(f_k) \end{vmatrix}}{(x - u_F - f_1) \cdots (x - u_F - f_k)},$$

are orthogonal with respect to $\mu_F(x - u_F)$, where $u_F = \sum_{f \in F} f - \binom{k}{2}$.

Using the self-duality (1.74) for Charlier polynomials, one can get, from the polynomials q_n above, the new polynomials

$$c_n^{a;F}(x) = \begin{vmatrix} c_{n-u_F}^a(x) & c_{n-u_F}^a(x+1) & \cdots & c_{n-u_F}^a(x+k) \\ c_{f_1}^a(x) & c_{f_1}^a(x+1) & \cdots & c_{f_1}^a(x+k) \\ \vdots & \vdots & \ddots & \vdots \\ c_{f_k}^a(x) & c_{f_k}^a(x+1) & \cdots & c_{f_1}^a(x+k) \end{vmatrix}. \tag{1.79}$$

Notice that we do not have a polynomial $c_n^{a;F}$ for each n. Indeed, first of all $n \geq u_F$. But also $n + u_F \notin \{u_F + f_1, \ldots, u_F + f_k\}$ because otherwise the determinant has two equal rows and collapses to 0. On the other hand, if we set

$$\sigma_F = \{u_F, u_F + 1, u_F + 2, \ldots\} \setminus \{u_F + f_1, \ldots, u_F + f_k\},$$

then, it is not difficult to see that $\deg c_n^{a;F} = n$ for $n \in \sigma_F$. It is also an easy computation (using (1.74)) to show that the polynomials $(q_n)_n$ and $(c_n^{a;F})_{n \in \sigma_F}$ are dual:

$$q_m(n) = \xi_m \zeta_n c_n^{a;F}(m),$$

where

$$\xi_m = \frac{(-a)^{(k+1)m}}{\prod_{i=0}^k (m+i)!}, \qquad \zeta_n = \frac{(-a)^{-n}(n-u_F)! \prod_{f \in F} f!}{\prod_{f \in F}(n-f-u_F)}.$$

Applying this duality, the three-term recurrence relation for the Krall–Charlier polynomials $(q_n)_n$ gives a second-order difference operator acting in the continuous variable x for the polynomials $(c_n^{a;F}(x))_{n \in \sigma_F}$. More precisely, the polynomials $(c_n^{a;F}(x))_{n \in \sigma_F}$ are common eigenfunctions of the second-order difference operator

$$D_F = h_{-1}(x)\mathsf{s}_{-1} + h_0(x)\mathsf{s}_0 + h_1(x)\mathsf{s}_1, \tag{1.80}$$

where

$$h_{-1}(x) = -x\frac{\Omega_F(x+1)}{\Omega_F^a(x)}, \qquad h_1(x) = -a\frac{\Omega_F^a(x)}{\Omega_F^a(x+1)},$$

$$h_0(x) = x + k + a + u_F - a\frac{\Lambda_F^a(x+1)}{\Omega_F^a(x+1)} + a\frac{\Lambda_F^a(x)}{\Omega_F^a(x)},$$

and $\Omega_F^a(x) = |c_{f_i}^a(x+j-1)|_{i,j=1}^k$,

$$\Lambda_F^a(x) = \begin{vmatrix} c_{f_1}^a(x) & c_{f_1}^a(x+1) & \cdots & c_{f_1}^a(x+k-2) & c_{f_1}^a(x+k) \\ \vdots & \vdots & \ddots & \vdots & \vdots \\ c_{f_k}^a(x) & c_{f_k}^a(x+1) & \cdots & c_{f_k}^a(x+k-2) & c_{f_k}^a(x+k) \end{vmatrix}.$$

Moreover, $D_F(c_n^{a;F}) = nc_n^{a;F}, n \in \sigma_F$.

1.6 Exceptional polynomials: continuous case. Second expansion of the Askey tableau.

1.6.1 Exceptional Charlier polynomials: admissibility

In the previous section we constructed from the Charlier polynomials a sequence of polynomials $(c_n^{a;F}(x))_{n \in \sigma_F}$ with some gaps in the degrees which are eigenfunctions of a second-order difference operator. In this section we discuss the question of when this sequence of polynomials is orthogonal with respect to a weight.

The starting point from the construction is the Krall–Charlier weight

$$\mu_a^F = \sum_{x=0}^{\infty} (x - f_1) \cdots (x - f_k) \frac{a^x}{x!} \delta_x. \tag{1.81}$$

where $F = \{f_1, \ldots, f_k\}$ is a finite set of positive integers. Notice that this measure is positive if and only if

$$(x - f_1) \cdots (x - f_k) \geq 0, \qquad x \in \mathbb{N}.$$

When this happens, we say that the finite set of positive integers $F = \{f_1, \ldots, f_k\}$ is admissible.

Exercise 1.29 Split up the set F in maximal segments, i.e., $F = \bigcup_{i=1}^{K} Y_i$, in such a way that $Y_i \cap Y_j = \emptyset, i \neq j$, the elements of each Y_i are consecutive integers and $1 + \max Y_i < \min Y_{i+1}, i = 1, \ldots, K - 1$. Then, F is admissible if and only if each $Y_i, i = 1, \ldots, K$, has an even number of elements.

This concept of admissibility has appeared several times in the literature. Relevant here because of the relationship with exceptional polynomials are [59] and [1], where the concept appears in connection with the zeros of certain Wronskian determinants associated with eigenfunctions of second-order differential operators of the form $-d^2/dx^2 + U$. Admissibility was also considered in [52] and [84].

Admissibility is the key to studying the orthogonality of the polynomials $(c_n^{a;F}(x))_{n \in \sigma_F}$. Indeed, let us consider again the determinants

$$\Phi_n^{a;F} = |c_{n+j-1}^a(f_i)|_{i,j=1,\ldots,k}, \quad \Omega_F^a(x) = |c_{f_i}^a(x + j - 1)|_{i,j=1}^k.$$

Using the self-duality (1.74), it is easy to find the following duality between $\Phi_n^{a;F}$ and $\Omega_F^a(x)$

$$\Omega_F^a(n) = \frac{\prod_{i=0}^{k-1} (n + i)!}{(-a)^{k(n-1)-u_F} \prod_{f \in F} f!} \Phi_n^{a;F}. \tag{1.82}$$

Lemma 1.30 *The following conditions are equivalent.*

(1) *The measure μ_a^F is positive.*

(2) *The finite set F is admissible.*

(3) $\Omega_F^a(n)\Omega_F^a(n+1) > 0$ *for all non-negative integer n.*

Proof We only provide a sketch (for a complete proof, see [15]).

The equivalence between (1) and (2) is just a consequence of the definition of admissibility.

Using the formula for the norm of a sequence of orthogonal polynomials with respect to the Christoffel transform of a weight (1.78) and the duality (1.82), we get

$$\langle q_n^{a;F}, q_n^{a;F} \rangle_{\mu_a^F} = (-1)^k \frac{n!}{(n+k)!} \langle c_n^a, c_n^a \rangle_{\mu_a} \Phi_n^{a;F} \Phi_{n+1}^{a;F}$$
$$= c(n,a,F)^2 \Omega_F^a(n)\Omega_F^a(n+1),$$

where $c(n,a,F)$ is certain real number depending on n, a and F. The equivalence between (2) and (3) can be obtained by exploiting this identity. □

Theorem 1.31 *Assume F is admissible. The exceptional Charlier polynomials $(c_n^{a;F})_{n \in \sigma_F}$ are orthogonal (and complete) with respect to the weight*

$$\omega_a^F = \sum_{x=0}^{\infty} \frac{a^x}{x! \Omega_F^a(x)\Omega_F^a(x+1)} \delta_x.$$

Proof The polynomials $(q_n^{a;F})_n$ are orthogonal and complete with respect to the weight μ_a^F (they are a basis of $L^2(\mu_a^F)$).

For $n \in \sigma_F$, the function

$$h_n(x) = \begin{cases} 1/\mu_a^F(n), & x = n, \\ 0, & x \neq n, \end{cases}$$

obviously belongs to $L^2(\mu_a^F)$ and its Fourier coefficients with respect to the orthonormal basis $(q_s^{a;F}/\|q_s^{a;F}\|_2)_s$ are $q_s^{a;F}(n)/\|q_s^{a;F}\|_2, s \geq 0$. Hence, the Parseval identity gives

$$\sum_{s=0}^{\infty} \frac{q_s^{a;F}(n)q_s^{a;F}(m)}{\|q_s^{a;F}\|_2^2} = \langle h_n, h_m \rangle_{\mu_a^F} = \frac{1}{\mu_a^F(n)} \delta_{n,m}.$$

Using the duality between $(q_s^{a;F})_s$ and $(c_n^{a;F})_{n \in \sigma_F}$, this gives

$$\sum_{s=0}^{\infty} \frac{\xi_s^2 \zeta_n \zeta_m c_n^{a;F}(s)c_m^{a;F}(s)}{\|q_s^{a;F}\|_2^2} = \langle h_n, h_m \rangle_{\mu_a^F} = \frac{1}{\mu_a^F(n)} \delta_{n,m}.$$

After some careful computations, this can be rewritten as

$$\sum_{s=0}^{\infty} \frac{c_n^{a;F}(s)c_m^{a;F}(s)a^s}{s!\Omega_F^a(s)\Omega_F^a(s+1)} = \frac{a^{n-u_F-k}e^a \prod_{f\in F}(n-f-u_F)}{(n-u_F)!}\delta_{n,m};$$

that is, $(c_n^{a;F})_n$ are orthogonal with respect to ω_a^F. For a proof of the completeness of $(c_n^{a;F})_{n\in\sigma_F}$ in $L^2(\omega_a^F)$ see [15]. $\square$

1.6.2 Exceptional Hermite polynomials by passing to the limit

In the previous section, we showed how to construct exceptional discrete polynomials using duality on the Krall discrete polynomials. We next show how to construct exceptional polynomials from exceptional discrete polynomials by passing to the limit as in the Askey tableau. Recall that one can navigate thorough the Askey tableau by passing to the limit in some of the parameters of the classical and classical discrete families (see Section 1.2.1). We are interested here in the limit (1.18)

$$\lim_{a\to\infty} \left(\frac{2}{a}\right)^{n/2} c_n^a(\sqrt{(2a)}x+a) = \frac{H_n(x)}{n!},\qquad(1.83)$$

because using it we construct exceptional Hermite polynomials from exceptional Charlier polynomials.

Indeed, using that $c_n^a(x+1) - c_n^a(x) = c_{n-1}^a(x)$, we can rewrite the exceptional Charlier polynomials (1.79) in the form

$$c_n^{a;F}(x) = \begin{vmatrix} c_{n-u_F}^a(x) & c_{n-u_F-1}^a(x) & \cdots & c_{n-u_F-k}^a(x) \\ c_{f_1}^a(x) & c_{f_1-1}^a(x) & \cdots & c_{f_1-k}^a(x) \\ \vdots & \vdots & \ddots & \vdots \\ c_{f_k}^a(x) & c_{f_k-1}^a(x) & \cdots & c_{f_k-k}^a(x) \end{vmatrix}.$$

We can now pass to the limit as in the Askey tableau using (1.83). After some easy computations and taking into account that $H_n'(x) = 2nH_{n-1}(x)$, we get a nice candidate for what we can call exceptional Hermite polynomials in the form of the following Wronskian of Hermite polynomials:

$$H_n^F(x) = \mathrm{Wr}(H_{n-u_F}, H_{f_1}, \ldots, H_{f_k}) = \begin{vmatrix} H_{n-u_F}(x) & H_{n-u_F}'(x) & \cdots & H_{n-u_F}^{(k)}(x) \\ H_{f_1}(x) & H_{f_1}'(x) & \cdots & H_{f_1}^{(k)}(x) \\ \vdots & \vdots & \ddots & \vdots \\ H_{f_k}(x) & H_{f_k}'(x) & \cdots & H_{f_k}^{(k)}(x) \end{vmatrix},$$

where $n \in \sigma_F = \{u_F, u_F + 1, u_F + 2, \dots\} \setminus \{u_F + f, f \in F\}$, and $u_F = \sum_{f \in F} f - \binom{k}{2}$.

More precisely,

$$\lim_{a \to +\infty} \left(\frac{2}{a}\right)^{n/2} c_n^{a;F}\left(\sqrt{2ax} + a\right) = \frac{1}{(n - u_F)! 2^{\binom{k+1}{2}} \prod_{f \in F} f!} H_n^F(x). \quad (1.84)$$

One can also work out the second-order difference operator for the exceptional Charlier polynomials $c_n^{a;F}$ (1.80) and passing to the limit we can prove that the polynomials H_n^F are eigenfunctions of the second-order differential operator

$$D_F = -\left(\frac{d}{dx}\right)^2 + 2\left(x + \frac{\Omega_F'(x)}{\Omega_F(x)}\right)\frac{d}{dx} + 2\left(k + u_F - x\frac{\Omega_F'(x)}{\Omega_F(x)}\right) - \frac{\Omega_F''(x)}{\Omega_F(x)},$$

where

$$\Omega_F(x) = \mathrm{Wr}(H_{f_1}, \dots, H_{f_k}).$$

Exceptional Hermite polynomials are *formally* orthogonal with respect to the weight

$$\omega_F = \frac{e^{-x^2}}{\Omega_F^2(x)}, \quad x \in \mathbb{R}, \qquad \Omega_F(x) = \mathrm{Wr}(H_{f_1}, \dots, H_{f_k}),$$

in the sense that this weight satisfies a Pearson equation $(a_2 \omega_F)' = a_1 \omega_F$ where a_2 and a_1 are the coefficients of the second-order differential operator D_F. More precisely

$$(\omega_F(x))' = -2\left(x + \frac{\Omega_F'(x)}{\Omega_F(x)}\right)\omega_F(x).$$

This can be used to prove the orthogonality of H_n^F with respect to ω_F through integration by parts in $\int H_n^F(x) H_k^F(x) d\omega_F(x)$ (in a similar form as the integration by parts needed to prove (2) $\Rightarrow$ (1) in Theorem 1.6 in Section 1.1). Of course, to perform this integration by parts, ω_F has to be an integrable function, that is, the polynomial Ω_F in the denominator of ω_F cannot vanish in the real line. This question is equivalent to the regularity of the operator D_F (no singular points in $\mathbb{R}$).

The problem of studying the real zeros of the Wronskian Ω_F is a very interesting one. Questions related to the real zeros of a Wronskian whose entries are orthogonal polynomials were first studied by Karlin and Szegő in 1960 in their famous paper [52] for the particular case when F is a segment (i.e., the elements of F are consecutive integers). A few years before that, the Ukrainian mathematician M.G. Krein had found a characterization for the finite sets F for which $\Omega_F(x) \neq 0, x \in \mathbb{R}$, [59]. This happens if and only if F is an admissible set. This was proved independently by Adler in 1994, [1].

Our approach to exceptional polynomials using duality provides an alternative proof for this result. The admissibility condition for the set F then appears in a very natural form. Indeed, as we have already explained, the starting point of our approach is the Krall–Charlier weight μ_F^a (1.81) and the admissibility condition for F is equivalent to the positivity of this weight. Using the admissibility condition, one can prove that $\Omega_F(x) \neq 0, x \in \mathbb{R}$, by taking limit in the weight ω_F^a for the exceptional Charlier polynomials. To work out this limit is much more complicated than the limit (1.84) for the polynomials (where only algebraic questions are involved). For details, the reader should consult [15]. Using this approach, we have discovered the corresponding admissibility condition for the Laguerre and Jacobi exceptional polynomials. This solved one of the most important open problem in this area; see [16, 21].

1.6.3 Exceptional Meixner and Laguerre polynomials

In this section we summarize without any proof the approach to exceptional Laguerre polynomials using duality (details can be seen in [16]). The starting point is the Krall–Meixner polynomials. They are constructed by applying a suitable Christoffel transform to the Meixner weight. In this case we consider a couple $\mathscr{F} = (F_1, F_2)$ of finite sets F_1 and F_2 of positive integers. We write $k_1 = |F_1|, k_2 = |F_2|$ and $k = k_1 + k_2$, and define the number

$$u_{\mathscr{F}} = \sum_{f \in F_1} f + \sum_{f \in F_2} f - \binom{k_1 + 1}{2} - \binom{k_2}{2}.$$

The Krall–Meixner weight is then defined by

$$\rho_{a,c}^{\mathscr{F}} = \sum_{x = u_{\mathscr{F}}}^{\infty} \prod_{f \in F_1} (x - f - u_{\mathscr{F}}) \prod_{f \in F_2} (x + c + f - u_{\mathscr{F}}) \frac{a^{x - u_{\mathscr{F}}} \Gamma(x + c - u_{\mathscr{F}})}{(x - u_{\mathscr{F}})!} \delta_x.$$

$$(1.85)$$

Orthogonal polynomials with respect to this weight are eigenfunctions of higher-order difference operators. This was first conjectured by the author in 2012, [13]. A year later, I proved the conjecture for the particular case when one of the sets is empty and the other is equal to $\{1, \ldots, k\}$, [14]. In collaboration with M.D. de la Iglesia, we proved the conjecture in [25]. In the last two papers, the method of $\mathscr{D}$-operators was used. Meixner poly-nomials have two $\mathscr{D}$-operators. Namely $\mathscr{D}_1 = \dfrac{1}{1-a}\nabla$ associated to the sequence of numbers defined by $\varepsilon_n = \dfrac{1}{1-a}$ and $\mathscr{D}_2 = \dfrac{a}{1-a}\Delta$ defined by $\varsigma_n = \dfrac{a}{1-a}$.

Krall–Meixner polynomials $q_n^{\mathscr{F}}$, orthogonal with respect to the weight $\rho_{a,c}^{\mathscr{F}}$, can be computed using (1.56); thus,

$$
q_n^{\mathscr{F}}(x) = \frac{
\begin{vmatrix}
m_n^{a,c}(x-u_{\mathscr{F}}) & m_{n+1}^{a,c}(x-u_{\mathscr{F}}) & \cdots & m_{n+k}^{a,c}(x-u_{\mathscr{F}}) \\
\begin{bmatrix} m_n^{a,c}(f) \\ f \in F_1 \end{bmatrix} & m_{n+1}^{a,c}(f) & \cdots & m_{n+k}^{a,c}(f) \\
\begin{bmatrix} m_n^{a,c}(-f-c) \\ f \in F_2 \end{bmatrix} & m_{n+1}^{a,c}(-f-c) & \cdots & m_{n+k}^{a,c}(-f-c) \end{bmatrix}
\end{vmatrix}
}{
\prod_{f \in F_1}(x-f-u_{\mathscr{F}}) \prod_{f \in F_2}(x+c+f-u_{\mathscr{F}})
},
$$

$$(1.86)$$

where we use the following notation. Given a finite set of positive integers $F = \{f_1, \ldots, f_m\}$, the expression

$$
\begin{bmatrix}
z_{f,1} & z_{f,2} & \cdots & z_{f,m} \\
f \in F
\end{bmatrix}
\tag{1.87}
$$

inside of a matrix or a determinant will mean the submatrix defined by

$$
\begin{pmatrix}
z_{f_1,1} & z_{f_1,2} & \cdots & z_{f_1,m} \\
\vdots & \vdots & \ddots & \vdots \\
z_{f_m,1} & z_{f_m,2} & \cdots & z_{f_m,m}
\end{pmatrix}.
$$

The determinant (1.86) should be understood in this form.

We can construct exceptional Meixner polynomials using the self-duality of Meixner polynomials given by (1.75). Note that there is a problem in the third block in (1.86), because the number $f - c$ in which Meixner polyno-mials are evaluated is not, in general, a non-negative integer. To fix this problem, we rewrite (1.86) using the identity $m_n^{a,c}(x) = (-1)^n m_n^{1/a,c}(-x-$

c) to get

$$
q_n^{\mathscr{F}}(x) = \frac{\begin{vmatrix} m_n^{a,c}(x-u_{\mathscr{F}}) & m_{n+1}^{a,c}(x-u_{\mathscr{F}}) & \cdots & m_{n+k}^{a,c}(x-u_{\mathscr{F}}) \\ \begin{bmatrix} m_n^{a,c}(f) & m_{n+1}^{a,c}(f) & \cdots & m_{n+k}^{a,c}(f) \\ f \in F_1 \end{bmatrix} \\ \begin{bmatrix} m_n^{1/a,c}(f) & -m_{n+1}^{1/a,c}(f) & \cdots & (-1)^k m_{n+k}^{1/a,c}(f) \\ f \in F_2 \end{bmatrix} \end{vmatrix}}{(-1)^{nk_2}\prod_{f\in F_1}(x-f-u_{\mathscr{F}})\prod_{f\in F_2}(x+c+f-u_{\mathscr{F}})}.
\tag{1.88}
$$

By dualizing (1.88), we construct exceptional Meixner polynomials

$$
m_n^{a,c;\mathscr{F}}(x) = \begin{vmatrix} m_{n-u_{\mathscr{F}}}^{a,c}(x) & m_{n-u_{\mathscr{F}}}^{a,c}(x+1) & \cdots & m_{n-u_{\mathscr{F}}}^{a,c}(x+k) \\ \begin{bmatrix} m_f^{a,c}(x) & m_f^{a,c}(x+1) & \cdots & m_f^{a,c}(x+k) \\ f \in F_1 \end{bmatrix} \\ \begin{bmatrix} m_f^{1/a,c}(x) & m_f^{1/a,c}(x+1)/a & \cdots & m_f^{1/a,c}(x+k)/a^k \\ f \in F_2 \end{bmatrix} \end{vmatrix},
\tag{1.89}
$$

where $n \in \sigma_{\mathscr{F}} = \{u_{\mathscr{F}}, u_{\mathscr{F}}+1, u_{\mathscr{F}}+2,\dots\} \setminus \{u_{\mathscr{F}}+f, f \in F_1\}$. Duality gives a second-order difference equation for $m_n^{a,c;\mathscr{F}}$, $n \in \sigma_{\mathscr{F}}$, from the three-term recurrence relation of $(q_n^{\mathscr{F}})_{n\geq 0}$.

As for the Charlier case, the admissibility condition appears by assuming the Krall–Meixner weight (1.85) to be positive. Hence, we say that the real number c and the couple $\mathscr{F} = (F_1, F_2)$ of finite sets of positive integers are admissible if

$$
\prod_{f\in F_1}(x-f)\prod_{f\in F_2}(x+c+f)\Gamma(x+c) \geq 0, \quad x \in \mathbb{N}.
$$

Using the properties of the gamma function, this can be simplified to

$$
(x+c)_{\hat{c}}\prod_{f\in F_1}(x-f)\prod_{f\in F_2}(x+c+f) \geq 0, \quad x \in \mathbb{N},
$$

where $\hat{c} = \max\{-[c], 0\}$.

Using the limit (1.17) from Meixner to Laguerre polynomials, we construct exceptional Laguerre polynomials from the exceptional Meixner polynomials (1.89)

$$L_n^{\alpha;\mathscr{F}}(x) = \begin{vmatrix} L_{n-u_{\mathscr{F}}}^{\alpha}(x) & (L_{n-u_{\mathscr{F}}}^{\alpha})'(x) & \cdots & (L_{n-u_{\mathscr{F}}}^{\alpha})^{(k)}(x) \\ \begin{bmatrix} L_f^{\alpha}(x) & (L_f^{\alpha})'(x) & \cdots & (L_f^{\alpha})^{(k)}(x) \\ f \in F_1 \end{bmatrix} \\ \begin{bmatrix} L_f^{\alpha}(-x) & L_f^{\alpha+1}(-x) & \cdots & L_f^{\alpha+k}(-x) \\ f \in F_2 \end{bmatrix} \end{vmatrix},$$

where $n \in \sigma_{\mathscr{F}}$. They are eigenfunctions of the second-order differential operator

$$D_{\mathscr{F}} = x\left(\frac{d}{dx}\right)^2 + h_1(x)\frac{d}{dx} + h_0(x),$$

where

$$h_1(x) = \alpha + 1 - x + k - 2x\frac{\Omega'_{\mathscr{F}}(x)}{\Omega_{\mathscr{F}}(x)},$$

$$h_0(x) = -k_1 - u_{\mathscr{F}} + (x - \alpha - k)\frac{\Omega'_{\mathscr{F}}(x)}{\Omega_{\mathscr{F}}(x)} + x\frac{\Omega''_{\mathscr{F}}(x)}{\Omega_{\mathscr{F}}(x)},$$

and

$$\Omega_{\mathscr{F}}^{\alpha}(x) = \begin{vmatrix} \begin{bmatrix} L_f^{\alpha}(x) & (L_f^{\alpha})'(x) & \cdots & (L_f^{\alpha})^{(k-1)}(x) \\ f \in F_1 \end{bmatrix} \\ \begin{bmatrix} L_f^{\alpha}(-x) & L_f^{\alpha+1}(-x) & \cdots & L_f^{\alpha+k-1}(-x) \\ f \in F_2 \end{bmatrix} \end{vmatrix}.$$

Exceptional Laguerre polynomials are formally orthogonal in $(0, +\infty)$ with respect to the weight function

$$\omega_{\alpha;\mathscr{F}}(x) = \frac{x^{\alpha+k}e^{-x}}{(\Omega_{\mathscr{F}}^{\alpha}(x))^2}$$

(in the sense that this weight and the coefficients a_2 and a_1 of $D_{\mathscr{F}}^{\alpha}$ satisfy the Pearson equation $(a_2\omega_F^{\alpha})' = a_1\omega_F^{\alpha}$).

A real proof of this orthogonality follows from the two conditions:

(1) $\alpha + k > -1$;
(2) $\Omega_{\mathscr{F}}^{\alpha}(x) \neq 0, x > 0$.

Using our approach, one can prove that these two conditions are equivalent to the admissibility of the $\alpha + 1$ and $\mathscr{F}$ (see [16] and [29] for details).

1.6.4 Second expansion of the Askey tableau

A good summary of these lectures is the expanded Askey tableau in Figure 1.8 by the inclusion of Krall and exceptional polynomials.

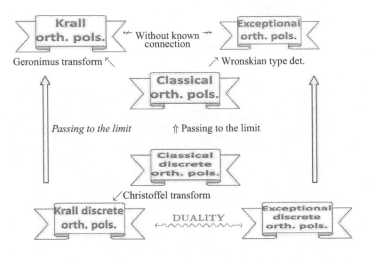

Figure 1.8

1.7 Appendix: Symmetries for Wronskian type determinants whose entries are classical and classical discrete orthogonal polynomials

In this appendix, we include some very nice symmetries and invariant properties we have found for Casoratian and Wronskian type determinants whose entries are classical and classical discrete orthogonal polynomials. These determinants satisfy some very impressive invariance properties (see [17, 20, 23, 24, 74, 75]). They have been found using different approaches. S. Odake and R. Sasaki in [74, 75] use the equivalence between eigenstate adding and deleting Darboux transformations for solvable (discrete) quantum mechanical systems. Instead I have used different approaches, from the one based in certain purely algebraic transformations of a Wronskian type determinant whose entries are orthogonal polynomials, [20], to the one using different determinantal expressions for Krall discrete polynomials, [24]. I consider here this last approach.

Let us start with the Charlier polynomials. We have obtained two determinantal identities for the Krall–Charlier polynomials. The first is coming from the Christoffel formula (1.56)

$$
q_n(x) = \frac{\begin{vmatrix} c_n^a(x) & c_{n+1}^a(x) & \cdots & c_{n+k}^a(x) \\ c_n^a(f_1) & c_{n+1}^a(f_1) & \cdots & c_{n+k}^a(f_1) \\ \vdots & \vdots & \ddots & \vdots \\ c_n^a(f_k) & c_{n+1}^a(f_k) & \cdots & c_{n+k}^a(f_k) \end{vmatrix}}{(x-f_1)\cdots(x-u_F-f_k)}.
$$

The second one is coming from the method of $\mathscr{D}$-operators (1.64)

$$
\tilde{q}_n(x) = \begin{vmatrix} c_n^a(x-f_k-1) & -c_{n-1}^a(x-f_k-1) & \cdots & (-1)^m c_{n-m}^a(x-f_k-1) \\ c_{g_1}^{-a}(-n-1) & c_{g_1}^{-a}(-n) & \cdots & c_{g_1}^{-a}(-n+m-1) \\ \vdots & \vdots & \ddots & \vdots \\ c_{g_m}^{-a}(-n-1) & c_{g_m}^{-a}(-n) & \cdots & c_{g_m}^{-a}(-n+m-1) \end{vmatrix},
$$

where $G = \{g_1, \ldots, g_m\} = I(F)$ and I is the involution (1.63).

This gives $q_n(x) = \alpha_n \tilde{q}_n(x)$, for a certain normalization constant α_n. This identity is highly non-trivial: it can not be proved by manipulating rows or columns in the determinants because the size of both determinants are, in general, very different. Indeed, the first determinant has size $(k+1) \times (k+1)$ while the second determinant has size $(m+1) \times (m+1)$, and $m = \max(F) - k + 1$.

There is a very nice invariant property of the polynomial

$$
\Omega_F^a(x) = |c_{f_i}^a(x+j-1)|_{i,j=1}^k,
$$

underlying the fact that the sequence of polynomials $(q_n)_n$ admit both determinantal definitions:

$$
\Omega_F^a(x) = (-1)^{k+u_F} \Omega_{I(F)}^{-a}(-x).
$$

For segments $F = \{n, n+1, \ldots, n+k-1\}$, I conjectured this identity in [17] and proved it in [20]. For arbitrary finite sets F, I conjectured it in [15] and proved it in [24].

This symmetry and the analogous one for the other classical discrete families are related to the celebrated Selberg integral, [33], and its extensions, and with some of the so-called constant-term identities. Here is an example of one of the latter:

$$
\mathrm{CT}_z \prod_{j=1}^n e^{-az_j} \frac{(1+z_j)^x}{z_j^{m+n-1}} \prod_{1 \le i < j \le n} (z_i - z_j)^2
$$
$$
= (-1)^{\binom{n}{2}+nm} n! \det \left(c_{n+i-1}^{-a}(-x+j-1) \right)_{i,j=1}^m,
$$

where $x \in \mathbb{N}$ (see [20]).

Passing to the limit, one can get a nice symmetry for a Wronskian whose entries are Hermite polynomials (see [24]). Write

$$\Omega_F(x) = \mathrm{Wr}(H_{f_1}, \ldots, H_{f_k}).$$

Then

$$\mathrm{Wr}(H_{f_1}(x), \ldots, H_{f_k}(x)) = \frac{i^{u_G + m} 2^{\binom{k}{2} - \binom{m}{2}} V_F}{V_G} \mathrm{Wr}(H_{g_1}(-ix), \ldots, H_{g_m}(-ix)).$$

The following example shows how different the size of both determinants in that identity can be: if $F = \{1, \ldots, k\}$, then $G = I(F) = \{k\}$ and so

$$\begin{vmatrix} H_1(x) & H_1'(x) & \cdots & H_1^{(k-1)}(x) \\ \vdots & \vdots & \ddots & \vdots \\ H_k(x) & H_k'(x) & \cdots & H_k^{(k-1)}(x) \end{vmatrix} = \left(i^k 2^{\binom{k}{2}} \prod_{j=1}^{k-1} j! \right) H_k(-ix).$$

For the Laguerre polynomials we have the following identities (see [24]). If

$$\Omega_{\mathscr{F}}^{\alpha}(x) = \left| \begin{bmatrix} L_f^{\alpha}(x) & (L_f^{\alpha})'(x) & \cdots & (L_f^{\alpha})^{(k-1)}(x) \\ f \in F_1 & & & \\ L_f^{\alpha}(-x) & L_f^{\alpha+1}(-x) & \cdots & L_f^{\alpha+k-1}(-x) \\ f \in F_2 & & & \end{bmatrix} \right|,$$

then

$$\Omega_{\mathscr{F}}^{\alpha}(x) = (-1)^{v_{\mathscr{F}}} \Omega_{\mathscr{G}}^{-\alpha - \max F_1 - \max F_2 - 2}(-x).$$

For instance, for $F_1 = \{1, \ldots, k_1\}, F_2 = \{1, \ldots, k_2\}$, we have $G_1 = I(F_1) = \{k_1\}, G_2 = I(F_2) = \{k_2\}$ and then

$$\left| \begin{bmatrix} L_f^{\alpha}(x) & (L_f^{\alpha})'(x) & \cdots & (L_f^{\alpha})^{(k-1)}(x) \\ f \in F_1 & & & \\ L_f^{\alpha}(-x) & L_f^{\alpha+1}(-x) & \cdots & L_f^{\alpha+k-1}(-x) \\ f \in F_2 & & & \end{bmatrix} \right|$$
$$= (-1)^{\binom{k}{2}} \begin{vmatrix} L_{k_1}^{-\alpha - 2k - 2}(-x) & -L_{k_1 - 1}^{-\alpha - 2k - 1}(-x) \\ L_{k_2}^{-\alpha - 2k - 2}(x) & L_{k_2}^{-\alpha - 2k - 1}(x) \end{vmatrix}.$$

References

[1] V.E. Adler, A modification of Crum's method, *Theor. Math. Phys.* **101** (1994), 1381–1386.

[2] R. Álvarez-Nodarse and A.J. Durán, Using $\mathscr{D}$-operators to construct orthogonal polynomials satisfying higher-order q-difference equations, *J. Math. Anal. Appl.* **424** (2015), 304–320.

[3] N.I. Akhiezer, *The Classical Moment Problem and Some Related Questions in Analysis* (Translated from the Russian) Oliver and Boyd, Edinburgh, 1965.

[4] W.A. Al-Salam, Characterization theorems for orthogonal polynomials. In *Orthogonal Polynomials: Theory and Practice*, P. Nevai, (ed.), Proc. Nato ASI (Columbus, Ohio) Series C: 294, Kluwer Academic Publishers, Dordrecht (1990), 1–24.

[5] H. Bavinck and H. van Haeringen, Difference equations for generalizations of Meixner polynomials, *J. Math. Anal. Appl.* **184** (1994), 453–463.

[6] H. Bavinck and R. Koekoek, On a difference equation for generalizations of Charlier polynomials, *J. Approx. Theory* **81** (1995), 195–206.

[7] S. Bochner, Über Sturm–Liouvillesche polynomsysteme, *Math. Z.* **29** (1929), 730–736.

[8] C. Brezinski, L. Gori and A. Ronveaux (eds.), *Orthogonal Polynomials and their Applications*, IMACS Annals on Computing and Applied Mathematics (No. 9), J.C. Baltzer AG, Basel, 1991.

[9] M.I. Bueno and F. Marcellán, Darboux transformations and perturbation of linear functionals. *Lin. Alg. Appl.* **384** (2004), 215–242.

[10] T.S. Chihara, *An Introduction to Orthogonal Polynomials*, Gordon and Breach, New York, 1978.

[11] J.J. Duistermaat and F.A. Grünbaum, Differential equations in the spectral parameter, *Comm. Math. Phys.* **103**, 177–240 (1986).

[12] A.J. Durán, A generalization of Favard's theorem for polynomials satisfying a recurrence relation. *J. Approx. Theory* **74** (1993), 83–109.

[13] A. J. Durán, Orthogonal polynomials satisfying higher-order difference equations. *Constr. Approx.* **36** (2012), 459-486.

[14] A.J. Durán, Using $\mathscr{D}$-operators to construct orthogonal polynomials satisfying higher-order difference or differential equations. *J. Approx. Theory* **174** (2013), 10–53.

[15] A.J. Durán, Exceptional Charlier and Hermite polynomials. *J. Approx. Theory* **182** (2014) 29–58.

[16] A.J. Durán. Exceptional Meixner and Laguerre polynomials. *J. Approx. Theory* **184** (2014) 176–208.

[17] A.J. Durán, Symmetries for Casorati determinants of classical discrete orthogonal polynomials, *Proc. Amer. Math. Soc.* **142** (2014), 915–930.

[18] A.J. Durán, Constructing bispectral dual Hahn polynomials, *J. Approx. Theory* **189** (2015) 1–28.

[19] A.J. Durán, Higher-order recurrence relation for exceptional Charlier, Meixner, Hermite and Laguerre orthogonal polynomials. *Integral Transforms Spec. Funct.* **26** (2015) 357–376.

[20] A.J. Durán, Wronskian-type determinants of orthogonal polynomials, Selberg-type formulas and constant term identities, *J. Combin. Theory Ser. A.* **124** (2014), 57–96.

[21] A.J. Durán. Exceptional Hahn and Jacobi polynomials. *J. Approx. Theory* **214** (2017) 9–48.

[22] A.J. Durán. From Krall discrete orthogonal polynomials to Krall polynomials. *J. Math. Anal. Appl.* **450** (2017), 888–900.

[23] A.J. Durán and J. Arvesú *q*-Casorati determinants of some *q*-classical orthogonal polynomials, *Proc. Amer. Math. Soc.* **144** (2016), 1655–1668.

[24] A.J. Durán, G. Curbera, Invariant properties for Wronskian type determinants of classical and classical discrete orthogonal polynomials, *J. Math. Anal. Appl.* **474** (2019), 748–764.

[25] A.J. Durán and M.D. de la Iglesia, Constructing bispectral orthogonal polynomials from the classical discrete families of Charlier, Meixner and Krawtchouk, *Constr. Approx.* **41** (2015), 49–91.

[26] A.J. Durán and M.D. de la Iglesia, Differential equations for discrete Laguerre–Sobolev orthogonal polynomials, *J. Approx. Theory* **195** (2015), 70–88.

[27] A.J. Durán and M.D. de la Iglesia, Constructing Krall–Hahn orthogonal polynomials, *J. Math. Anal. Appl.* **424** (2015), 361–384.

[28] A.J. Durán and M.D. de la Iglesia, Differential equations for discrete Jacobi–Sobolev orthogonal polynomials, *J. Spectral Theory* **8** (2018), 191–234.

[29] A.J. Durán and M. Pérez, Admissibility condition for exceptional Laguerre polynomials. *J. Math. Anal. Appl.* **424**, (2015) 1042–1053.

[30] S.Y. Dubov, V.M. Eleonskii, and N.E. Kulagin, Equidistant spectra of anharmonic oscillators, *Sov. Phys. JETP* **75** (1992), 47–53; Chaos 4 (1994) 47–53.

[31] A. Erdélyi, W. Magnus, F. Oberhettinger, F.G. Tricomi, *Higher Trascendental Functions, Volume II*. McGraw Hill, New York, 1953.

[32] J. Favard, Sur les polynomes de Tchebicheff, *C.R. Acad. Sci. Paris* **200** (1935), 2052–2053.

[33] P.J. Forrester and S. Ole Warnaar, The importance of the Selberg integral, *Bull. Amer. Math. Soc.* **45** (2008), 489–534.

[34] Ya.L. Geronimus, On the polynomials orthogonal with respect to a given number sequence and a theorem by W. Hahn, *Izv. Akad. Nauk. SSSR* **4** (1940), 215–228.

[35] D. Gómez-Ullate, Y. Grandati and R. Milson, Rational extensions of the quantum harmonic oscillator and exceptional Hermite polynomials, *J. Phys. A* **47** (2014), 015203.

[36] D. Gómez-Ullate, N. Kamran and R. Milson, An extended class of orthogonal polynomials defined by a Sturm–Liouville problem, *J. Math. Anal. Appl.* **359** (2009), 352–367.

[37] D. Gómez-Ullate, N. Kamran and R. Milson, An extension of Bochner's problem: exceptional invariant subspaces, *J. Approx. Theory* **162** (2010), 987–1006.

[38] D. Gómez-Ullate, N. Kamran and R. Milson, Exceptional orthogonal polynomials and the Darboux transformation, *J. Phys. A* **43** (2010), 434016.

[39] D. Gómez-Ullate, F. Marcellán, and R. Milson, Asymptotic and interlacing properties of zeros of exceptional Jacobi and Laguerre polynomials, *J. Math. Anal. Appl.* **399** (2013), 480–495.

[40] Y. Grandati, Multistep DBT and regular rational extensions of the isotonic oscillator, *Ann. Physics* **327** (2012), 2411-2431.

[41] Y. Grandati, Rational extensions of solvable potentials and exceptional orthogonal polynomials. In *7th International Conference on Quantum Theory and Symmetries, Journal of Physics: Conference Series* **343** (2012), 012041, 11 pp.

[42] F.A. Grünbaum, Random walks and orthogonal polynomials: some challenges. In *Probability, Geometry and Integrable Systems*, M. Pinsky and B. Birnir (eds.), Math. Sci. Res. Inst. Publ., no. 55, Cambridge University Press, 241–260 (2008).

[43] F.A. Grünbaum and L. Haine, Orthogonal polynomials satisfying differential equations: the role of the Darboux transformation. In *Symmetries and Integrability of Differential Equations*, D. Levi, L. Vinet, P. Winternitz (eds.), CRM Proc. Lecture Notes, vol. 9, Amer. Math. Soc. Providence, RI, 1996, 143–154.

[44] F.A. Grünbaum, L. Haine and E. Horozov, Some functions that generalize the Krall–Laguerre polynomials, *J. Comput. Appl. Math.* **106**, 271–297 (1999).

[45] F.A. Grünbaum and M. Yakimov, Discrete bispectral Darboux transformations from Jacobi operators, *Pac. J. Math.* **204**, 395–431 (2002).

[46] L. Haine, Beyond the classical orthogonal polynomials. In *The Bispectral Problem (Montreal, PQ, 1997)*, CRM Proc. Lecture Notes, vol. 14, Amer. Math. Soc., Providence, RI, 1998, pp. 47–65.

[47] C.-L. Ho, Prepotential approach to solvable rational extensions of harmonic oscillator and Morse potentials, *J. Math. Phys.* **52** (2011), 122107.

[48] P. Iliev, Krall–Jacobi commutative algebras of partial differential operators, *J. Math. Pures Appl.* **96** (2011) 446-461.

[49] P. Iliev, Krall–Laguerre commutative algebras of ordinary differential operators, *Ann. Mat. Pur. Appl.* **192** (2013), 203–224.

[50] M.E.H. Ismail, *Classical and Quantum Orthogonal Polynomials in One Variable*, Encyclopedia of Mathematics and its Applications volume 98, Cambridge University Press, 2009.

[51] S. Karlin and J. McGregor, Ehrenfest urn models, *J. Appl. Prob.* **2** (1965), 352–376.

[52] S. Karlin and G. Szegő, On certain determinants whose elements are orthogonal polynomials, *J. Analyse Math.* **8** (1961), 1–157.

[53] R. Koekoek, P.A. Lesky and L.F. Swarttouw, *Hypergeometric Orthogonal Polynomials and their q-Analogues*, Springer Verlag, Berlin, 2008.

[54] J. Koekoek and R. Koekoek, On a differential equation for Koornwinder's generalized Laguerre polynomials, *Proc. Amer. Math. Soc.* **112** (1991), 1045–1054.

[55] J. Koekoek and R. Koekoek, Differential equations for generalized Jacobi polynomials, *J. Comput. Appl. Math.* **126** (2000), 1–31.

[56] R. Koekoek, Differential equations for symmetric generalized ultraspherical polynomials, *Trans. Amer. Math. Soc.* **345** (1994), 47–72.

[57] H.L. Krall, Certain differential equations for Tchebycheff polynomials, *Duke Math. J.* **4** (1938), 705–718.

[58] H.L. Krall, *On orthogonal polynomials satisfying a certain fourth-order differential equation*. The Pennsylvania State College Studies, No. 6, 1940.

[59] M.G. Krein, A continual analogue of a Christoffel formula from the theory of orthogonal polynomials, *Dokl. Akad. Nauk. SSSR* **113** (1957), 970–973.

[60] K.H. Kwon and D.W. Lee, Characterizations of Bochner–Krall orthogonal polynomials of Jacobi type, *Constr. Approx.* **19** (2003), 599–619.

[61] O.E. Lancaster, Orthogonal polynomials defined by difference equations, *Amer. J. Math.* **63** (1941), 185–207.

[62] D. Leonard, Orthogonal polynomials, duality, and association schemes, *SIAM J. Math. Anal.* **13** (1982), 656–663.

[63] L.L. Littlejohn, The Krall polynomials: a new class of orthogonal polynomials, *Quaest. Math.* **5** (1982), 255–265.

[64] L. L. Littlejohn, An application of a new theorem on orthogonal polynomials and differential equations, *Quaest. Math.* **10** (1986), 49–61.

[65] P. Maroni, Tchebychev forms and their perturbed forms as second degree forms, *Ann. Numer. Math.* **2** (1995), 123–143.

[66] B. Midya and B. Roy, Exceptional orthogonal polynomials and exactly solvable potentials in position-dependent mass Schrödinger Hamiltonians, *Phys. Lett. A* **373** (2009), 4117–4122.

[67] A.F. Nikiforov and V.B. Uvarov, *Special Functions of Mathematical Physics*, Birkhäuser, Basel, 1988.

[68] A.F. Nikiforov, S.K. Suslov and V. B. Uvarov, *Classical Orthogonal Polynomials of a Discrete Variable*, Springer Verlag, Berlin, 1991.

[69] S. Odake and R. Sasaki, Infinitely many shape invariant potentials and new orthogonal polynomials, *Phys. Lett. B* **679** (2009), 414–417.

[70] S. Odake and R. Sasaki, Infinitely many shape invariant discrete quantum mechanical systems and new exceptional orthogonal polynomials related to the Wilson and Askey–Wilson polynomials, *Phys. Lett. B* **682** (2009), 130–136.

[71] S. Odake and R. Sasaki, The exceptional (X_ℓ) (q)-Racah polynomials, *Prog. Theor. Phys.* **125** (2011), 851–870.

[72] S. Odake and R. Sasaki, Dual Christoffel transformations, *Prog. Theor. Phys.* **126** (2011), 1–34.

[73] S. Odake and R. Sasaki, Multi-indexed q-Racah polynomials, *J. Phys. A* **45** (2012) 385201 (21pp).

[74] S. Odake and R. Sasaki, Krein–Adler transformations for shape-invariant potentials and pseudo virtual states, *J. Phys. A* **46** (2013) 245201 (24pp).

[75] S. Odake and R. Sasaki, Casoratian identities for the Wilson and Askey–Wilson polynomials, *J. Approx. Theory.* **193** (2015), 184–209.

[76] C. Quesne, Exceptional orthogonal polynomials, exactly solvable potentials and supersymmetry, *J. Phys. A* **41**, (2008), no. 39, 392001, 6pp.

[77] E.J. Routh, On some properties of certain solutions of a differential equation of the second order, *Proc. London Math. Soc.* **16** (1884) 245–261.

[78] R. Sasaki, S. Tsujimoto, and A. Zhedanov, Exceptional Laguerre and Jacobi polynomials and the corresponding potentials through Darboux–Crum transformations, *J. Phys. A* **43** (2010) n. 31, 315204, 20 pp.

[79] E. Schrödinger Quantisierung als Eigenwertproblem, *Annalen der Physik* **384**(4) (1926) 361–377

[80] V. Spiridonov and A. Zhedanov, Discrete Darboux transformations, discrete time Toda lattice and the Askey–Wilson polynomials, *Methods Appl. Anal.* **2** (1995), no. 4, 369398.

[81] M.H. Stone, *Linear Transformations in Hilbert Space and their Applications to Analysis*, American Mathematical Society, Providence, R.I., 1932.

[82] G. Szegő, *Orthogonal Polynomials*, Fourth Edition, American Mathematical Society, Providence, R.I., 1975.

[83] T. Tanaka, N-fold supersymmetry and quasi-solvability associated with X2-Laguerre polynomials, *J. Math. Phys.* **51** (2010), 032101.

[84] O. Yermolayeva and A. Zhedanov, Spectral transformations and generalized Pollaczek polynomials, *Methods and Appl. Anal.* **6** (1999), 261–280.

[85] G.J. Yoon, Darboux Transforms and orthogonal polynomials, *Bull. Korean Math. Soc.* **39**, (2002) 359–376.

[86] A. Zhedanov, A method of constructing Krall's polynomials, *J. Comput. Appl. Math.* **107** (1999), 1–20.

2

A Brief Review of q-Series

Mourad E.H. Ismail

Abstract: We sample certain results from the theory of q-series including summation and transformation formulas, as well as some recent results which are not available in book form. Our approach is systematic and uses the Askey–Wilson calculus and Rodrigues-type formulas.

Dedicated to the memory of Richard Askey in friendship and gratitude

2.1 Introduction

The subject of q-series is very diverse and the present lecture notes represent the author's philosophy on how to approach the subject. Other points of view are available in [4, 16, 35]. We have used our own approach, which was jointly developed over the years with several coauthors, and we owe a great deal to our coauthors, especially Dennis Stanton.

Central to our approach is the use of a calculus for the Askey–Wilson operator, and the identity theorem for analytic functions.

We built our approach to q-series on the theory of the Askey–Wilson polynomials and related topics. Other approaches use the theory of q-series to treat the Askey–Wilson polynomials but we develop both theories simultaneously and we believe it is this interaction that best motivates and explains the subject. We hope we have succeeded in demonstrating the power of the Askey–Wilson calculus which started in [7] and later developed in a series of papers by the present author and his collaborators.

One essential tool is the theory of q-Taylor series for the Askey–Wilson operator. The first attempt to use this tool was in [18]. This was followed by [24] and [25] where the theory was applied and a q-Taylor series was developed for expansions of entire functions. The latter theory requires further development. Another ingredient is the Rodrigues formulas and the usage of an explicit expression for the powers of the Askey–Wilson operator.

Recall the identity theorem for analytic functions.

Theorem 2.1   [30] *Let $f(z)$ and $g(z)$ be analytic in a domain Ω and assume that $f(z_n) = g(z_n)$ for a sequence $\{z_n\}$ converging to an interior point of Ω. Then $f(z) = g(z)$ at all points of Ω.*

Theorem 2.1 will be used to derive summation theorems and identities throughout this work by identifying a special parameter in the identity to be proven. Let b be the special parameter. Usually one can prove the desired identity when $b = \lambda q^n, n = 0, 1, \ldots$. When both sides of the desired formula are analytic in the special parameter b, we invoke Theorem 2.1 and conclude its validity in the domain of analyticity in b. We must note that one has to be careful with the use of this technique as George Gasper explains some of the pitfalls of misapplying it in his interesting article [15].

2.2 Notation and q-operators

This section contains all the notation used in the rest of these notes. The q-shifted factorials are

$$(a;q)_0 := 1, \quad (a;q)_n := \prod_{k=1}^{n}(1 - aq^{k-1}), \quad n = 1, 2, \ldots, \text{ or } \infty,$$

$$(a_1, a_2, \ldots, a_k; q)_n := \prod_{j=1}^{k}(a_j; q)_n. \tag{2.1}$$

The q-analogue of the binomial coefficient is

$$\begin{bmatrix} n \\ k \end{bmatrix}_q := \frac{(q;q)_n}{(q;q)_k (q;q)_{n-k}}. \tag{2.2}$$

Usually it is refered to as the q-binomial coefficient or the Gaussian binomial coefficient. It is clear from (2.1) that

$$(a;q)_n = (a;q)_\infty / (aq^n; q)_\infty, \qquad n = 0, 1, \ldots,$$

which suggests the following definition for q-shifted factorials of any n

$$(a;q)_z := (a;q)_\infty / (aq^z; q)_\infty. \tag{2.3}$$

It is easy to see that

$$(a;q)_{-n} = 1/(aq^{-n};q)_n, \quad (a;q)_m (aq^m; q)_n = (a;q)_{m+n}, \left.\vphantom{\begin{matrix}1\\1\end{matrix}}\right\}$$
$$m, n = 0, \pm 1, \pm 2, \ldots. \qquad\qquad (2.4)$$

Some useful identities which follow from the definitions (2.1) and (2.3) are

$$
\begin{aligned}
(aq^{-n};q)_k &= \frac{(a;q)_k(q/a;q)_n}{(q^{1-k}/a;q)_n}q^{-nk}, \\
(a;q^{-1})_n &= (1/a;q)_n(-a)^n q^{-\binom{n}{2}},
\end{aligned}
\qquad (2.5)
$$

$$
\left.
\begin{aligned}
(aq^{-n};q)_n &= (q/a;q)_n(-a)^n q^{-\binom{n+1}{2}}, \\
(a;q)_{-n} &= \frac{(-1)^n q^{\binom{n+1}{2}}}{(q/a;q)_n a^n}, \quad n=0,1,2,\ldots,
\end{aligned}
\right\}
\qquad (2.6)
$$

$$
\left.
\begin{aligned}
(a;q)_{n-k} &= \frac{(a;q)_n}{(q^{1-n}/a;q)_k}(-a)^{-k} q^{\binom{k+1}{2}-nk}, \\
\frac{(a;q)_{n-k}}{(b;q)_{n-k}} &= \frac{(a;q)_n(q^{1-n}/b;q)_k}{(b;q)_n(q^{1-n}/a;q)_k}\left(\frac{b}{a}\right)^k.
\end{aligned}
\right\}
\qquad (2.7)
$$

Unless we say otherwise we shall always assume that

$$
0 < q < 1. \qquad (2.8)
$$

The q-gamma function is

$$
\Gamma_q(z) = \frac{(q;q)_\infty}{(q^z;q)_\infty}(1-q)^{1-z}. \qquad (2.9)
$$

One can show that [3] $\lim_{q\to 1^-}\Gamma_q(z) = \Gamma(z)$. It is easy to see that

$$
\Gamma_q(z+1) = \frac{1-q^z}{1-q}\Gamma_q(z), \quad \Gamma_q(1) = \Gamma_q(2) = 1. \qquad (2.10)
$$

A **basic hypergeometric series** is

$$
\begin{aligned}
{}_r\phi_s\left(\begin{array}{c} a_1,\ldots,a_r \\ b_1,\ldots,b_s \end{array} \middle| \; q,\,z\right) &= {}_r\phi_s(a_1,\ldots,a_r;b_1,\ldots,b_s;q,z) \\
&= \sum_{n=0}^{\infty}\frac{(a_1,\ldots,a_r;q)_n}{(q,b_1,\ldots,b_s;q)_n}z^n\left(-q^{(n-1)/2}\right)^{n(s+1-r)}.
\end{aligned}
\qquad (2.11)
$$

Note that $(q^{-k};q)_n = 0$ for $n = k+1, k+2, \ldots$, when k is a non-negative integer. If one of the numerator parameters is of the form q^{-k} then the sum on the right-hand side of (2.11) is a finite sum and we say that the series in (2.11) is **terminating**. A series that does not terminate is called **nonterminating**.

The radius of convergence of the series in (2.11) is $1, 0$ or ∞ when $r = s+1, r > s+1$ or $r < s+1$, respectively, as can be seen from the ratio test. These notions extend the familiar notions of shifted and multishifted factorials

$$(a)_0 := 1, \qquad (a)_n := a(a+1)\dots(a+n-1),$$
$$\left.\begin{array}{c} \\ (a_1,\dots,a_k)_n = \prod_{j=1}^{k}(a_j)_n, \quad n=1,2,\dots \end{array}\right\} \tag{2.12}$$

and the **generalized hypergeometric functions**

$$_rF_s\left(\begin{array}{c} a_1,\dots,a_r \\ b_1,\dots,b_s \end{array} \middle| z\right) = {}_rF_s(a_1,\dots,a_r;b_1,\dots,b_s;z)$$
$$= \sum_{n=0}^{\infty} \frac{(a_1,\dots,a_r)_n}{(1,b_1,\dots,b_s)_n} z^n. \tag{2.13}$$

It is clear that $\lim_{q\to 1^-}(q^a;q)_n/(1-q)^n = (a)_n$; therefore

$$\left.\begin{array}{c} \lim_{q\to 1^-}{}_r\phi_s\left(\begin{array}{c} q^{a_1},\dots,q^{a_r} \\ q^{b_1},\dots,q^{b_s} \end{array} \middle| q,\, z(1-q)^{s+1-r}\right) \\[3mm] = {}_rF_s\left(\begin{array}{c} a_1,\dots,a_r \\ b_1,\dots,b_s \end{array} \middle| (-1)^{s+1-r}z\right), \quad r \le s+1. \end{array}\right\} \tag{2.14}$$

We shall also use the Bailey notation

$$\left.\begin{array}{c} {}_{r+3}W_{r+2}(a^2;a_1,\dots,a_r;q,z) := \\[3mm] {}_{r+3}\phi_{r+2}\left(\begin{array}{c} a^2,qa,-qa,a_1,\dots,a_r \\ a,-a,qa^2/a_1,\dots,qa^2/a_r \end{array} \middle| q,\, z\right). \end{array}\right\} \tag{2.15}$$

The ϕ function in (2.15) is called **very well-poised**.

An important tool in our treatment is the Askey–Wilson operator $\mathscr{D}_q$, which will be defined below. Given a polynomial f we set $\check{f}(e^{i\theta}) := f(x)$, with $x = \cos\theta$; that is

$$\check{f}(z) = f((z+1/z)/2), \quad z = e^{i\theta}. \tag{2.16}$$

So we think of $f(\cos\theta)$ as a function of $e^{i\theta}$. Now the Askey–Wilson divided difference operator $\mathscr{D}_q$ is

$$(\mathscr{D}_q f)(x) := \frac{\check{f}(q^{1/2}z) - \check{f}(q^{-1/2}z)}{\check{e}(q^{1/2}z) - \check{e}(q^{-1/2}z)}, \quad x = (z+1/z)/2, \tag{2.17}$$

with

$$e(x) = x. \tag{2.18}$$

The definition (2.17) easily reduces to

$$(\mathscr{D}_q f)(x) = \frac{\check{f}(q^{1/2}z) - \check{f}(q^{-1/2}z)}{(q^{1/2}-q^{-1/2})\,[z-1/z]/2}, \quad x = [z+1/z]/2. \tag{2.19}$$

Note that $\mathscr{D}_q = \mathscr{D}_{q^{-1}}$.

It is important to note that although we use $x = \cos\theta$, θ is not necessarily real. In fact $e^{i\theta}$ is defined as

$$e^{i\theta} = x + \sqrt{x^2 - 1},$$

and the branch of the square root is taken such that $\sqrt{x^2 - 1} \approx x$ as $x \to \infty$. It is clear that the Askey–Wilson operator is linear.

The Chebyshev polynomials of the first and second kind are

$$T_n(\cos\theta) = \cos(n\theta), \qquad U_n(\cos\theta) = \frac{\sin((n+1)\theta)}{\sin\theta}. \qquad (2.20)$$

It is clear that both T_n and U_n have degree n.

As an example we apply $\mathscr{D}_q$ to a Chebyshev polynomial. It is clear that $\check{T}_n(z) = (z^n + z^{-n})/2$. It is easy to see that

$$\mathscr{D}_q T_n(x) = \frac{q^{n/2} - q^{-n/2}}{q^{1/2} - q^{-1/2}} U_{n-1}(x). \qquad (2.21)$$

Therefore

$$\lim_{q\to 1}(\mathscr{D}_q f)(x) = f'(x), \qquad (2.22)$$

holds for $f = T_n$, therefore for all polynomials, since $\{T_n(x)\}_0^\infty$ is a basis for the vector space of all polynomials and $\mathscr{D}_q$ is a linear operator. Note that the Askey–Wilson operator maps theta functions (defined for instance in [43]) to theta functions. Even though this is the case, we could not find the Askey–Wilson operator in any of Jacobi's works on theta functions.

For $v > 0$ we define the function space

$$H_v = \{f : \check{f}(z) \text{ is analytic for } q^v \le |z| \le q^{-v}\},$$

and for $f, g \in H_{1/2}$ we define the inner product

$$\langle f, g \rangle = \int_{-1}^{1} f(x)\overline{g(x)}\,\frac{dx}{\sqrt{1-x^2}}. \qquad (2.23)$$

We have the following integration by parts formula for the Askey–Wilson operator [10, 19].

Theorem 2.2 ([19, Theorem 16.1.1]) *Suppose that $f, g \in H_{1/2}$. Then*

$$\langle \mathscr{D}_q f, g \rangle = \frac{\pi q^{1/2}}{1-q}\left[\check{f}(q^{1/2})\check{g}(1) - \check{f}(-q^{1/2})\check{g}(-1)\right]$$
$$- \langle f(x), (1-x^2)^{1/2}\mathscr{D}_q((1-x^2)^{-1/2}g(x))\rangle. \qquad (2.24)$$

Let η_q denote the shift operator

$$(\eta_q f)(x) = (\eta_q \check{f})(z) = \check{f}(q^{1/2}z). \qquad (2.25)$$

The Askey–Wilson operator satisfies the product rule

$$(\mathscr{D}_q(fg))(x) = (\eta_q g)(x)(\mathscr{D}_q f)(x) + (\eta_{q^{-1}} f)(x)(\mathscr{D}_q g)(x). \qquad (2.26)$$

The Leibniz rule for the operator $\mathscr{D}_q$ is, [18, (1.22)],

$$\mathscr{D}_q^n(fg) = \sum_{k=0}^{n} q^{k(k-n)/2} \begin{bmatrix} n \\ k \end{bmatrix}_q (\eta_q^k \mathscr{D}_q^{n-k} f)(\eta_q^{k-n} \mathscr{D}_q^k g), \qquad (2.27)$$

which follows from (2.26) by induction. A more symmetric form of (2.26) is

$$\left. \begin{aligned} (\mathscr{D}_q(fg))(x) &= (\mathscr{A}_q g)(x)(\mathscr{D}_q f)(x) + (\mathscr{A}_q f)(x)(\mathscr{D}_q g)(x), \\ \mathscr{A}_q &:= \frac{1}{2}[\eta_q + \eta_{q^{-1}}]. \end{aligned} \right\} \qquad (2.28)$$

An interesting formula due to S. Cooper [11] is

$$\left. \begin{aligned} (\mathscr{D}_q^n f)(x) = \\ \frac{2^n q^{n(1-n)/4}}{(q^{1/2} - q^{-1/2})^n} \sum_{k=0}^{n} \begin{bmatrix} n \\ k \end{bmatrix}_q \frac{q^{k(n-k)} z^{2k-n} \check{f}(q^{n/2-k} z)}{(q^{n-2k+1} z^2; q)_k (q^{2k-n+1} z^{-2}; q)_{n-k}}. \end{aligned} \right\} \qquad (2.29)$$

This can also be proved by induction.

2.3 q-Taylor series

We shall base our development of the theory of q-series on an Askey–Wilson calculus. In the Askey–Wilson calculus we will have a product rule, a Leibniz rule, and q-analogues of integration by parts, as well as q-Taylor series. The Askey–Wilson calculus has natural polynomial bases which play the role of $\{(x-a)^n : n = 0, 1, \ldots\}$ in calculus. With $x = \cos\theta$ they are

$$\phi_n(x; a) = (ae^{i\theta}, ae^{-i\theta}; q)_n = \prod_{k=0}^{n-1} [1 - 2axq^k + a^2 q^{2k}], \qquad (2.30)$$

$$\phi_n(x) = (q^{1/4} e^{i\theta}, q^{1/4} e^{-i\theta}; q^{1/2})_n = \prod_{k=0}^{n-1} [1 - 2xq^{1/4+k/2} + q^{1/2+k}] \qquad (2.31)$$

$$p_n(x) = (1 + e^{2i\theta}) e^{-in\theta} (-q^{2-n} e^{2i\theta}; q^2)_{n-1}, \quad n > 0, \quad p_0(x) := 1. \qquad (2.32)$$

It is worth noting that

$$\left.\begin{array}{l} \rho_{2n}(x) = q^{n(1-n)}(-e^{2i\theta}, -e^{-2i\theta}; q^2)_n, \\ \rho_{2n+1}(x) = 2q^{-n^2}\cos\theta(-qe^{2i\theta}, -qe^{-2i\theta}; q^2)_n. \end{array}\right\} \qquad (2.33)$$

Theorem 2.3 *The action of the Askey–Wilson operator on the bases* $\{\phi_n(x;a)\}$, $\{\phi_n(x)\}$, *and* $\{\rho_n(x;a)\}$ *is given by*

$$\mathscr{D}_q\phi_n(x;a) = -\frac{2a(1-q^n)}{1-q}\phi_{n-1}(x;aq^{1/2}), \qquad (2.34)$$

$$\mathscr{D}_q\phi_n(x) = -2q^{1/4}\frac{1-q^n}{1-q}\phi_{n-1}(x), \qquad (2.35)$$

$$\mathscr{D}_q\rho_n(x) = 2q^{(1-n)/2}\frac{1-q^n}{1-q}\rho_{n-1}(x). \qquad (2.36)$$

The proof follows from the definitions (2.30)–(2.32).

We next present a q-Taylor series for polynomials.

Theorem 2.4 (Ismail [18]) *Let* f *be a polynomial of degree n. Then*

$$f(x) = \sum_{k=0}^n f_k(ae^{i\theta}, ae^{-i\theta}; q)_k, \qquad (2.37)$$

where

$$f_k = \frac{(q-1)^k}{(2a)^k(q;q)_k}q^{-k(k-1)/4}(\mathscr{D}_q^k f)(x_k) \qquad (2.38)$$

with

$$x_k := \frac{1}{2}\left(aq^{k/2}+q^{-k/2}/a\right). \qquad (2.39)$$

Proof It is clear that the expansion (2.37) exists, so we now compute the f_k. Formula (2.34) yields

$$\mathscr{D}_q^k(ae^{i\theta}, ae^{-i\theta}; q)_n|_{x=x_k}$$

$$= (2a)^k\frac{q^{(0+1+\cdots+k-1)/2}(q;q)_n}{(q-1)^k(q;q)_{n-k}}(aq^{k/2}e^{i\theta}, aq^{k/2}e^{-i\theta}; q)_{n-k}|_{e^{i\theta}=aq^{k/2}}$$

$$= \frac{(q;q)_k}{(q-1)^k}(2a)^kq^{k(k-1)/4}\delta_{k,n}.$$

The theorem now follows by applying $\mathscr{D}_q^j$ to both sides of (2.37) then setting $x = x_j$. $\qquad\qquad\square$

Similarly one can prove the following q-Taylor expansions from [24].

Theorem 2.5 *If f is a polynomial of degree n then*

$$f(x) = \sum_{k=0}^{n} f_k(\phi)\phi_k(x), \; f_k(\phi) := \frac{(q-1)^k q^{-k/4}}{2^k (q;q)_k}(\mathscr{D}_q^k f)(\zeta_0), \quad (2.40)$$

$$f(x) = \sum_{k=0}^{n} f_k(\rho)\rho_k(x), \; f_k(\rho) := \frac{(1-q)^k q^{k(k-1)/4}}{2^k (q;q)_k}(\mathscr{D}_q^k f)(0), \quad (2.41)$$

where

$$\zeta_0 := \frac{1}{2}(q^{1/4} + q^{-1/4}). \quad (2.42)$$

Elliptic analogues of the q-Taylor series have been introduced by Schlosser and Schlosser and Yoo in the very interesting papers [39, 40].

Another q-analogue of the derivative is the q-difference operator

$$(D_q f)(x) = (D_{q,x} f)(x) = \frac{f(x) - f(qx)}{(1-q)x}. \quad (2.43)$$

It is clear that

$$D_{q,x}(x^n) = \frac{1-q^n}{1-q} x^{n-1}, \quad (2.44)$$

and for differentiable functions

$$\lim_{q \to 1^-} (D_q f)(x) = f'(x).$$

The operator D_q was employed systematically by Jackson who was not the first to use it. Interestingly, while the q-derivative was apparently anticipated by Euler and by Heine, it does not seem to appear explicitly in their work. The first explicit occurrence of the q-derivative, together with applications such as q-Taylor series expansions, probably appeared in the 1877 work of Leopold Schendel [38].

The evaluation of a q-integral is a restatement of a series identity. However its interpretation as a q-integral may give insight to its origin and where it fits conceptually.

For finite a and b the q-integral is

$$\left. \begin{array}{rcl} \int_0^a f(x)\,d_q x & := & \sum_{n=0}^{\infty} [aq^n - aq^{n+1}] f(aq^n), \\ \int_a^b f(x)\,d_q x & := & \int_0^b f(x)\,d_q x - \int_0^a f(x)\,d_q x. \end{array} \right\} \quad (2.45)$$

It is clear from (2.45) that the q-integral is an infinite Riemann sum with the division points in a geometric progression. We would then expect

$$\int_a^b f(x)\,d_qx \to \int_a^b f(x)\,dx \quad \text{as } q \to 1$$

for continuous functions. The q-integral over $[0,\infty)$ uses the division points $\{q^n : -\infty < n < \infty\}$ and is

$$\int_0^\infty f(x)\,d_qx := (1-q)\sum_{n=-\infty}^\infty q^n f(q^n). \tag{2.46}$$

The relationship

$$\int_a^b f(x)g(qx)\,d_qx = q^{-1}\int_a^b g(x)f(x/q)\,d_qx$$
$$+ q^{-1}(1-q)[ag(a)f(a/q) - bg(b)f(b/q)] \tag{2.47}$$

follows from series rearrangements. The proof is straightforward and will be omitted.

Consider the weighted inner product

$$\langle f,g\rangle_q := \int_a^b f(t)\overline{g(t)}\,w(t)\,d_qt$$
$$= (1-q)\sum_{k=0}^\infty f(y_k)\overline{g(y_k)}\,y_k w(y_k)$$
$$- (1-q)\sum_{k=0}^\infty f(x_k)\overline{g(x_k)}\,x_k w(x_k), \tag{2.48}$$

where

$$x_k := aq^k, \quad y_k := bq^k, \tag{2.49}$$

and $w(x_k) > 0$ and $w(y_k) > 0$ for $k = 0,1,\ldots$. We will take $a \le 0 \le b$.

Theorem 2.6 *An analogue of integration by parts for $D_{q,x}$ is*

$$\langle D_{q,x}f,g\rangle_q = -f(x_0)\overline{g(x_{-1})}\,w(x_{-1}) + f(y_0)\overline{g(y_{-1})}\,w(y_{-1})$$
$$- q^{-1}\left\langle f, \frac{1}{w(x)}D_{q^{-1},x}(g(x)w(x))\right\rangle_q, \tag{2.50}$$

provided that the series on both sides of (2.50) converge absolutely and

$$\lim_{n\to\infty} w(x_n)f(x_{n+1})\overline{g(x_n)} = \lim_{n\to\infty} w(y_n)f(y_{n+1})\overline{g(y_n)} = 0. \tag{2.51}$$

The proof is left as an exercise.

The product rule for D_q is

$$(D_q(fg))(x) = f(x)(D_q g)(x) + g(qx)(D_q f)(x). \qquad (2.52)$$

In this work we do not develop applications of q-integrals for two reasons. The first is that they are treated in some detail in the books [16] and [4]. The second is that we have concentrated on integrals with respect to measures which are absolutely continuous with respect to the Lebesgue measure in order to include new material not available in book form. We note that infinite q-Taylor series expansions involving D_q has been rigorousley developed by Annaby and Mansour in [5]. For earlier or formal treatments see the references in [5].

When studying an operator it is helpful to describe its null space. For differentiation the null space is the constant functions. For D_q, the null space is the set of functions satisfying $f(x) = f(qx)$. If we write $x = q^y$, then the null space is the set of functions periodic in y with period 1. For the Askey–Wilson operator we use $x = \cos\theta$. As functions of θ the null space of the Askey–Wilson operator is clearly the doubly periodic functions with periods 2π and $i\log q$.

2.4 Summation theorems

We will use the q-Taylor expansion for polynomials to derive some of the summation theorems in the q-calculus. The idea is to expand polynomials in one of the bases in (2.30), (2.31), or (2.32). The explicit q-Taylor coefficients are found either by computing the action of $\mathscr{D}_q^k$ on the function to be expanded, or using the Cooper formula (2.29). It must be noted that Zhigo Liu [32]–[34] used the operator $D_{q,x}$ and his characterization of functions satisfying

$$D_{q,x}f(x,y) = D_{q,y}f(x,y)$$

to obtain summation theorems and q-series identities. His approach is noteworthy.

A basic hypergeometric function (2.11) is called **balanced** if

$$r = s + 1 \quad \text{and} \quad qa_1 a_2 \cdots a_{s+1} = b_1 b_2 \cdots b_s. \qquad (2.53)$$

Theorem 2.7 (q-Pfaff–Saalschütz). *The sum of a terminating balanced*

$_3\phi_2$ *is given by*

$$_3\phi_2 \left(\begin{matrix} q^{-n}, a, b \\ c, d \end{matrix} \middle| q, q \right) = \frac{(d/a, d/b; q)_n}{(d, d/ab; q)_n}, \quad \text{with } cd = abq^{1-n}. \quad (2.54)$$

Proof Apply Theorem 2.4 to $f(\cos\theta) = (be^{i\theta}, be^{-i\theta}; q)_n$ and use (2.37) and (2.38) to obtain

$$f_k = \frac{(q; q)_n (b/a)^k}{(q; q)_k (q; q)_{n-k}} (bq^{k/2} e^{i\theta}, bq^{k/2} e^{-i\theta}; q)_{n-k} \big|_{e^{i\theta} = aq^{k/2}}$$

$$= \frac{(q; q)_n (b/a)^k}{(q; q)_k (q; q)_{n-k}} (abq^k, b/a; q)_{n-k}.$$

Therefore (2.37) becomes

$$\frac{(be^{i\theta}, be^{-i\theta}; q)_n}{(q; q)_n} = \sum_{k=0}^{n} \frac{b^k (ae^{i\theta}, ae^{-i\theta}; q)_k (abq^k, b/a; q)_{n-k}}{a^k (q; q)_k (q; q)_{n-k}}. \quad (2.55)$$

Using (2.7) we can rewrite the above equation in the form

$$\frac{(be^{i\theta}, be^{-i\theta}; q)_n}{(ab, b/a; q)_n} = {}_3\phi_2 \left(\begin{matrix} q^{-n}, ae^{i\theta}, ae^{-i\theta} \\ ab, q^{1-n} a/b \end{matrix} \middle| q, q \right), \quad (2.56)$$

which is equivalent to the desired result. □

Theorem 2.8 *We have the q-analogue of the Chu–Vandermonde sum*

$$_2\phi_1(q^{-n}, a; c; q, q) = \frac{(c/a; q)_n}{(c; q)_n} a^n, \quad (2.57)$$

and the q-analogue of Gauss' theorem

$$_2\phi_1(a, b; c; q, c/ab) = \frac{(c/a, c/b; q)_\infty}{(c, c/ab; q)_\infty}, \quad |c/ab| < 1. \quad (2.58)$$

Proof Let $d = abq^{1-n}/c$ in (2.54) then let $n \to \infty$. Taking the limit inside the sum is justified since $(a, b; q)_k/(q, c; q)_k$ is bounded, [9]. The result is (2.58). When $b = q^{-n}$ then (2.58) becomes

$$_2\phi_1(q^{-n}, a; c; q, cq^n/a) = \frac{(c/a; q)_n}{(c; q)_n}. \quad (2.59)$$

To prove (2.57) express the left-hand side of the above equation as a sum, over k, say, replace k by $n - k$, then apply (2.7) and arrive at (2.57) after some simplifications and substitutions. This completes the proof. □

Replace a, b, c by q^a, q^b, q^c, respectively, in (2.58), then let $q \to 1^-$ to see that (2.58) reduces to Gauss's theorem, [4, 36]:

$$_2F_1(a,b;c;1) = \frac{\Gamma(c)\Gamma(c-a-b)}{\Gamma(c-a)\Gamma(c-b)}, \qquad \Re(c-a-b) > 0. \quad (2.60)$$

Theorem 2.9 *If $|z| < 1$ or $a = q^{-n}$ then the q-binomial theorem*

$$_1\phi_0(a;-;q,z) = \frac{(az;q)_\infty}{(z;q)_\infty}, \qquad (2.61)$$

holds.

Proof Let $c = abz$ in (2.58) then let $b \to 0$. $\qquad \square$

As $q \to 1^-$, $_1\phi_0(a;-;q,z)$ tends to $\sum_{n=0}^\infty (a)_n z^n/n! = (1-z)^{-a}$, by the binomial theorem. Thus the right-hand side of (2.61) is a q-analogue of $(1-z)^{-a}$.

Theorem 2.10 (Euler). *We have*

$$e_q(z) := \sum_{n=0}^\infty \frac{z^n}{(q;q)_n} = \frac{1}{(z;q)_\infty}, \qquad |z| < 1, \qquad (2.62)$$

$$E_q(z) := \sum_{n=0}^\infty \frac{z^n}{(q;q)_n} q^{n(n-1)/2} = (-z;q)_\infty. \qquad (2.63)$$

Proof Formula (2.62) is the special case $a = 0$ of (2.61). To get (2.63), we replace z by $-z/a$ in (2.61) and let $a \to \infty$. This implies (2.63) and the proof is complete. $\qquad \square$

The left-hand sides of (2.62) and (2.63) are q-analogues of the exponential function.

The terminating version of the q-binomial theorem may be written as

$$(z;q)_n = \sum_{k=0}^n \begin{bmatrix} n \\ k \end{bmatrix}_q q^{\binom{k}{2}} (-z)^k. \qquad (2.64)$$

Theorem 2.11 *The following summation theorem holds*

$$\frac{(ae^{i\theta}, ae^{-i\theta};q^2)_n}{(aq^{-1/2};q)_{2n}} = {}_4\phi_3\left(\begin{array}{c} q^{-n}, -q^{-n}, q^{1/2}e^{i\theta}, q^{1/2}e^{-i\theta} \\ -q, aq^{-1/2}, q^{-2n+3/2}/a \end{array} \middle| q, q \right). \qquad (2.65)$$

Proof Let $f(x) = (ae^{i\theta}, ae^{-i\theta};q)_n$ in (2.40). Using (2.34) we see that the coefficient of $\phi_n(x)$ is

$$\frac{(q-1)^k q^{-k/4}}{2^k (q;q)_k} \frac{(2a)^k (q;q)_n}{(q-1)^k (q;q)_{n-k}} q^{k(k-1)/4} (aq^{(2k+1)/4}, aq^{(2k-1)/4}; q)_{n-k}$$

$$= \frac{a^k q^{-k/4}}{(q;q)_k (q;q)_{n-k}} q^{k(k-2)/4} (aq^{(2k-1)/4}; q^{1/2})_{2n-2k}.$$

Replace q by q^2 and after some simplification we establish (2.65). $\quad\square$

Ismail and Stanton proved Theorem 2.11 in [24]. Schlosser and Yoo [40] observed that Theorem 2.11 follows from (1.4) in [14].

For completeness we mention, without proof, a non-polynomial version of q-Taylor series. The maximum modulus of an entire function f, $M(r; f)$, is [8]

$$M(r; f) = \sup \{|f(z)| : |z| \le r\}. \tag{2.66}$$

Theorem 2.12 ([25]) *Let f be analytic in a bounded domain D and let C be a contour within D and x belonging to the interior of C. If the distance between C and the set of zeros of $\phi_\infty(x; a)$ is positive then*

$$f(x) = \frac{\phi_\infty(x; a)}{2\pi i} \oint_C \frac{f(y)}{y - x} \frac{dy}{\phi_\infty(y; a)} - \frac{a}{\pi i} \sum_{n=0}^{\infty} q^n \phi_n(x; a) \oint_C \frac{f(y) \, dy}{\phi_{n+1}(y; a)}, \tag{2.67}$$

where $\phi_\infty(\cos\theta; a) := (ae^{i\theta}, ae^{-i\theta}; q)_\infty$.

This is true for any entire function, but if we let all points on C tend to ∞, the first integral may or may not converge to 0. We now restrict ourselves to entire functions satisfying

$$\limsup_{r \to +\infty} \frac{\ln M(r; f)}{\ln^2 r} = c, \tag{2.68}$$

for a particular c which, of course, may depend upon q.

Theorem 2.13 ([25]) *Any entire function f satisfying (2.68) with $c < 1/(2\ln q^{-1})$ has a convergent expansion*

$$f(x) = \sum_{k=0}^{\infty} f_{k,\phi} \, \phi_k(x; a), \tag{2.69}$$

with $\{f_{k,\phi}\}$ as defined in Theorem 2.4. Moreover any such f is uniquely determined by its values at the points $\{x_n : n \ge 0\}$ defined in (2.39).

The q-shifted factorial $(a;q)_n$ for $n < 0$ has been stated in (2.6). A **bilateral basic hypergeometric function** is

$$
{}_m\psi_m\left(\begin{array}{c} a_1,\ldots,a_m \\ b_1,\ldots,b_m \end{array}\bigg|\, q,\, z\right) = \sum_{n=-\infty}^{\infty} \frac{(a_1,\ldots,a_m;q)_n}{(b_1,\ldots,b_m;q)_n} z^n. \tag{2.70}
$$

It is easy to see that the series in (2.70) converges if

$$
\left|\frac{b_1 b_2 \cdots b_m}{a_1 a_2 \cdots a_m}\right| < |z| < 1. \tag{2.71}
$$

Our next result is the Ramanujan ${}_1\psi_1$ sum.

Theorem 2.14 *For $|b/a| < |z| < 1$ we have*

$$
{}_1\psi_1(a;b;q,z) = \frac{(b/a, q, q/az, az; q)_\infty}{(b, b/az, q/a, z; q)_\infty}. \tag{2.72}
$$

Proof ([17]). Observe that both sides of (2.72) are analytic functions of b for $|b| < |az|$ since, by (2.4), we have

$$
{}_1\psi_1(a;b;q,z) = \sum_{n=0}^{\infty} \frac{(a;q)_n}{(b;q)_n} z^n + \sum_{n=0}^{\infty} \frac{(q/b;q)_n}{(q/a;q)_n}\left(\frac{b}{az}\right)^n.
$$

Moreover when $b = q^{m+1}$, m a positive integer, it follows that $1/(b;q)_n = (q^n b; q)_{-n} = 0$ for $n < -m$, see (2.4). Therefore the q-binomial theorem (2.61) gives

$$
\begin{aligned}
&{}_1\psi_1(a;q^{m+1};q,z) \\
&= \sum_{n=-m}^{\infty} \frac{(a;q)_n}{(q^{m+1};q)_n} z^n = z^{-m} \frac{(a;q)_{-m}}{(q^{m+1};q)_{-m}} \sum_{n=0}^{\infty} \frac{(aq^{-m};q)_n}{(q;q)_n} z^n \\
&= z^{-m} \frac{(a;q)_{-m}}{(q^{m+1};q)_{-m}} \frac{(azq^{-m};q)_\infty}{(z;q)_\infty} = \frac{z^{-m}(az;q)_\infty (q, azq^{-m};q)_m}{(z;q)_\infty (aq^{-m};q)_m}.
\end{aligned}
$$

Using (2.4)–(2.5) we simplify the above formula to

$$
{}_1\psi_1(a;q^{m+1};q,z) = \frac{(q^{m+1}/a, q, q/az, az; q)_\infty}{(q^{m+1}, q^{m+1}/az, q/a, z; q)_\infty},
$$

which is (2.72) with $b = q^{m+1}$. The identity theorem for analytic functions then establishes the theorem. $\qquad\square$

For other proofs and references see [4] and [16].

Theorem 2.15 (Jacobi Triple Product Identity). *We have*

$$
\sum_{n=-\infty}^{\infty} q^{n^2} z^n = (q^2, -qz, -q/z; q^2)_\infty. \tag{2.73}
$$

Proof It readily follows from (2.72) that

$$\sum_{n=-\infty}^{\infty} q^{n^2} z^n = \lim_{c \to 0} {}_1\psi_1(-1/c;0;q^2,qzc) = \lim_{c \to 0} \frac{(q^2,-qz,-q/z;q^2)_\infty}{(-q^2c,qcz;q^2)_\infty}$$

$$= (q^2,-qz,-q/z;q^2)_\infty,$$

which is (2.73). □

2.5 Transformations

The Euler transformation for a hypergeometric function is, [4, 36],

$$_2F_1(a,b;c;z) = (1-z)^{c-a-b}\,{}_2F_1(c-a,c-b;c;z). \tag{2.74}$$

The next theorem is a *q*-analogue of (2.74).

Theorem 2.16 *We have the q-analogue of the Euler transformation*

$$_2\phi_1\left(\begin{matrix} A,B \\ C \end{matrix}\,\bigg|\, q, Z\right) = \frac{(ABZ/C;q)_\infty}{(Z;q)_\infty}\,{}_2\phi_1\left(\begin{matrix} C/A,C/B \\ C \end{matrix}\,\bigg|\, q, ABZ/C\right)$$

$$\tag{2.75}$$

and the symmetric form of the Sears transformation

$$(ab,b\beta,b\gamma;q)_m b^{-m}\,{}_4\phi_3\left(\begin{matrix} q^{-m},q^{m-1}ab\beta\gamma,bz,b/z \\ ab,b\beta,b\gamma \end{matrix}\,\bigg|\, q, q\right)$$

$$= (ab,a\beta,a\gamma;q)_m a^{-m}\,{}_4\phi_3\left(\begin{matrix} q^{-m},q^{m-1}ab\beta\gamma,az,a/z \\ ab,a\beta,a\gamma \end{matrix}\,\bigg|\, q, q\right). \tag{2.76}$$

Proof As usual we use $z = e^{i\theta}$. Multiply (2.55) by $t^n/(ab;q)_n$ and add for $n \geq 0$. The result is

$$_2\phi_1\left(\begin{matrix} bz,b/z \\ ab \end{matrix}\,\bigg|\, q, t\right) = {}_1\phi_0(b/a;-;q,t)\,{}_2\phi_1\left(\begin{matrix} az,a/z \\ ab \end{matrix}\,\bigg|\, q, \frac{bt}{a}\right).$$

We sum the $_1\phi_0$ by the *q*-binomial theorem (2.61) and establish (2.75). The proof of (2.76) is similar. Multiply (2.55) by

$$\frac{(q^{-m},q^{m-1}ab\beta\gamma;q)_n}{(ab,b\beta,b\gamma;q)_n}\,q^n$$

and sum for $0 \leq n \leq m$. After interchanging the k and n sums we get

$$
{}_4\phi_3 \left(\begin{array}{c} q^{-m}, q^{m-1}ab\beta\gamma, bz, b/z \\ ab, b\beta, b\gamma \end{array} \bigg| q, q \right)
$$

$$
= \sum_{k=0}^{m} \frac{(q^{-m}, q^{m-1}ab\beta\gamma; q)_k}{(q;q)_k} (az, a/z; q)_k
$$

$$
\times \left(\frac{b}{a} \right)^k \sum_{n=k}^{m} \frac{(q^{k-m}, q^{k+m-1}ab\beta\gamma, b/a; q)_{n-k}}{(ab;q)_k (b\beta, b\gamma; q)_n (q;q)_{n-k}} q^n
$$

$$
= \sum_{k=0}^{m} \frac{(q^{-m}, q^{m-1}ab\beta\gamma; q)_k}{(q, ab, b\beta, b\gamma; q)_k} (az, a/z; q)_k \left(\frac{qb}{a} \right)^k
$$

$$
\times {}_3\phi_2 \left(\begin{array}{c} q^{k-m}, q^{m+k-1}ab\beta\gamma, b/a \\ b\beta q^k, b\gamma q^k \end{array} \bigg| q, q \right).
$$

By the q-Pfaff–Saalschütz theorem the ${}_3\phi_2$ is

$$
\frac{(a\beta q^k, q^{1-m}/a\gamma; q)_{m-k}}{(b\beta q^k, q^{1-m}/b\gamma; q)_{m-k}} = \frac{(a\beta, a\gamma; q)_m (b\beta, b\gamma; q)_k}{(b\beta, b\gamma; q)_m (a\beta, a\gamma; q)_k} \left(\frac{b}{a} \right)^{m-k}.
$$

This establishes (2.76). □

The transformation (2.75) is one of three known as Heine transformations.

The Sears transformation is usually written, [16], as

$$
{}_4\phi_3 \left(\begin{array}{c} q^{-n}, A, B, C \\ D, E, F \end{array} \bigg| q, q \right)
$$

$$
= \frac{(E/A, F/A; q)_n}{(E, F; q)_n} A^n {}_4\phi_3 \left(\begin{array}{c} q^{-n}, A, D/B, D/C \\ D, Aq^{1-n}/E, Aq^{1-n}/F \end{array} \bigg| q, q \right), \quad (2.77)
$$

where $DEF = ABCq^{1-n}$. We leave as an exercise for the reader to show the equivalence of (2.77) and (2.76).

An iterate of the Sears transformation is

$$
{}_4\phi_3 \left(\begin{array}{c} q^{-n}, A, B, C \\ D, E, F \end{array} \bigg| q, q \right) = \frac{(A, EF/AB, EF/AC; q)_n}{(E, F, EF/ABC; q)_n}
$$

$$
\times {}_4\phi_3 \left(\begin{array}{c} q^{-n}, E/A, F/A, EF/ABC \\ EF/AB, EF/AC, q^{1-n}/A \end{array} \bigg| q, q \right),
$$

$$
(2.78)
$$

where $DEF = ABCq^{1-n}$.

Note that the form which we proved of the q-analogue of the Euler transformation is the symmetric form

$$(t;q)_\infty \, {}_2\phi_1 \left(\begin{array}{c} bz, b/z \\ ab \end{array} \middle| q, \, t \right) = (bt/a;q)_\infty \, {}_2\phi_1 \left(\begin{array}{c} az, a/z \\ ab \end{array} \middle| q, \, \frac{bt}{a} \right).$$

$$(2.79)$$

A limiting case of the Sears transformation is the useful transformation stated below.

Theorem 2.17   *The following ${}_3\phi_2$ transformation holds*

$${}_3\phi_2 \left(\begin{array}{c} q^{-n}, a, b \\ c, d \end{array} \middle| q, q \right) = \frac{b^n (d/b;q)_n}{(d;q)_n} {}_3\phi_2 \left(\begin{array}{c} q^{-n}, b, c/a \\ c, q^{1-n} b/d \end{array} \middle| q, \frac{aq}{d} \right).$$

$$(2.80)$$

Proof   In (2.77) set $F = ABCq^{1-n}/DE$ then let $C \to 0$ while all the other parameters remain constant. The result is

$${}_3\phi_2 \left(\begin{array}{c} q^{-n}, A, B \\ D, E \end{array} \middle| q, q \right) = \frac{A^n (E/A;q)_n}{(E;q)_n} {}_3\phi_2 \left(\begin{array}{c} q^{-n}, A, D/B \\ D, q^{1-n} A/E \end{array} \middle| q, \frac{Bq}{E} \right).$$

The result now follows with the parameter identification $A = b, B = a, D = c, E = d$.   □

An interesting application of (2.80) follows by taking $b = \lambda d$ and letting $d \to \infty$. The result is

$${}_2\phi_1 \left(\begin{array}{c} q^{-n}, a \\ c \end{array} \middle| q, q\lambda \right) = (\lambda q^{1-n};q)_n \sum_{j=0}^{n} \frac{(q^{-n}, c/a;q)_j q^{j(j-1)/2}}{(q, c, \lambda q^{1-n};q)_j} (-\lambda aq)^j.$$

Now replace λ by λq^{n-1} and observe that the above identity becomes the special case $\gamma = q^n$ of

$${}_2\phi_1 \left(\begin{array}{c} a, 1/\gamma \\ c \end{array} \middle| q, \gamma\lambda \right) = \frac{(\lambda;q)_\infty}{(\lambda\gamma;q)_\infty} \sum_{j=0}^{\infty} \frac{(1/\gamma, c/a;q)_j q^{j(j-1)/2}}{(q, c, \lambda;q)_j} (-\lambda a\gamma)^j.$$

$$(2.81)$$

Since both sides of the relationship (2.81) are analytic functions of γ when $|\gamma| < 1$ and they are equal when $\gamma = q^n$ then they must be identical for all γ if $|\gamma| < 1, |\lambda\gamma| < 1$.

It is more convenient to write the identity (2.81) in the form

$$\sum_{n=0}^{\infty} \frac{(A, C/B;q)_n}{(q, C, Az;q)_n} q^{n(n-1)/2} (-Bz)^n = \frac{(z;q)_\infty}{(Az;q)_\infty} {}_2\phi_1 (A, B; C; q, z). \quad (2.82)$$

In terms of basic hypergeometric functions (2.82) takes the form

$$_2\phi_2(A,C/B;C,Az;q,Bz) = \frac{(z;q)_\infty}{(Az;q)_\infty} {}_2\phi_1(A,B;C;q,z). \qquad (2.83)$$

Observe that (2.82) or (2.83) is the q-analogue of the Pfaff–Kummer transformation, [36],

$$_2F_1(a,b;c;z) = (1-z)^{-a}{}_2F_1(a,c-b;c;z/(z-1)), \qquad (2.84)$$

which holds when $|z| < 1$ and $|z/(z-1)| < 1$.

Later on we will come across the following $_3\phi_2$ transformation

$$_3\phi_2 \left(\begin{array}{c} q^{-n},a,b \\ c,0 \end{array} \middle| q,\ q \right) = \frac{(b;q)_n a^n}{(c;q)_n} {}_2\phi_1 \left(\begin{array}{c} q^{-n},c/b \\ q^{1-n}/b \end{array} \middle| q,\ q/a \right). (2.85)$$

Proof of (2.85)   Let $c \to 0$ in Theorem 2.17 to get

$$_3\phi_2 \left(\begin{array}{c} q^{-n},a,b \\ d,0 \end{array} \middle| q,\ q \right) = \frac{(d/b;q)_n b^n}{(d;q)_n} {}_2\phi_1 \left(\begin{array}{c} q^{-n},b \\ q^{1-n}b/d \end{array} \middle| q,\ qa/d \right).$$

On the $_2\phi_1$ side replace the summation index, say k, by $n-k$ then apply (2.7) to obtain a result equivalent to (2.85).   □

The transformation (2.85) has an interesting consequence. Since the left-hand side is symmetric in a, b then

$$(b;q)_n a^n\, {}_2\phi_1 \left(\begin{array}{c} q^{-n},c/b \\ q^{1-n}/b \end{array} \middle| q,\ q/a \right) = (a;q)_n b^n\, {}_2\phi_1 \left(\begin{array}{c} q^{-n},c/a \\ q^{1-n}/a \end{array} \middle| q,\ q/b \right).$$
$$\qquad (2.86)$$

Theorem 2.18   *The following two Heine transformations hold when both sides are well defined:*

$$_2\phi_1 \left(\begin{array}{c} a,b \\ c \end{array} \middle| q,\ z \right) = \frac{(az,c/a;q)_\infty}{(c,z;q)_\infty} {}_2\phi_1 \left(\begin{array}{c} a,abz/c \\ az \end{array} \middle| q,\ c/a \right), \quad (2.87)$$

$$_2\phi_1 \left(\begin{array}{c} a,b \\ c \end{array} \middle| q,\ z \right) = \frac{(b,az;q)_\infty}{(c,z;q)_\infty} {}_2\phi_1 \left(\begin{array}{c} c/b,z \\ az \end{array} \middle| q,\ b \right). \qquad (2.88)$$

Proof   In (2.86) replace a and b by q^{1-n}/a and q^{1-n}/b, respectively and conclude that

$$
{}_2\phi_1\left(\begin{array}{c} q^{-n}, q^{n-1}bc \\ b \end{array} \middle| q,\, q^n a\right)
$$

$$
= \frac{(q^{1-n}/a;q)_n a^n}{(q^{1-n}/b;q)_n b^n}{}_2\phi_1\left(\begin{array}{c} q^{-n}, q^{n-1}ac \\ a \end{array} \middle| q,\, q^n b\right)
$$

$$
= \frac{(a, bq^n; q)_\infty}{(b, aq^n; q)_\infty}{}_2\phi_1\left(\begin{array}{c} q^{-n}, q^{n-1}ac \\ a \end{array} \middle| q,\, q^n b\right).
$$

Now observe that the above equation, with c replaced by cq, is the case $\gamma = q^n$ of the transformation

$$
{}_2\phi_1\left(\begin{array}{c} 1/\gamma, bc\gamma \\ b \end{array} \middle| q,\, \gamma a\right) = \frac{(a, b\gamma; q)_\infty}{(b, a\gamma; q)_\infty}{}_2\phi_1\left(\begin{array}{c} 1/\gamma, ac\gamma \\ a \end{array} \middle| q,\, \gamma b\right).
$$

With $|a| \le 1, |b| \le 1$, both sides of the above identity are analytic functions of γ in the open unit disc; therefore its validity for the sequence $\gamma = q^n$ implies its validity for $|\gamma| < 1$. This is an equivalent form of (2.87). Apply (2.75) to (2.87) to derive (2.88). $\qquad\square$

2.6 q-Hermite polynomials

Our approach to the theory of q-series is based on the theory of Askey–Wilson operators and polynomials. The first step in this program is to develop the continuous q-Hermite polynomials $\{H_n(x|q)\}$. They are generated by the recursion relation

$$
2xH_n(x|q) = H_{n+1}(x|q) + (1 - q^n)H_{n-1}(x|q), \qquad (2.89)
$$

and the initial conditions

$$
H_0(x|q) = 1, \quad H_1(x|q) = 2x. \qquad (2.90)
$$

We solve this recurrence relation using the generating function technique. Set

$$
H(x,t) := \sum_{n=0}^{\infty} H_n(x|q) \frac{t^n}{(q;q)_n}. \qquad (2.91)
$$

Multiply (2.89) by $t^n/(q;q)_n$, and sum over $n = 1, 2, \ldots$. After the use of the initial conditions (2.90) we obtain the q-difference equation

$$
H(x,t) = \frac{H(x,qt)}{1 - 2xt + t^2} = \frac{H(x,qt)}{(te^{i\theta};q)_1(te^{-i\theta};q)_1}, \qquad x = \cos\theta.
$$

After iterating the above functional equation n times we get

$$H(\cos\theta, t) = \frac{H(\cos\theta, q^n t)}{(te^{i\theta}, te^{-i\theta}; q)_n}.$$

As $n \to \infty$, $H(x, q^n t) \to H(x, 0) = 1$. This establishes the generating function

$$\sum_{n=0}^{\infty} H_n(\cos\theta \,|\, q) \frac{t^n}{(q;q)_n} = \frac{1}{(te^{i\theta}, te^{-i\theta}; q)_\infty}. \tag{2.92}$$

To obtain an explicit formula for the H_n we expand $1/(te^{\pm i\theta}; q)_\infty$ by (2.62), then multiply the resulting series. This gives

$$H_n(\cos\theta \,|\, q) = \sum_{k=0}^{n} \begin{bmatrix} n \\ k \end{bmatrix}_q e^{i(n-2k)\theta} = \sum_{k=0}^{n} \begin{bmatrix} n \\ k \end{bmatrix}_q \cos(n-2k)\theta. \tag{2.93}$$

The representation (2.93) reflects the polynomial character of $H_n(x\,|\,q)$. The generating function (2.92) implies the symmetry property

$$H_n(-x\,|\,q) = (-1)^n H_n(x\,|\,q). \tag{2.94}$$

It is worth noting that

$$H_n(x\,|\,q) = (2x)^n + \text{lower order terms}, \tag{2.95}$$

which follows from (2.89) and (2.90).

Our next goal is to prove the orthogonality relation

$$\int_{-1}^{1} H_m(x\,|\,q) H_n(x\,|\,q) w(x\,|\,q)\,dx = \frac{2\pi(q;q)_n}{(q;q)_\infty} \delta_{m,n}, \tag{2.96}$$

where

$$w(x\,|\,q) = \frac{(e^{2i\theta}, e^{-2i\theta}; q)_\infty}{\sqrt{1-x^2}}, \qquad x = \cos\theta, \ 0 \le \theta \le \pi. \tag{2.97}$$

The proof of (2.96) is based on the following lemma.

Lemma 2.19 *Let $j \in \mathbb{Z}$. Then we have the following evaluation*

$$\int_0^\pi e^{2ij\theta} (e^{2i\theta}, e^{-2i\theta}; q)_\infty \, d\theta = \frac{\pi(-1)^j}{(q;q)_\infty} (1+q^j) q^{j(j-1)/2}. \tag{2.98}$$

Proof Let I_j denote the left side of (2.98). The Jacobi triple product identity (2.73) gives

$$I_j = \int_0^\pi e^{2ij\theta}(1 - e^{2i\theta})(qe^{2i\theta}, e^{-2i\theta}; q)_\infty d\theta$$

$$= \int_0^\pi \frac{e^{2ij\theta}(1 - e^{2i\theta})}{(q;q)_\infty} \sum_{n=-\infty}^\infty (-1)^n q^{n(n+1)/2} e^{2in\theta} d\theta$$

$$= \sum_{n=-\infty}^\infty \frac{(-1)^n q^{n(n+1)/2}}{2(q;q)_\infty} \int_{-\pi}^\pi (1 - e^{i\theta}) e^{i(j+n)\theta} d\theta.$$

The result now follows from the orthogonality of the trigonometric functions on $[-\pi, \pi]$.   □

Proof of (2.96)   Since the weight function $w(x|q)$ is an even function of x, it follows that (2.96) trivially holds if $|m - n|$ is odd. Thus there is no loss of generality in assuming $m \le n$ and $n - m$ is even. It is clear that we can replace $n - 2k$ by $|n - 2k|$ in (2.93). Therefore we evaluate the following integrals only for $0 \le j \le n/2$. We now have

$$\int_0^\pi e^{i(n-2j)\theta} H_n(\cos\theta \,|\, q)(e^{2i\theta}, e^{-2i\theta}; q)_\infty d\theta$$

$$= \sum_{k=0}^n \frac{(q;q)_n}{(q;q)_k(q;q)_{n-k}} \int_0^\pi e^{2i(n-j-k)\theta}(e^{2i\theta}, e^{-2i\theta}; q)_\infty d\theta$$

$$= \frac{\pi}{(q;q)_\infty} \sum_{k=0}^n \frac{(-1)^{j+k+n}(q;q)_n}{(q;q)_k(q;q)_{n-k}}(1 + q^{n-j-k}) q^{(n-j-k)(n-j-k-1)/2}$$

$$= \frac{(-1)^{n+j}\pi}{(q;q)_\infty} q^{(n-j)(n-j-1)/2}[{}_1\phi_0(q^{-n}; -; q, q^{j+1}) + q^{n-j}{}_1\phi_0(q^{-n}; -; q, q^j)].$$

We evaluate the ${}_1\phi_0$ functions by the q-binomial theorem and after some simplification we arrive at

$$\int_0^\pi e^{i(n-2j)\theta} H_n(\cos\theta \,|\, q)(e^{2i\theta}, e^{-2i\theta}; q)_\infty d\theta$$

$$= \frac{(-1)^{n+j}\pi}{(q;q)_\infty} q^{(n-j)(n-j-1)/2}[(q^{-n+j+1}; q)_n + q^{n-j}(q^{-n+j}; q)_n]. \quad (2.99)$$

For $0 < j < n$ it is clear that the right-hand side of (2.99) vanishes. When $j = 0$, the right-hand side of (2.99) is

$$\frac{\pi}{(q;q)_\infty} q^{n(n-1)/2}(-1)^n q^n (q^{-n}; q)_n.$$

Thus

$$\int_0^\pi e^{i(n-2j)\theta} H_n(\cos\theta \,|\, q)(e^{2i\theta}, e^{-2i\theta}; q)_\infty d\theta = \frac{\pi(q;q)_n}{(q;q)_\infty} \delta_{j,0}, \quad 0 \le j < n.$$

$$(2.100)$$

This calculation establishes (2.96) when $m < n$. When $m = n$ we use (2.99) and (2.95) to obtain

$$\int_0^\pi H_m(\cos\theta\,|\,q)H_n(\cos\theta\,|\,q)(e^{2i\theta}, e^{-2i\theta}; q)_\infty d\theta$$

$$= 2\int_0^\pi e^{in\theta}H_n(\cos\theta\,|\,q)(e^{2i\theta}, e^{-2i\theta}; q)_\infty d\theta = \frac{2\pi(q;q)_n}{(q;q)_\infty}. \qquad (2.101)$$

This proves (2.96). $\qquad\qquad\qquad\qquad\qquad\qquad\qquad\qquad\square$

It is a fact that

$$\lim_{q\to 1^-} \left(\frac{2}{1-q}\right)^{n/2} H_n(x\sqrt{(1-q)/2}\,|\,q) = H_n(x). \qquad (2.102)$$

This can be verified using (2.89) and (2.90) and comparing them with the recursive definition of the Hermite polynomials, namely

$$H_0(x) = 1, \quad H_1(x) = 2x, \quad H_{n+1}(x) = 2xH_n(x) - 2nH_{n-1}(x), \quad (2.103)$$

[19, 36, 41]. Note that it is an interesting exercise to see how the orthogonality relation for $\{H_n(x\,|\,q)\}$ tends to the orthogonality relation for $\{H_n(x)\}$.

Theorem 2.20 *The linearization of products of continuous q-Hermite polynomials is given by*

$$H_m(x\,|\,q)H_n(x\,|\,q) = \sum_{k=0}^{m\wedge n} \frac{(q;q)_m(q;q)_n}{(q;q)_k(q;q)_{m-k}(q;q)_{n-k}} H_{m+n-2k}(x\,|\,q), \quad (2.104)$$

where $m \wedge n := \min\{m, n\}$.

Proof It is clear from (2.94) that $H_m(x\,|\,q)H_n(x\,|\,q)$ has the same parity as $H_{m+n}(x\,|\,q)$. Therefore there exists a sequence $\{a_{m,n,k} : 0 \le k \le m\wedge n\}$ such that

$$H_m(x\,|\,q)H_n(x\,|\,q) = \sum_{k=0}^{m\wedge n} a_{m,n,k}H_{m+n-2k}(x\,|\,q) \qquad (2.105)$$

and $a_{m,n,k}$ is symmetric in m and n. Furthermore

$$a_{m,0,k} = a_{0,n,k} = \delta_{k,0} \qquad (2.106)$$

holds and (2.95) implies

$$a_{m,n,0} = 1. \qquad (2.107)$$

Multiply (2.105) by $2x$ and use the three-term recurrence relation (2.89) to obtain

$$\sum_{k=0}^{(m+1)\wedge n} a_{m+1,n,k}H_{m+n+1-2k}(x\,|\,q)+$$

$$+(1-q^m)\sum_{k=0}^{(m-1)\wedge n} a_{m-1,n,k}H_{m+n-1-2k}(x\,|\,q)$$

$$=\sum_{k=0}^{m\wedge n} a_{m,n,k}[H_{m+n+1-2k}(x\,|\,q)+(1-q^{m+n-2k})H_{m+n-1-2k}(x\,|\,q)],$$

with $H_{-1}(x\,|\,q):=0$. This leads us to the system of difference equations

$$a_{m+1,n,k+1}-a_{m,n,k+1}=(1-q^{m+n-2k})a_{m,n,k}-(1-q^m)a_{m-1,n,k}, \quad (2.108)$$

subject to the initial conditions (2.106) and (2.107). When $k=0$, equations (2.108) and (2.107) imply

$$a_{m+1,n,1}=a_{m,n,1}+q^m(1-q^n),$$

which leads to

$$a_{m,n,1}=(1-q^n)\sum_{k=0}^{m-1}q^k=\frac{(1-q^m)(1-q^n)}{1-q}. \quad (2.109)$$

Setting $k=1$ in (2.108) and applying (2.109) we find

$$a_{m+1,n,2}=a_{m,n,2}+q^{m-1}(1-q^m)(1-q^n)(1-q^{n-1})/(1-q),$$

whose solution is

$$a_{m,n,2}=\frac{(1-q^n)(1-q^{n-1})(1-q^m)(1-q^{m-1})}{(1-q)(1-q^2)}.$$

From here we suspect the pattern

$$a_{m,n,k}=\frac{(q;q)_m(q;q)_n}{(q;q)_{m-k}(q;q)_{n-k}(q;q)_k},$$

which can be proved from (2.108) by a straightforward induction. □

The next result is the Poisson kernel

$$\sum_{n=0}^{\infty}\frac{H_n(\cos\theta\,|\,q)H_n(\cos\phi\,|\,q)}{(q;q)_n}t^n$$

$$=\frac{(t^2;q)_\infty}{(te^{i(\theta+\phi)},te^{i(\theta-\phi)},te^{-i(\theta+\phi)},te^{-i(\theta-\phi)};q)_\infty}, \quad (2.110)$$

To prove (2.110) multiply (2.104) by $t^m s^n / (q;q)_m (q;q)_n$ and add for $m, n \geq 0$. The result is

$$\frac{1}{(te^{i\theta}, te^{-i\theta}, se^{i\theta}, se^{-i\theta}; q)_\infty} = \sum_{k=0}^{\infty} \sum_{m \geq k, n \geq k} \frac{t^m s^n H_{m+n-2k}(\cos\theta | q)}{(q;q)_{m-k}(q;q)_{n-k}(q;q)_k}$$

$$= \sum_{k=0}^{\infty} \frac{(ts)^k}{(q;q)_k} \sum_{m,n=0}^{\infty} \frac{t^m s^n H_{m+n}(\cos\theta | q)}{(q;q)_m (q;q)_n}$$

$$= \frac{1}{(st;q)_\infty} \sum_{p=0}^{\infty} \frac{H_p(\cos\theta | q)}{(q;q)_p} \sum_{n=0}^{p} \begin{bmatrix} p \\ n \end{bmatrix}_q t^{p-n} s^n.$$

Now let $t = ue^{i\phi}, s = ue^{-i\phi}$ and use (2.93) to write the last equation as

$$\frac{(u^2;q)_\infty}{(ue^{i(\phi+\theta)}, ue^{i(\phi-\theta)}, ue^{i(\theta-\phi)}, ue^{-i(\phi+\theta)}; q)_\infty}$$

$$= \sum_{p=0}^{\infty} \frac{u^p}{(q;q)_p} H_p(\cos\theta | q) H_p(\cos\phi | q).$$

This proves (2.110).

The Askey–Wilson operator acts on $H_n(x | q)$ in a natural way.

Theorem 2.21   *The polynomials $\{H_n(x | q)\}$ have the ladder operators*

$$\mathscr{D}_q H_n(x | q) = \frac{2(1 - q^n)}{1 - q} q^{(1-n)/2} H_{n-1}(x | q) \qquad (2.111)$$

and

$$\frac{1}{w(x | q)} \mathscr{D}_q \{w(x | q) H_n(x | q)\} = -\frac{2q^{-n/2}}{1 - q} H_{n+1}(x | q), \qquad (2.112)$$

where $w(x | q)$ is as defined in (2.97).

Proof   Apply $\mathscr{D}_q$ to (2.92) and get

$$\sum_{n=0}^{\infty} \frac{t^n}{(q;q)_n} \mathscr{D}_q H_n(x | q) = \frac{2t/(1-q)}{(tq^{-1/2}e^{i\theta}, tq^{-1/2}e^{-i\theta}; q)_\infty}.$$

The above and (2.92) imply (2.111). Similarly one can prove (2.112).   □

Since $(q^{-1}; q^{-1})_n = (-1)^n (q;q)_n q^{-n(n+1)/2}$, we derive

$$H_n(\cos\theta | q^{-1}) = \sum_{k=0}^{n} \frac{(q;q)_n}{(q;q)_k (q;q)_{n-k}} q^{k(k-n)} e^{i(n-2k)\theta} \qquad (2.113)$$

from (2.92).

Theorem 2.22 *The polynomials $\{H_n(x|q^{-1})\}$ have the generating function*

$$\sum_{n=0}^{\infty} \frac{H_n(\cos\theta\,|q^{-1})}{(q;q)_n}(-1)^n t^n q^{\binom{n}{2}} = (te^{i\theta}, te^{-i\theta}; q)_\infty. \qquad (2.114)$$

Proof Insert $H_n(\cos\theta\,|q^{-1})$ from (2.113) into the left-hand side of (2.114) to see that

$$\sum_{n=0}^{\infty} \frac{H_n(\cos\theta\,|q^{-1})}{(q;q)_n}(-1)^n t^n q^{\binom{n}{2}}$$

$$= \sum_{n\geq k\geq 0} \frac{q^{k(k-n)+n(n-1)/2}}{(q;q)_k(q;q)_{n-k}}(-t)^n e^{i(n-2k)\theta}$$

$$= \sum_{k=0}^{\infty} \frac{(-t)^k}{(q;q)_k} q^{k(k-1)/2} e^{-ik\theta} \sum_{n=0}^{\infty} \frac{q^{n(n-1)/2}(-t)^n}{(q;q)_n} e^{in\theta}$$

and the result follows from Euler's theorem (2.63). $\qquad\qquad\square$

Exercises

Exercise 2.23 Prove that the Poisson kernel (2.110) is equivalent to the linearization formula (2.104).

Exercise 2.24 Prove that the linearization formula (2.104) has the inverse relation

$$\frac{H_{n+m}(x|q)}{(q;q)_m(q;q)_n} = \sum_{k=0}^{m\wedge n} \frac{(-1)^k q^{k(k-1)/2}}{(q;q)_k} \frac{H_{n-k}(x|q)}{(q;q)_{n-k}} \frac{H_{m-k}(x|q)}{(q;q)_{m-k}}. \qquad (2.115)$$

Exercise 2.25 Show that if $g_n(z) = \sum_{k=0}^{n} \begin{bmatrix} n \\ k \end{bmatrix}_q z^k$ then

$$\sum_{n=0}^{\infty} t^n g_n(z)/(q;q)_n = 1/(t, tz; q)_\infty.$$

Exercise 2.26 Prove that

$$H_{2n+1}(0|q) = 0, \quad \text{and} \quad H_{2n}(0|q) = (-1)^n (q; q^2)_n. \qquad (2.116)$$

Exercise 2.27 Show that

$$H_n((q^{1/4} + q^{-1/4})/2\,|q) = (-1)^n H_n(-(q^{1/4} + q^{-1/4})/2\,|q)$$
$$= q^{-n/4}(-q^{1/2}; q^{1/2})_n. \qquad (2.117)$$

Exercise 2.28 Prove (2.112) using the orthogonality relation, (2.111), Theorem 2.2, and the completeness of $\{H_n(x|q)\}$ in $L_2[-1, 1, w(x|q)]$.

Exercise 2.29 Prove that

$$\frac{H_n(x|q)}{(q;q)_n} = \sum_{k=0}^{\lfloor n/2 \rfloor} \frac{(-1)^k q^{k(3k-2n-1)/2}}{(q;q)_k (q;q)_{n-2k}} H_{n-2k}(x|q^{-1}). \tag{2.118}$$

Exercise 2.30 Prove the inverse relation

$$H_n(x|q^{-1}) = \sum_{s=0}^{\lfloor n/2 \rfloor} \frac{q^{-s(n-s)}(q;q)_n}{(q;q)_s (q;q)_{n-2s}} H_{n-2s}(x|q), \tag{2.119}$$

2.7 The Askey–Wilson polynomials

Theorem 2.31 *For* $\max\{|t_j| : 1 \le j \le 4\} < 1$, *we have the Askey–Wilson* *q-beta integral evaluation,*

$$\int_0^\pi \frac{(e^{2i\theta}, e^{-2i\theta}; q)_\infty}{\prod_{j=1}^4 (t_j e^{i\theta}, t_j e^{-i\theta}; q)_\infty} d\theta = \frac{2\pi (t_1 t_2 t_3 t_4; q)_\infty}{(q;q)_\infty \prod_{1 \le j < k \le 4} (t_j t_k; q)_\infty}. \tag{2.120}$$

Proof Note that each factor $1/(t_j e^{i\theta}, t_j e^{-i\theta}; q)_\infty$ is a generating function for q-Hermite polynomials. Expanding these factors and using the linearization formula (2.104) we find that the left-hand side of (2.120) is

$$\frac{2\pi}{(q;q)_\infty} \int_0^\pi \sum_{k_1,k_2,n_1,n_2,n_3,n_4=0}^\infty \frac{(e^{2i\theta}, e^{-2i\theta}; q)_\infty \prod_{j=1}^4 t_j^{n_j}}{\prod_{r=1}^2 (q;q)_{n_r-k_1} (q;q)_{n_{r+2}-k_2} (q;q)_{k_r}}$$

$$\times H_{n_1+n_2-2k_1}(\cos\theta|q) H_{n_3+n_4-2k_2}(\cos\theta|q) d\theta.$$

We now put $n_1 = k_1 + m_1, n_2 = k_1 + m_2, n_3 = k_2 + m_3, n_4 = k_2 + m_4$. Thus $m_1 + m_2 = m_3 + m_4 = m$, say, since the other terms when $m_1 + m_2 \ne m_3 + m_4$ vanish sue to the orthogonality relation (2.96). With this notation the above expression becomes

$$\frac{2\pi}{(q;q)_\infty} \sum \frac{(q;q)_m t_1^{k_1+m_1} t_2^{k_1+m-m_1} t_3^{k_2+m_3} t_4^{k_2+m-m_3}}{(q;q)_{m_1} (q;q)_{m-m_1} (q;q)_{m_3} (q;q)_{m-m_3} (q;q)_{k_1} (q;q)_{k_2}} =$$

$$\frac{2\pi}{(q, t_1 t_2, t_3 t_4; q)_\infty} \sum_{m=0}^\infty \frac{1}{(q;q)_m} \sum_{m_1=0}^m \begin{bmatrix} m \\ m_1 \end{bmatrix}_q t_1^{m_1} t_2^{m-m_1} \sum_{m_3=0}^m \begin{bmatrix} m \\ m_3 \end{bmatrix}_q t_3^{m_3} t_4^{m-m_3}.$$

We set $t_1 = re^{-i\theta}, t_2 = re^{i\theta}, t_3 = \rho e^{-i\phi}$, and $t_4 = \rho e^{i\phi}$ then use (2.93) and the Poisson kernel (2.110) to see that the above expression equals

$$\frac{2\pi (r^2 \rho^2; q)_\infty}{(q, t_1 t_2, t_3 t_4; q)_\infty (r\rho e^{i(\theta+\phi)}, r\rho e^{i(\theta-\phi)}, r\rho e^{i(\phi-\theta)}, r\rho e^{i(-\theta-\phi)}; q)_\infty},$$

which is the desired result. □

The above evaluation is due to Ismail and Stanton [23]. Other proofs are in [4] and [16]. The original proof by R. Askey and J. Wilson in [7] is rather lengthy.

Askey–Wilson polynomials are orthogonal with respect to the weight function whose total mass is given by (2.120). To save space we shall use the vector notation $\mathbf{t}$ to denote the ordered tuple (t_1, t_2, t_3, t_4). Their weight function is, with $x = \cos\theta$,

$$w(x;\mathbf{t}|q) = w(x;t_1,t_2,t_3,t_4|q) = \frac{(e^{2i\theta}, e^{-2i\theta}; q)_\infty}{\prod_{j=1}^{4}(t_j e^{i\theta}, t_j e^{-i\theta}; q)_\infty} \frac{1}{\sqrt{1-x^2}}. \quad (2.121)$$

Theorem 2.32 *The Askey–Wilson polynomials are given by*

$$p_n(x;\mathbf{t}|q) =$$

$$t_1^{-n}(t_1 t_2, t_1 t_3, t_1 t_4; q)_n \, {}_4\phi_3\left(\begin{array}{c} q^{-n}, t_1 t_2 t_3 t_4 q^{n-1}, t_1 e^{i\theta}, t_1 e^{-i\theta} \\ t_1 t_2, \ t_1 t_3, \ t_1 t_4 \end{array}\middle| q, q\right). \quad (2.122)$$

They satisfy the orthogonality relation

$$\int_{-1}^{1} p_m(x;\mathbf{t}|q)\, p_n(x;\mathbf{t}|q)\, w(x;\mathbf{t}|q)\, dx$$

$$= \frac{2\pi(t_1 t_2 t_3 t_4 q^{2n}; q)_\infty (t_1 t_2 t_3 t_4 q^{n-1}; q)_n}{(q^{n+1}; q)_\infty \prod_{1\le j<k\le 4}(t_j t_k q^n; q)_\infty} \, \delta_{m,n}. \quad (2.123)$$

Proof We evaluate the integral of p_n as given by (2.122) times any polynomial of degree $\le n$ against the weight function in (2.121). From (2.120) it is clear that for $j \le n$ we have

$$\frac{t_1^n}{\prod_{s=2}^{4}(t_1 t_s; q)_n} \int_0^{\pi} (t_2 e^{i\theta}, t_2 e^{-i\theta}; q)_j \, p_n(\cos\theta; \mathbf{t}|q) w(\cos\theta; \mathbf{t}|q) \sin\theta \, d\theta$$

$$= \sum_{k=0}^{n} \frac{(q^{-n}, t_1 t_2 t_3 t_4 q^{n-1}; q)_k}{(q, t_1 t_2, \ t_1 t_3, \ t_1 t_4; q)_k} q^k \int_{-1}^{1} w(x; t_1 q^k, t_2 q^j; t_3, t_4)\, dx$$

$$= \sum_{k=0}^{n} \frac{(q^{-n}, t_1 t_2 t_3 t_4 q^{n-1}; q)_k}{(q, t_1 t_2, \ t_1 t_3, \ t_1 t_4; q)_k} q^k \frac{2\pi(q^{j+k} t_1 t_2 t_3 t_4; q)_\infty / (q; q)_\infty}{(q^{j+k} t_1 t_2, q^k t_1 t_3, q^k t_1 t_4, q^j t_2 t_3, q^j t_2 t_4, t_3 t_4; q)_\infty}$$

$$= \frac{2\pi(q^j t_1 t_2 t_3 t_4; q)_\infty / (q; q)_\infty}{(q^j t_1 t_2, q^j t_2 t_3, q^j t_2 t_4, t_1 t_3, t_1 t_4, t_3 t_4; q)_\infty}$$

$$\times {}_3\phi_2\left(\begin{array}{c} q^{-n}, t_1 t_2 t_3 t_4 q^{n-1}, t_1 t_2 q^j \\ t_1 t_2, \ t_1 t_2 t_3 t_4 q^j \end{array}\middle| q, q\right).$$

We apply (2.54) to see that the sum of the ${}_3\phi_2$ is

$$\frac{(t_3 t_4, q^{j-n+1}; q)_n}{(t_1 t_2 t_3 t_4 q^j, q^{1-n}/t_1 t_2; q)_n},$$

which vanishes if $j < n$. Therefore

$$\frac{t_1^n}{\prod_{s=2}^4(t_1t_s;q)_n}\int_0^\pi (t_2e^{i\theta},t_2e^{-i\theta};q)_j p_n(\cos\theta;\mathbf{t}|q)w(\cos\theta;\mathbf{t}|q)\sin\theta\,d\theta$$

$$-\frac{2\pi(q^nt_1t_2t_3t_4;q)_\infty/(q;q)_\infty}{(q^nt_1t_2,q^nt_2t_3,q^nt_2t_4,t_1t_3,t_1t_4,t_3t_4;q)_\infty}\frac{(-t_1t_2)^nq^{\binom{n}{2}}(t_3t_4,q;q)_n}{(t_1t_2t_3t_4q^n,t_1t_2;q)_n}\delta_{j,n}.$$

Using

$$(ae^{i\theta},ae^{-i\theta};q)_n = (-2a)^nq^{\binom{n}{2}}x^n + \text{lower order terms}$$

and the explicit representation (2.122), we conclude that

$$\int_0^\pi (t_1e^{i\theta},t_1e^{-i\theta};q)_n p_n(\cos\theta;\mathbf{t}|q)w(\cos\theta;\mathbf{t}|q)\sin\theta\,d\theta$$

$$=\frac{2\pi(q^nt_1t_2t_3t_4;q)_\infty/(q;q)_\infty}{(q^nt_1t_2,q^nt_2t_3,q^nt_2t_4,t_1t_3,t_1t_4,t_3t_4;q)_\infty}\frac{(-t_1)^nq^{\binom{n}{2}}(t_3t_4,q;q)_n}{(t_1t_2t_3t_4q^n,t_1t_2;q)_n}.$$

We then use the explicit form (2.122) and the above evaluation to establish the desired orthogonality relation. □

Observe that the weight function in (2.121) and the right-hand side of (2.123) are symmetric functions of t_1,t_2,t_3,t_4. The uniqueness of the polynomials orthogonal with respect to a positive measure shows that the Askey–Wilson polynomials are symmetric in the four parameters t_1,t_2,t_3,t_4. This symmetry is the Sears transformation in the form

$$t_1^{-n}(t_1t_2,t_1t_3,t_1t_4;q)_n {}_4\phi_3\left(\begin{matrix}q^{-n},t_1t_2t_3t_4q^{n-1},t_1e^{i\theta},t_1e^{-i\theta}\\t_1t_2,\ t_1t_3,\ t_1t_4\end{matrix}\bigg| q,q\right)$$

$$= t_2^{-n}(t_2t_1,t_2t_3,t_2t_4;q)_n {}_4\phi_3\left(\begin{matrix}q^{-n},t_1t_2t_3t_4q^{n-1},t_2e^{i\theta},t_2e^{-i\theta}\\t_2t_1,\ t_2t_3,\ t_2t_4\end{matrix}\bigg| q,q\right),$$

which we saw earlier in (2.76).

We now establish a generating function for the Askey–Wilson polynomials following a technique due to Ismail and Wilson [27].

Theorem 2.33 *The Askey–Wilson polynomials have the generating function*

$$\sum_{n=0}^\infty \frac{p_n(\cos\theta;\mathbf{t}|q)}{(q,t_1t_2,t_3t_4;q)_n}t^n =$$

$${}_2\phi_1\left(\begin{matrix}t_1e^{i\theta},t_2e^{i\theta}\\t_1t_2\end{matrix}\bigg| q,te^{-i\theta}\right){}_2\phi_1\left(\begin{matrix}t_3e^{-i\theta},t_4e^{-i\theta}\\t_3t_4\end{matrix}\bigg| q,te^{i\theta}\right).\qquad(2.124)$$

Proof   Apply (2.77) with

$$A = t_1 e^{i\theta}, \ B = t_1 e^{-i\theta}, \ C = t_1 t_2 t_3 t_4 q^{n-1}, \ D = t_1 t_2, \ E = t_1 t_3, \ F = t_1 t_4,$$

to obtain

$$p_n(x; \mathbf{t} \mid q) = (t_1 t_2, q^{1-n} e^{i\theta} / t_3, q^{1-n} e^{i\theta} / t_4; q)_n (t_3 t_4 q^{n-1} e^{-i\theta})^n$$
$$\times {}_4\phi_3 \left(\begin{array}{c} q^{-n}, t_1 e^{i\theta}, t_2 e^{i\theta}, q^{1-n} / t_3 t_4 \\ t_1 t_2, \ q^{1-n} e^{i\theta} / t_3, \ q^{1-n} e^{i\theta} / t_4 \end{array} \middle| q, q \right). \qquad (2.125)$$

Using (2.6) we write

$$(q^{1-n} e^{i\theta} / t_3, q^{1-n} e^{i\theta} / t_4; q)_n = e^{2in\theta} q^{-n(n-1)} (t_3 t_4)^{-n} (t_3 e^{-i\theta}, t_4 e^{-i\theta}; q)_n.$$

Furthermore, if the summation index of the ${}_4\phi_3$ in (2.125) is k then we may use (2.5) to get

$$\frac{(q^{-n}, q^{1-n} / t_3 t_4; q)_k}{(q^{1-n} e^{i\theta} / t_3, q^{1-n} e^{i\theta} / t_4; q)_k}$$
$$= \frac{(q, t_3 t_4; q)_n}{(q, t_3 t_4; q)_{n-k}} \frac{(t_3 e^{-i\theta}, t_4 e^{-i\theta}; q)_{n-k}}{(t_3 e^{-i\theta}, t_4 e^{-i\theta}; q)_n} (q e^{2i\theta})^{-k}.$$

Therefore

$$\frac{p_n(x; t_1, t_2, t_3, t_4 \mid q)}{(q, t_1 t_2, t_3 t_4; q)_n}$$
$$= \sum_{k=0}^{n} \frac{(t_1 e^{i\theta}, t_2 e^{i\theta}; q)_k}{(q, t_1 t_2; q)_k} e^{-ik\theta} \frac{(t_3 e^{-i\theta}, t_4 e^{-i\theta}; q)_{n-k}}{(q, t_3 t_4; q)_{n-k}} e^{i(n-k)\theta}. \qquad (2.126)$$

It is clear that (2.126) implies (2.124) and the proof is complete. $\qquad \square$

The Askey–Wilson polynomials are orthogonal polynomials; therefore they satisfy a three-term recurrence relation of the form

$$2x p_n(x; \mathbf{t} \mid q) = A_n p_{n+1}(x; \mathbf{t} \mid q) + B_n p_n(x; \mathbf{t} \mid q) + C_n p_{n-1}(x; \mathbf{t} \mid q). \qquad (2.127)$$

The coefficient of x^n in p_n is $2^n (t_1 t_2 t_3 t_4 q^{n-1}; q)_n$. Equating the coefficients of x^{n+1} on both sides of (2.127) we get

$$A_n = \frac{1 - t_1 t_2 t_3 t_4 q^{n-1}}{(1 - t_1 t_2 t_3 t_4 q^{2n-1})(1 - t_1 t_2 t_3 t_4 q^{2n})}. \qquad (2.128)$$

We next choose the special values $e^{-i\theta} = t_1, t_2$ in (2.122) and obtain

$$p_n((t_1 + 1/t_1)/2; \mathbf{t} \mid q) = (t_1 t_2, t_1 t_3, t_1 t_4; q)_n t_1^{-n},$$
$$p_n((t_2 + 1/t_2)/2; \mathbf{t} \mid q) = (t_2 t_1, t_2 t_3, t_2 t_4; q)_n t_2^{-n}.$$

With A_n given by (2.128), we substitute $x = (t_j + 1/t_j)/2$, $j = 1, 2$ in (2.127) and solve for B_n and C_n. The result is

$$
\left.
\begin{aligned}
C_n &= \frac{(1-q^n)\prod_{1\leq j<k\leq 4}(1-t_jt_kq^{n-1})}{(1-t_1t_2t_3t_4q^{2n-2})(1 \quad t_1t_2t_3t_4q^{2n-1})}, \\
B_n &= t_1 + t_1^{-1} - A_nt_1^{-1}\prod_{j=2}^{4}(1-t_1t_jq^n) - \frac{t_1C_n}{\prod_{2\leq k\leq 4}(1-t_1t_kq^{n-1})}.
\end{aligned}
\right\}
$$

$$(2.129)$$

Theorem 2.34 *With $z = e^{i\theta}$ and $x = \cos\theta$ we have*

$$
\frac{p_n(\cos\theta;\mathbf{t}\,|\,q)}{(q,t_1t_2,t_3t_4;q)_n}
$$
$$
= z^n\frac{(t_1/z,t_3/z;q)_\infty}{(z^{-2},q;q)_\infty}\sum_{m=0}^{\infty}\frac{(qz/t_1,qz/t_3;q)_m}{(q,qz^2;q)_m}(t_1t_3q^n)^m
$$
$$
\times {}_2\phi_2\left(\begin{array}{c}t_1z,t_1/z\\t_1t_2,t_1q^{-m}/z\end{array}\middle|\,q,t_2q^{-m}/z\right) {}_2\phi_2\left(\begin{array}{c}t_3z,t_3/z\\t_3t_4,t_3q^{-m}/z\end{array}\middle|\,q,t_4q^{-m}/z\right)
$$
$$
+ \text{a similar term with } z \text{ and } 1/z \text{ interchanged.} \tag{2.130}
$$

Proof Let $F(x,t)$ denote the right-hand side of (2.124). Apply the q-analogue of the Pfaff–Kummer transformation, (2.82) to the ${}_2\phi_1$ in $F(x,t)$. Thus

$$
F(x,t) = \frac{(tt_1,tt_3;q)_\infty}{(tz,t/z;q)_\infty}\sum_{k,j=0}^{\infty}\frac{(t_1z,t_1/z;q)_k(t_3z,t_3/z;q)_j}{(q,t_1t_2,tt_1;q)_k(q,t_3t_4,tt_3;q)_j}
$$
$$
\times q^{\binom{j}{2}+\binom{k}{2}}(-tt_4)^j(-tt_2)^k.
$$

Cauchy's theorem shows that

$$
\frac{p_n(\cos\theta;\mathbf{t}\,|\,q)}{(q,t_1t_2,t_3t_4;q)_n} = \frac{1}{2\pi i}\int_C F(x,t)t^{-n-1}\,dt, \tag{2.131}
$$

where C is the contour $\{t : |t| = r\}$, with $r < \left|e^{-i\theta}\right| = 1/|z|$. Now think of the contour C as a contour around the point $t = \infty$ with the wrong orientation, so it encloses all the poles of $F(x,t)$. Therefore the right-hand side of (2.131) is $-\sum$ Residues. Now $t = 0$ is outside the contour and the singularities of F inside the contour are $t = q^{-m}z^{\pm1}$, $m = 0, 1, \ldots$. It is straightforward to see that

$$
\begin{aligned}
\operatorname{Res}&\{F(x,t) : t = zq^{-m}\}\\
&= -\frac{(q^{-m}zt_1, q^{-m}zt_3; q)_\infty}{(q^{-m}; q)_m (q, q^{-m}z^2; q)_\infty}(zq^{-m})^{-n}\\
&\quad \times \sum_{k,j=0}^{\infty} \frac{(t_1z, t_1/z; q)_k (t_3z, t_3/z; q)_j}{(q, t_1t_2, t_1zq^{-m}; q)_k (q, t_3t_4, t_3zq^{-m}; q)_j}\\
&\quad \times q^{\binom{j}{2}+\binom{k}{2}}(-t_4)^j(-t_2)^k(q^{-m}z)^{j+k}\\
&= -z^{-n}\frac{(zt_1, zt_3; q)_\infty (q/zt_1, q/zt_3; q)_m}{(q, z^2; q)_\infty (q, q/z^2; q)_m}(t_1t_3q^n)^m\\
&\quad \times {}_2\phi_2\left(\begin{matrix} t_1z, t_1/z\\ t_1t_2, t_1zq^{-m}\end{matrix}\,\middle|\, q, t_2zq^{-m}\right) {}_2\phi_2\left(\begin{matrix} t_3z, t_3/z\\ t_3t_4, t_3zq^{-m}\end{matrix}\,\middle|\, q, t_4zq^{-m}\right).
\end{aligned}
$$

For the residue at $t = q^{-m}/z$ replace z by $1/z$, and we establish the theorem.

$\square$

Observe that the series (2.130) is both an explicit formula and an asymptotic series for large n.

2.8 Ladder operators and Rodrigues formulas

Theorem 2.35 *The Askey–Wilson polynomials satisfy the lowering and raising relations*

$$
\mathscr{D}_q p_n(x; \mathbf{t} | q) = 2\frac{(1-q^n)(1-t_1t_2t_3t_4q^{n-1})}{(1-q)q^{(n-1)/2}} p_{n-1}(x; q^{1/2}\mathbf{t} | q) \quad (2.132)
$$

$$
\frac{2q^{(1-n)/2}}{q-1}p_n(x; \mathbf{t} | q) = \\
\frac{1}{w(x; \mathbf{t} | q)}\mathscr{D}_q\left[w(x; q^{1/2}\mathbf{t} | q)p_{n-1}(x; q^{1/2}\mathbf{t} | q)\right], \quad (2.133)
$$

respectively, where $w(x; \mathbf{t} | q)$ is defined in (2.121).

Proof The lowering relation (2.132) follows from (2.34) and the representation (2.122). To prove (2.133) we use (2.132) and Theorem 2.2 to write the orthogonality relation (2.123) in the form

$$
2\frac{2\pi\,(t_1t_2t_3t_4q^{2n+2};q)_\infty\,(t_1t_2t_3t_4q^n;q)_{n+1}}{(q^{n+2};q)_\infty\,\prod_{1\le j<k\le 4}(t_jt_kq^{n+1};q)_\infty}\frac{q^{-n/2}}{1-q}\delta_{m,n},
$$
$$
-\int_{-1}^{1}p_m(x;q^{1/2}\mathbf{t}|q)\,w(x,q^{1/2}\mathbf{t}|q)\,\mathscr{D}_q p_{n+1}(x;\mathbf{t}|q)\,dx
$$
$$
=-\int_{-1}^{1}p_{n+1}(x;\mathbf{t}|q)\,\mathscr{D}_q\left[w(x;q^{1/2}\mathbf{t}|q)p_m(x;q^{1/2}\mathbf{t}|q)\right]\,dx.
$$

In the last step we used integration by parts for the Askey–Wilson operator (Theorem 2.2) and the fact that the boundary term vanishes since the weight function, as a function of z, vanishes when $z = q^{\pm 1/2}$. Therefore

$$
\frac{1}{w(x;\mathbf{t}|q)}\,\mathscr{D}_q\left[w(x;q^{1/2}\mathbf{t}|q)p_m(x;q^{1/2}\mathbf{t}|q)\right]
$$

must be a constant multiple of $p_{m+1}(x;\mathbf{t}|q)$ because the Askey–Wilson polynomials are complete in $L_2[-1,1,w(x;\mathbf{t}|q)]$. From the orthogonality relation we find that the sought multiple is $2q^{-m/2}/(q-1)$ and (2.133) follows. □

By iterating (2.133) one derive the Rodrigues-type formula

$$
w(x;\mathbf{t}\,|\,q)\,p_n(x;\mathbf{t}\,|\,q) = \left(\frac{q-1}{2}\right)^n q^{n(n-1)/4}\mathscr{D}_q^n\left[w(x;q^{n/2}\mathbf{t}|q)\right].\quad (2.134)
$$

Combining (2.132) and (2.133) we conclude that the Askey–Wilson polynomials solve the second order Sturm–Liouville equation

$$
\frac{1}{w(x;\mathbf{t}|q)}\,\mathscr{D}_q\left[w(x;q^{1/2}\mathbf{t}|q)\mathscr{D}_q y\right]
$$
$$
= \frac{4q}{(1-q)^2}(1-q^{-n})(1-t_1t_2t_3t_4q^{n-1})\,y.\quad (2.135)
$$

Theorem 2.36 *The Askey–Wilson polynomials have the $_8W_7$ representation*

$$
p_n(\cos\theta;\mathbf{t}|q) =
$$
$$
\frac{z^n\prod_{j=1}^4(t_j/z;q)_n}{(1/z^2;q)_n}{}_8W_7(q^{-n}z^2;q^{-n},t_1z,t_2z,t_3z,t_4z;q,q^{2-n}/t_1t_2t_3t_4).
$$
$$
(2.136)
$$

Proof From the Rodrigues formula (2.134) and Cooper's formula (2.29) and with $z = e^{i\theta}$ we see that

$$p_n(x;\mathbf{t}|q)$$

$$= \sum_{k=0}^{n} \begin{bmatrix} n \\ k \end{bmatrix}_q \frac{1-z^2 q^{n-2k}}{1-z^2} \frac{q^{k(1+n-k)} z^{2k-n}}{(q/z^2;q)_k (qz^2;q)_{n-k}} \prod_{j=1}^{4} (t_j z;q)_{n-k}(t_j/z;q)_k$$

$$= \sum_{k=0}^{n} \begin{bmatrix} n \\ k \end{bmatrix}_q \frac{1-z^2 q^{-n} q^{2k}}{1-z^2} \frac{q^{(n-k)(1+k)} z^{n-2k}}{(q/z^2;q)_{n-k}(qz^2;q)_k} \prod_{j=1}^{4} (t_j z;q)_k (t_j/z;q)_{n-k}.$$

After routine manipulations we arrive at the representation (2.136). □

Note that the representation (2.136) exhibits the symmetry of the Askey–Wilson polynomials under permutations of (t_1, t_2, t_3, t_4) but does not show the polynomial nature of $p_n(x;\mathbf{t}|q)$.

Theorem 2.37 *We have the Watson transformation*

$$_8\phi_7 \left(\begin{array}{c} a, qa^{1/2}, -qa^{1/2}, b, c, d, e, q^{-n} \\ a^{1/2}, -a^{1/2}, qa/b, qa/c, qa/d, qa/e, aq^{n+1} \end{array} \middle| q, \frac{a^2 q^{n+2}}{bcde} \right)$$

$$= \frac{(qa, qa/de;q)_n}{(qa/d, qa/e;q)_n} {}_4\phi_3 \left(\begin{array}{c} qa/bc, d, e, q^{-n} \\ qa/b, qa/c, deq^{-n}/a \end{array} \middle| q, q \right). \qquad (2.137)$$

Proof We equate the right-hand sides of the representations of $p_n(x;\mathbf{t}|q)$ in (2.122) and (2.136). After routine manipulations we arrive at the representation

$$p_n(\cos\theta;\mathbf{t}|q)$$

$$= t_1^{-n}(t_1 t_2, t_1 t_3, t_1 t_4;q)_n \, {}_4\phi_3 \left(\begin{array}{c} q^{-n}, t_1 t_2 t_3 t_4 q^{n-1}, t_1 e^{i\theta}, t_1 e^{-i\theta} \\ t_1 t_2, \, t_1 t_3, \, t_1 t_4 \end{array} \middle| q, q \right)$$

$$= \frac{z^n \prod_{j=1}^{4}(t_j/z;q)_n}{(1/z^2;q)_n} {}_8 W_7(q^{-n}z^2; q^{-n}, t_1 z, t_2 z, t_3 z, t_4 z; q, q^{2-n}/t_1 t_2 t_3 t_4).$$

$$(2.138)$$

To reach the form (2.137), we apply the Sears transformation (2.77) with $A = t_1 z, D = t_1 t_4$. □

We now evaluate certain integrals involving q-functions. Our first result is an integral representation for a ${}_6\phi_5$ function.

Theorem 2.38 *If $|t_j| < 1$ for $1 \le j \le 6$ then*

$$
\int_0^\pi \prod_{j=5}^6 {}_2\phi_1 \left(\begin{array}{c} t_1 e^{i\theta}, t_2 e^{i\theta} \\ t_1 t_2 \end{array} \bigg| q, \, t_j e^{-i\theta} \right) {}_2\phi_1 \left(\begin{array}{c} t_3 e^{-i\theta}, t_4 e^{-i\theta} \\ t_3 t_4 \end{array} \bigg| q, \, t_j e^{i\theta} \right)
$$

$$
\times \frac{(e^{2i\theta}, e^{-2i\theta}; q)_\infty}{\prod_{j=1}^4 (t_j e^{i\theta}, t_j e^{-i\theta}; q)_\infty} \, d\theta
$$

$$
= \frac{2\pi (t_1 t_2 t_3 t_4; q)_\infty}{(q;q)_\infty \prod_{1 \le j < k \le 4}(t_j t_k; q)_\infty}
$$

$$
\times {}_6\phi_5 \left(\begin{array}{c} \sqrt{t_1 t_2 t_3 t_4/q}, -\sqrt{t_1 t_2 t_3 t_4/q}, t_1 t_3, t_1 t_4, t_2 t_3, t_2 t_4 \\ \sqrt{t_1 t_2 t_3 t_4 q}, -\sqrt{t_1 t_2 t_3 t_4 q}, t_1 t_2, t_3 t_4, t_1 t_2 t_3 t_4/q \end{array} \bigg| q, \, t_5 t_6 \right),
$$
(2.139)

Proof Multiply the orthogonality relation (2.123) by

$$
\frac{t_5^m t_6^n}{(q, t_1 t_2, t_3 t_4; q)_m \, (q, t_1 t_2, t_3 t_4; q)_n}
$$

and add for all $m, n \ge 0$ and then make use of (2.124). $\square$

The next integral to consider is the Nassrallah–Rahman integral which we do in the following theorem.

Theorem 2.39 *We have the Nassrallah–Rahman integral evaluation*

$$
\int_0^\pi \frac{(e^{2i\theta}, e^{-2i\theta}; q)_\infty (\alpha e^{i\theta}, \alpha e^{-i\theta}; q)_\infty \, d\theta}{\prod_{j=1}^5 (t_j e^{i\theta}, t_j e^{-i\theta}; q)_\infty}
$$

$$
= \frac{2\pi (t_1 t_5, t_1 t_2 t_3 t_4, t_2 t_3 t_4 t_5; q)_\infty \prod_{j=2}^4 (\alpha t_j; q)_\infty}{(q, \alpha t_2 t_3 t_4; q)_\infty \prod_{1 \le j < k \le 5}(t_j t_k; q)_\infty}
$$

$$
\times {}_8 W_7 (\alpha t_2 t_3 t_4/q; t_2 t_3, t_2 t_4, t_3 t_4, \alpha/t_1, \alpha/t_5; q, t_1 t_5), \tag{2.140}
$$

when $|t_j| < 1; \, 1 \le j \le 5$.

The proof is based on the following lemma.

Lemma 2.40   *We have the evaluation*

$$\int_0^\pi \frac{(e^{2i\theta}, e^{-2i\theta}; q)_\infty (\alpha e^{i\theta}, \alpha e^{-i\theta}; q)_n \, d\theta}{\prod_{j=1}^4 (t_j e^{i\theta}, t_j e^{-i\theta}; q)_\infty}$$

$$= \frac{2\pi(\alpha/t_4, \alpha t_4; q)_n (t_1 t_2 t_3 t_4; q)_\infty}{(q;q)_\infty \prod_{1 \le j < k \le 4} (t_j t_k; q)_\infty} {}_4\phi_3 \left(\begin{array}{c} q^{-n}, t_1 t_4, t_2 t_4, t_3 t_4 \\ \alpha t_4, t_1 t_2 t_3 t_4, q^{1-n} t_4/\alpha \end{array} \middle| q, q \right)$$

$$= \frac{2\pi(t_1 t_2 t_3 t_4; q)_\infty (\alpha t_4; q)_n}{(q;q)_\infty \prod_{1 \le j < k \le 4} (t_j t_k; q)_\infty}$$

$$\times \sum_{k=0}^n \frac{(q;q)_n}{(q;q)_k} \frac{(t_1 t_4, t_2 t_4, t_3 t_4; q)_k (\alpha/t_4; q)_{n-k}}{(\alpha t_4, t_1 t_2 t_3 t_4; q)_k (q;q)_{n-k}} \left(\frac{\alpha}{t_4} \right)^k, \qquad (2.141)$$

where $|t_j| < 1$, $1 \le j \le 4$.

Proof   Denote the Askey–Wilson integral in (2.120) by $I(t_1, t_2, t_3, t_4)$. Apply (2.55) with $b = \alpha$ and $a = t_4$ to see that the extreme left-hand side of (2.141) is

$$\sum_{k=0}^n \frac{(q, \alpha t_4; q)_n (\alpha/t_4)^k}{(\alpha t_4, q; q)_k (q;q)_{n-k}} (\alpha/t_4; q)_{n-k} I(t_1, t_2, t_3, q^k t_4)$$

$$= \frac{2\pi(q, \alpha t_4; q)_n (t_1 t_2 t_3 t_4; q)_\infty}{(q;q)_\infty \prod_{1 \le j < k \le 4} (t_j t_k; q)_\infty} \sum_{k=0}^n \frac{(t_1 t_4, t_2 t_4, t_3 t_4; q)_k (\alpha/t_4; q)_{n-k}}{(q, \alpha t_4, t_1 t_2 t_3 t_4; q)_k (q;q)_{n-k}} (\alpha/t_4)^k,$$

and the lemma follows.   □

Proof of (2.140)   We first take $t_5 = \alpha q^n$ and apply Lemma 2.40. Apply the Watson transformation (2.137) to the ${}_4\phi_3$ in Lemma 2.40 with the choices:

$$aq = \alpha t_2 t_3 t_4, \quad b = t_2 t_3, \quad c = \alpha/t_1, \quad d = t_2 t_4, \quad e = t_3 t_4.$$

This establishes the theorem when $\alpha = t_5 q^n$. Since both sides of (2.140) are analytic functions of α the identity theorem for analytic functions establishes the result.   □

Note that if we did not know the right-hand side of (2.140) we would have discovered it by replacing q^n in our calculations by t_5/α, then continue the argument given in the proof. It is also important to note that the left-hand side of (2.140) is symmetric under interchanging t_j and t_k for any $1 \le j, k \le 5$. The right-hand side is obviously symmetric under $t_1 \leftrightarrow t_5$. The symmetry under $t_i \leftrightarrow t_j$ gives transformation formulas for the ${}_8 W_7$ functions.

We next solve the connection coefficient problem for the Askey–Wilson polynomials.

Theorem 2.41 *The Askey–Wilson polynomials have the connection relation*

$$p_n(x;\mathbf{b}|q) = \sum_{k=0}^{n} c_{n,k}(\mathbf{a},\mathbf{b})\, p_k(x;\mathbf{a}|q), \qquad (2.142)$$

where

$$
\begin{aligned}
& c_{n,k}(\mathbf{a},\mathbf{b}) \\
&= \frac{(q;q)_n\,(b_4^{k-n}b_1b_2b_3b_4q^{n-1};q)_k (b_1b_4,b_2b_4,b_3b_4;q)_n}{(q;q)_{n-k}(q,t_1t_2t_3t_4q^{k-1};q)_k (b_1b_4,b_2b_4,b_3b_4;q)_k} \\
&\quad \times q^{k(k-n)} \sum_{j,l\geq 0} \frac{(q^{k-n},b_1b_2b_3b_4q^{n+k-1},t_4b_4q^k;q)_{j+l}\,q^{j+l}}{(b_1b_4q^k,b_2b_4q^k,b_3b_4q^k;q)_{j+l}(q;q)_j(q;q)_l} \\
&\quad \times \frac{(a_1a_4q^k,a_2a_4q^k,a_3a_4q^k;q)_l(b_4/a_4;q)_j}{(a_4b_4q^k,a_1a_2a_3a_4q^{2k};q)_l}\left(\frac{b_4}{t_4}\right)^l. \qquad (2.143)
\end{aligned}
$$

Proof Denote the coefficient of $\delta_{m,n}$ in (2.123) by $h_n(\mathbf{t})$. The coefficients $c_{n,k}$ are given by

$$h_k(\mathbf{a})c_{n,k}(\mathbf{a},\mathbf{b}) = \left\langle \sqrt{1-x^2}\, p_n(x;\mathbf{b}|q), w(x;\mathbf{a}|q)p_k(x;\mathbf{a}|q) \right\rangle, \qquad (2.144)$$

where $\langle f,g\rangle$ is the inner product $\int_{-1}^{1} f(x)\overline{g(x)}(1-x^2)^{-1/2}\,dx$, see (2.23). We use the Rodrigues formula (2.134) and the integration by parts formula (2.24) to find

$$
\begin{aligned}
h_k(\mathbf{a})c_{n,k} &= \left[\frac{q-1}{2}\right]^k q^{k(k-1)/4} \left\langle \sqrt{1-x^2}\, p_n(x;\mathbf{b}|q), \mathscr{D}_q^k w(x;q^{k/2}\mathbf{a}|q) \right\rangle \\
&= \left[\frac{1-q}{2}\right]^k q^{k(k-1)/4} \left\langle \mathscr{D}_q^k p_n(x;\mathbf{b}|q), \sqrt{1-x^2}\, w(x;q^{k/2}\mathbf{a}|q) \right\rangle \\
&= q^{k(k-n)/2}(b_1b_2b_3b_4q^{n-1};q)_k \frac{(q;q)_n}{(q;q)_{n-k}} \\
&\quad \times \left\langle p_{n-k}(x;q^{k/2}\mathbf{b}|q), \sqrt{1-x^2}\, w(x;q^{k/2}\mathbf{a}|q) \right\rangle \\
&= b_4^{k-n}(b_1b_2b_3b_4q^{n-1};q)_k(b_1b_4q^k,b_2b_4q^k,b_3b_4q^k;q)_{n-k} \\
&\quad \times q^{k(k-n)}\frac{(q;q)_n}{(q;q)_{n-k}} \sum_{j=0}^{n-k} \frac{(q^{k-n},b_1b_2b_3b_4q^{n+k-1};q)_j}{(q,b_1b_4q^k,b_2b_4q^k,b_3b_4q^k;q)_j}q^j \\
&\quad \times \left\langle \phi_j(x;b_4q^{k/2}), \sqrt{1-x^2}\, w(x;q^{k/2}\mathbf{a}|q) \right\rangle.
\end{aligned}
$$

In the above steps we applied Lemma 2.2 repeatedly and used the fact that the boundary terms vanish since the weight function vanished when $z = q^{\pm j/2}$ for any non-negative integer j. Using Lemma 2.40 we see that the j-sum is

$$\frac{2\pi(t_1t_2t_3t_4q^{2k};q)_\infty}{(q;q)_\infty \prod_{1\leq r<s\leq 4}(t_rt_sq^k;q)_\infty} \sum_{j=0}^{n-k} \frac{(q^{k-n},b_1b_2b_3b_4q^{n+k-1},t_4b_4q^k;q)_j}{(b_1b_4q^k,b_2b_4q^k,b_3b_4q^k;q)_j}$$

$$\times \sum_{l=0}^{j} \frac{(a_1a_4q^k,a_2a_4q^k,a_3a_4q^k;q)_l}{(q,a_4b_4q^k,a_1a_2a_3a_4q^{2k};q)_l} \frac{(b_4/a_4;q)_{j-l}}{(q;q)_{j-l}} \left(\frac{b_4}{t_4}\right)^l,$$

and after some manipulations one completes the proof.   □

The special case $a_4 = b_4$ is worth noting.

Corollary 2.42   (Askey and Wilson [7]) *We have the connection relation*

$$p_n(x;b_1,b_2,b_3,a_4|q) = \sum_{k=0}^{n} d_{n,k}\, p_k(x;a_1,a_2,a_3,a_4|q), \qquad (2.145)$$

where

$$d_{n,k} = \frac{a_4^{k-n}(b_1b_2b_3a_4q^{n-1};q)_k(q,b_1a_4,b_2a_4,b_3a_4;q)_n}{(q;q)_{n-k}(q,a_1a_2a_3a_4q^{k-1};q)_k(b_1a_4,b_2a_4,b_3a_4;q)_k}q^{k(k-n)}$$

$$\times {}_5\phi_4\left(\begin{array}{c} q^{k-n},b_1b_2b_3a_4q^{n+k-1},a_1a_4q^k,a_2a_4q^k,a_3a_4q^k \\ b_1a_4q^k,b_2a_4q^k,b_3a_4q^k,a_1a_2a_3a_4q^{2k} \end{array}\Bigg|\, q,q\right).$$
$$(2.146)$$

The proof follows because the terms in the double series in (2.142) vanish unless $j = 0$ so the double sum reduces to a ${}_5\phi_4$.

The Askey–Wilson polynomials contain many special and limiting cases. For details see the Askey scheme in [31]. We will only mention the Al-Salam–Chihara polynomials introduced by W. Al-Salam and T. S. Chihara in [1]. Their weight function was first found in [6]. Al-Salam–Chihara polynomials correspond to the case $t_3 = t_4 = 0$ in the Askey–Wilson polynomials. They are defined by

$$p_n(x;t_1,t_2|q) = p_n(x;t_1,t_2,0,0|q)$$

$$= t_1^{-n}(t_1t_2;q)_n{}_3\phi_2\left(\begin{array}{c} q^{-n},t_1e^{i\theta},t_1e^{-i\theta} \\ t_1t_2,\ 0 \end{array}\Bigg|\, q,q\right). \qquad (2.147)$$

Their generating function is

$$\sum_{n=0}^{\infty} \frac{p_n(\cos\theta;t_1,t_2\,|\,q)}{(q;q)_n}t^n = \frac{(t_1t,t_2t;q)_\infty}{(te^{i\theta},te^{-i\theta};q)_\infty}. \qquad (2.148)$$

2.9 Identities and summation theorems

We next derive the sum of a very well-poised $_6\phi_5$ series.

Theorem 2.43 (Rogers) *The sum of a very well-poised $_6\phi_5$ series is*

$$
_6\phi_5\left(\begin{array}{c} A, qA^{1/2}, -qA^{1/2}, B, C, D \\ A^{1/2}, -A^{1/2}, qA/B, qA/C, qA/D \end{array} \;\middle|\; q, \; \frac{qA}{BCD}\right)
$$
$$
= \frac{(qA, qA/BC, qA/BD, qA/CD; q)_\infty}{(qA/B, qA/C, qA/D, qA/BCD; q)_\infty},
\tag{2.149}
$$

and when $D = q^{-n}$ is

$$
_6\phi_5\left(\begin{array}{c} A, qA^{1/2}, -qA^{1/2}, B, C, q^{-n} \\ A^{1/2}, -A^{1/2}, qA/B, qA/C, q^{n+1}A \end{array} \;\middle|\; q, \; \frac{q^{n+1}A}{BC}\right)
$$
$$
= \frac{(qA, qA/BC; q)_n}{(qA/B, qA/C; q)_n}.
\tag{2.150}
$$

Proof With $z = e^{i\theta}$ is straightforward to check that

$$
\mathscr{D}_q \frac{(az, a/z; q)_\infty}{(bz, b/z; q)_\infty} = \frac{2b(1-a/b)}{[q^{1/2} - q^{-1/2}]} \frac{(aq^{1/2}z, aq^{1/2}/z; q)_\infty}{(bq^{-1/2}z, bq^{-1/2}/z; q)_\infty}.
$$

Therefore

$$
\mathscr{D}_q^n \frac{(az, a/z; q)_\infty}{(bz, b/z; q)_\infty} = \frac{2^n b^n (a/b; q)_n q^{-n(n-1)/4}}{(q^{1/2} - q^{-1/2})^n} \frac{(aq^{n/2}z, aq^{n/2}/z; q)_\infty}{(bq^{-n/2}z, bq^{-n/2}/z; q)_\infty}.
$$

Using (2.29) we see that the left-hand side of the above equation is

$$
\frac{2^n q^{n(1-n)/4}}{(q^{1/2} - q^{-1/2})^n} \times
$$
$$
\sum_{k=0}^{n} \begin{bmatrix} n \\ k \end{bmatrix}_q \frac{q^{k(n-k)}z^{2k-n}}{(q^{1+n-2k}z^2; q)_k (q^{2k-n+1}z^{-2}; q)_{n-k}} \frac{(aq^{-k+n/2}z, aq^{k-n/2}/z; q)_\infty}{(bq^{-k+n/2}z, bq^{k-n/2}/z; q)_\infty}.
$$

We note that

$$
\frac{(bq^{-n/2}z, bq^{-n/2}/z; q)_\infty}{(aq^{n/2}z, aq^{n/2}/z; q)_\infty} \frac{(aq^{-k+n/2}z, aq^{k-n/2}/z; q)_\infty}{(bq^{-k+n/2}z, bq^{k-n/2}/z; q)_\infty}
$$
$$
= (bq^{-n/2}z, bq^{-n/2}/z; q)_n \frac{(aq^{-k+n/2}z; q)_k (aq^{k-n/2}/z; q)_{n-k}}{(bq^{-k+n/2}z; q)_k (bq^{k-n/2}/z; q)_{n-k}}.
$$

After some simplifications we find that

$$
\frac{b^n (a/b, q^{1-n}/z^2; q)_n}{(bq^{-n/2}z, bq^{-n/2}/z; q)_n} =
$$
$$
_6\phi_5\left(\begin{array}{c} q^{-n}/z^2, q^{1-n/2}/z, -q^{1-n/2}/z, bq^{-n/2}/z, q^{1-n/2}/az, q^{-n} \\ q^{-n/2}/z, -q^{-n/2}/z, q^{1-n/2}/bz, aq^{-n/2}/z, q/z^2 \end{array} \;\middle|\; q, \; \frac{q^n a}{b}\right)
$$

which is equivalent to (2.150). Now both sides of (2.149) are analytic in $1/D$ for $|1/D| < 1$ and are equal when $1/D = q^n$ by (2.150); therefore they are equal for all D, $|1/D| < 1$. We then analytically continue the result for all D for which both sides are well defined.   □

Another proof of (2.150) follows by choosing $d = \lambda/c$ in Theorem 2.37 and then letting $c \to 0$.

Theorem 2.44   (Bailey) *We have the* $_6\psi_6$ *sum*

$$_6\psi_6 \left(\begin{array}{c} qa^{1/2}, -qa^{1/2}, b, c, d, e \\ a^{1/2}, -a^{1/2}, aq/b, aq/c, aq/d, aq/e \end{array} \middle| q, \frac{qa^2}{bcde} \right)$$

$$= \frac{(qa, qa/bc, qa/bd, qa/be, qa/cd, qa/ce, qa/de, q, q/a; q)_\infty}{(qa/b, qa/c, qa/d, qa/e, q/b, q/c, q/d, q/e, qa^2/bcde; q)_\infty}. \quad (2.151)$$

Proof   Note that both sides of (2.151) are analytic in $z := qa/e$. If $z = q^{m+1}$, then $e = aq^{-m}$. Using (2.4) we see that the sum in the $_6\psi_6$ is now over all n, $n \geq -m$. Thus the $_6\psi_6$ becomes

$$\sum_{n=-m}^\infty \frac{(qa^{1/2}, -qa^{1/2}, b, c, d, aq^{-m}; q)_n}{(a^{1/2}, -a^{1/2}, aq/b, aq/c, aq/d, q^{m+1}; q)_n} \left(\frac{q^{m+1}a}{bcd} \right)^n$$

$$= \sum_{n=0}^\infty \frac{(qa^{1/2}, -qa^{1/2}, b, c, d, aq^{-m}; q)_{n-m}}{(a^{1/2}, -a^{1/2}, aq/b, aq/c, aq/d, q^{m+1}; q)_{n-m}} \left(\frac{q^{m+1}a}{bcd} \right)^{n-m}$$

$$= \frac{(qa^{1/2}, -qa^{1/2}, b, c, d, aq^{-m}; q)_{-m}}{(a^{1/2}, -a^{1/2}, aq/b, aq/c, aq/d, q^{m+1}; q)_{-m}} \left(\frac{q^{m+1}a}{bcd} \right)^{-m}$$

$$\times \, _6\phi_5 \left(\begin{array}{c} aq^{-2m}, q^{1-m}a^{1/2}, -q^{1-m}a^{1/2}, bq^{-m}, cq^{-m}, dq^{-m} \\ q^{-m}a^{1/2}, -q^{-m}a^{1/2}, aq^{1-m}/b, aq^{1-m}/c, aq^{1-m}/d \end{array} \middle| q, \frac{q^{m+1}a}{bcd} \right)$$

$$= \left(\frac{bcd}{q^{m+1}a} \right)^m \frac{(q^{1-2m}a, qa/bc, qa/bd, qa/cd; q)_\infty}{(qa/b, qa/c, qa/d, q^{m+1}a/bcd; q)_\infty}$$

$$\times \frac{(b, c, d, q^{-m}a, qa^{1/2}, -qa^{-1/2}; q)_{-m}}{(a^{1/2}, -a^{1/2}, q^{m+1}; q)_{-m}},$$

where we used (2.149). Note that

$$\frac{(q^{1-2m}a; q)_\infty (q^{-m}a, qa^{1/2}, -qa^{1/2}; q)_{-m}}{(a^{1/2}, -a^{1/2}, q^{m+1}; q)_{-m}} = (q^{-m}a, q; q)_m (qa; q)_\infty.$$

This shows that the $_6\psi_6$ is

$$\frac{(qa/bc, qa/bd, qa/cd; q)_\infty}{(qa/b, qa/c, qa/d, q^{m+1}a/bcd; q)_\infty} \frac{(q/a; q)_m (qa, q; q)_\infty}{(q/b, q/c, q/d; q)_m},$$

which, with $e = aq^{-m}$, takes the form

$$\frac{(q, qa/bc, qa/bd, qa/cd; q)_\infty}{(qa/b, qa/c, qa/d, qa/e, qa^2/bcde; q)_\infty} \cdot \frac{(qa, q/a, qa/be, qa/ce, qa/de; q)_\infty}{(q/b, q/c, q/d, q/e; q)_\infty}.$$

This completes the proof. $\square$

2.10 Expansions

This section is mostly based on [20] and [26]. We first expand an Askey–Wilson basis in terms of Askey–Wilson polynomials.

Proposition 2.45 *For any non-negative n,*

$$(be^{i\theta}, be^{-i\theta}; q)_n = \sum_{k=0}^{n} f_{n,k}(b, \mathbf{t}) p_k(x; \mathbf{t}|q), \tag{2.152}$$

where

$$f_{n,k}(b, \mathbf{t}) = \frac{(-b)^k q^{\binom{k}{2}} (q; q)_n (b/t_4, bt_4 q^k; q)_{n-k}}{(q, t_1 t_2 t_3 t_4 q^{k-1}; q)_k (q; q)_{n-k}}$$

$$\times {}_4\phi_3 \left(\begin{array}{c} q^{k-n}, \; t_1 t_4 q^k, \; t_2 t_4 q^k, \; t_3 t_4 q^k \\ bt_4 q^k, t_1 t_2 t_3 t_4 q^{2k}, q^{1+k-n} t_4/b \end{array} \bigg| q, q \right). \tag{2.153}$$

Proof Denote the coefficient of $\delta_{m,n}$ in (2.123) by $h_k(\mathbf{t})$. It is clear that

$$f_{n,k} h_k(\mathbf{t}) = \langle p_k(x; \mathbf{t}|q) w(x; \mathbf{t}|q), \sqrt{1-x^2} (be^{i\theta}, be^{-i\theta}; q)_n \rangle$$

$$= \left(\frac{q-1}{2} \right)^k q^{k(k-1)/4} \langle \mathscr{D}_q^k w(x; q^{k/2} \mathbf{t}|q), \sqrt{1-x^2} (be^{i\theta}, be^{-i\theta}; q)_n \rangle$$

$$= \left(\frac{1-q}{2} \right)^k q^{k(k-1)/4} \int_{-1}^{1} w(x; q^{k/2} \mathbf{t}|q) \mathscr{D}_q^k (be^{i\theta}, be^{-i\theta}; q)_n dx$$

$$= \frac{(-b)^k (q; q)_n}{(q; q)_{n-k}} q^{\binom{k}{2}} \int_{-1}^{1} (bq^{k/2} e^{i\theta}, bq^{k/2} e^{-i\theta}; q)_{n-k} w(x; q^{k/2} \mathbf{t}|q) dx.$$

In the above steps we used the Rodrigues formula (2.134), Theorem 2.2, and (2.34). The result follows from Lemma 2.40. $\square$

The special case $b = t_1$ of Proposition 2.45 is interesting.

$$(t_1 e^{i\theta}, t_1 e^{-i\theta}; q)_n = \sum_{k=0}^{n} \begin{bmatrix} n \\ k \end{bmatrix}_q (-t_1)^k q^{\binom{k}{2}} \frac{(t_1 t_2, t_1 t_3, t_1 t_4; q)_n}{(t_1 t_2, t_1 t_3, t_1 t_4; q)_k} \frac{1 - t_1 t_2 t_3 t_4 q^{2k-1}}{1 - t_1 t_2 t_3 t_4/q}$$

$$\times \frac{(t_1 t_2 t_3 t_4/q; q)_k}{(t_1 t_2 t_3 t_4; q)_{n+k}} p_k(x; \mathbf{t}|q). \tag{2.154}$$

For completeness we mention without a proof the following generalization of Proposition 2.45 due to Ismail and Simeonov [22].

Theorem 2.46 *The Askey–Wilson polynomials $\{p_n(x;\mathbf{a}|q)\}$ have the two parameter generating function*

$$\sum_{n=0}^{\infty} c_n(u,t,\mathbf{a})p_n(x;\mathbf{a}|q) = \frac{(ue^{i\theta}, ue^{-i\theta};q)_\infty}{(te^{i\theta}, te^{-i\theta};q)_\infty}, \qquad (2.155)$$

when $\max\{|a_1|,\ldots,|a_4|,|t|\} < 1$, where

$$c_n(u,t,\mathbf{a}) = \frac{t^n(u/t;q)_n(q^n a_1 u, q^n a_2 u, q^n a_3 u, q^n a_1 a_2 a_3 t;q)_\infty}{(q,q^{n-1}a_1 a_2 a_3 a_4;q)_n(a_1 t, a_2 t, a_3 t, q b_1;q)_\infty}$$

$$\times {}_8\phi_7 \left(\begin{matrix} b_1, qb_1^{1/2}, -qb_1^{1/2}, q^n a_1 a_2, q^n a_1 a_3, q^n a_2 a_3, u/a_4, q^n u/t \\ b_1^{1/2}, -b_1^{1/2}, q^n a_1 u, q^n a_2 u, q^n a_3 u, q^{2n}a_1 a_2 a_3 a_4, q^n a_1 a_2 a_3 t \end{matrix} \,\middle|\, q, a_4 t \right),$$

$$(2.156)$$

with $b_1 = q^{2n-1}a_1 a_2 a_3 u$. In particular,

$$c_n(0,t,\mathbf{a}) = \frac{t^n(q^n a_1 a_2 a_3 t;q)_\infty}{(q,q^{n-1}a_1 a_2 a_3 a_4;q)_n(a_1 t, a_2 t, a_3 t;q)_\infty}$$

$$\times {}_3\phi_2 \left(\begin{matrix} q^n a_1 a_2, q^n a_1 a_3, q^n a_2 a_3 \\ q^{2n}a_1 a_2 a_3 a_4, q^n a_1 a_2 a_3 t \end{matrix} \,\middle|\, q, a_4 t \right). \qquad (2.157)$$

Theorem 2.47 *We have the expansion*

$$\sum_{n=0}^{\infty} \frac{(t_4 z, t_4/z;q)_n}{(q;q)_n} \Lambda_n \zeta^n = \sum_{k=0}^{\infty} p_k(x;\mathbf{t}|q) \frac{(-t_4\zeta)^k q^{\binom{k}{2}}}{(q, t_1 t_2 t_3 t_4 q^{k-1};q)_k}$$

$$\times \sum_{n=0}^{\infty} \Lambda_{n+k} \frac{(t_1 t_4 q^k, t_2 t_4 q^k, t_3 t_4 q^k)_n}{(q, t_1 t_2 t_3 t_4 q^{2k})_n} \zeta^n.$$

$$(2.158)$$

Proof Interchange t_1 and t_4 in (2.154), multiply the result by $\Lambda_n \zeta^n/(q;q)_n$ and then sum over n. □

When the Λ_n is a quotient of products of q-shifted factorials we establish the following corollary.

Corollary 2.48 *We have the following expansion*

$${}_{p+1}\phi_p \left(\begin{matrix} a_1, \ldots, a_{p-1}, t_4 z, t_4/z \\ t_1 t_4, t_2 t_4, t_3 t_4, b_1, \ldots, b_{p-3} \end{matrix} \,\middle|\, q, \zeta \right)$$

$$= \sum_{k=0}^{\infty} p_k(x;\mathbf{t}|q) \frac{(a_1, \ldots, a_{p-1};q)_k}{(t_1 t_4, t_2 t_4, t_3 t_4, b_1, \ldots, b_{p-3};q)_k}$$

$$\times \frac{(-t_4\zeta)^k q^{\binom{k}{2}}}{(q, t_1 t_2 t_3 t_4 q^{k-1};q)_k} {}_{p-1}\phi_{p-2} \left(\begin{matrix} q^k a_1, \cdots, q^k a_{p-1} \\ q^k b_1, \cdots, q^k b_{p-3}, t_1 t_2 t_3 t_4 q^{2k} \end{matrix} \,\middle|\, q, \zeta \right).$$

The special case $p = 2$ and a special choice of the parameters lead to the following theorem.

Theorem 2.49 *We have the expansion formula*

$$
{}_3\phi_2\left(\begin{matrix} A, bz, b/z \\ bt_4, B \end{matrix}\,\middle|\, q, \frac{Bt_4}{bA}\right)
$$

$$
= \frac{(B/A, Bt_4/b; q)_\infty}{(B, Bt_4/bA; q)_\infty} \sum_{k=0}^{\infty} \frac{(A; q)_k}{(bt_4, Bt_4/b; q)_k} \left(\frac{Bt_4}{bA}\right)^k \frac{(-b)^k q^{\binom{k}{2}}}{(q, t_1 t_2 t_3 t_4 q^{k-1}; q)_k}
$$

$$
\times {}_4\phi_3\left(\begin{matrix} Aq^k, t_1 t_4 q^k, t_2 t_4 q^k, t_3 t_4 q^k \\ bt_4 q^k, t_1 t_2 t_3 t_4 q^{2k}, q^k Bt_4/b \end{matrix}\,\middle|\, q, \frac{B}{A}\right) p_k(x; \mathbf{t}|q).
$$

Theorem 2.49 appeared as Theorem 2.9 in [26]. Another result from [26] follows directly from (2.154).

Proposition 2.50 *We have the expansion*

$$
\sum_{n=0}^{\infty} \frac{\Lambda_n}{(q, bt_4; q)_n}(bz, b/z; q)_n
$$

$$
= \sum_{k=0}^{\infty} p_k(x; \mathbf{t}|q) \frac{(-b)^k q^{\binom{k}{2}}}{(q, bt_4, t_1 t_2 t_3 t_4 q^{k-1}; q)_k}
$$

$$
\times \sum_{s=0}^{\infty} \frac{(t_1 t_4 q^k, t_2 t_4 q^k, t_3 t_4 q^k; q)_s}{(q, bt_4 q^k, t_1 t_2 t_3 t_4 q^{2k}; q)_s} \left(\frac{b}{t_4}\right)^s \sum_{n=0}^{\infty} \frac{(b/t_4; q)_n}{(q; q)_n} \Lambda_{n+k+s}.
$$

Another choice for Λ_n in Proposition 2.50 is

$$
\Lambda_n = \frac{(q^{-N}, A; q)_n}{(B, q^{1-N} Ab/Bt_4; q)_n} q^n.
$$

This time the n-sum is evaluatable by the q-Pfaff–Saalschütz theorem. The result is the following.

Theorem 2.51 *The expansion of a general terminating* ${}_4\phi_3$ *in the Askey–Wilson polynomials is given by*

$$
{}_4\phi_3\left(\begin{matrix} q^{-N}, A, bz, b/z \\ bt_4, B, bAq^{1-N}/Bt_4 \end{matrix}\,\middle|\, q, q\right)
$$

$$
= \frac{(B/A, Bt_4/b; q)_N}{(B, Bt_4/Ab; q)_N} \sum_{k=0}^{N} \frac{(-t_4)^k q^{\binom{k+1}{2}}(q^{-N}, A; q)_k}{(q, bt_4, t_1 t_2 t_3 t_4 q^{k-1}, Bt_4/b, q^{1-N} A/B; q)_k} p_k(x; \mathbf{t}|q)
$$

$$
\times {}_5\phi_4\left(\begin{matrix} q^{-N+k}, Aq^k, t_1 t_4 q^k, t_2 t_4 q^k, t_3 t_4 q^k \\ bt_4 q^k, Bt_4 q^k/b, Aq^{k+1-N}/B, t_1 t_2 t_3 t_4 q^{2k}, \end{matrix}\,\middle|\, q, q\right).
$$

In Theorem 2.51 if we replace A by Aq^{N-1}, we can then identify parameters a_2, a_3 such that the ${}_4\phi_3$ therein is a multiple of $p_N(x; b, a_2, a_3, t_4 | q)$. As such Theorem 2.51 is equivalent to a connection coefficient problem solved in [7]. We also note that although Theorem 2.49 is the limiting case $N \to \infty$ of Theorem 2.51, Theorem 2.49 is not available in the literature.

Remark 2.52 If we specialize Theorem 2.51 to

$$b = t_2, \quad B = t_1 t_2, \quad z = t_3.$$

the ${}_5\phi_4$ in Theorem 2.51 reduces to a balanced ${}_3\phi_2$ which is again evaluable by the q-Pfaff–Saalschütz theorem [16, (II.12)]. The resulting identity is the terminating case of the Watson transformation [16, (III.18)]. The nonterminating case Watson transformation [16, (III.18)] follows by analytic continuation in the variable $d = q^N$.

We now go back to (2.158) and observe that $\{(t_4 z, t_4/z; q)_n\}$ is a basis for the space of polynomials; therefore we can replace $(t_4 z, t_4/z; q)_n$ by $A_n(t_4 z, t_4/z; q)_n$ and (2.158) will remain valid as long as the series on both sides converge. This establishes the following expansion theorem.

Proposition 2.53 *We have the general expansion*

$$\sum_{n=0}^{\infty} \frac{(az, a/z; q)_n}{(q; q)_n} A_n B_n \zeta^n$$

$$= \sum_{k=0}^{\infty} \frac{(-\zeta)^k q^{\binom{k}{2}}}{(q, Cq^{k-1}; q)_k} \left[\sum_{j=0}^{k} \frac{(q^{-k}, Cq^{k-1}; q)_j}{(q; q)_j} A_j (az, a/z; q)_j q^j \right]$$

$$\times \left[\sum_{n=0}^{\infty} \frac{B_{n+k} \zeta^n}{(q, Cq^{2k}; q)_n} \right]. \tag{2.159}$$

Proposition 2.53 writes a triple sum as a single sum. Fields and Ismail [12] pointed out that identities like (2.159) follow from a matrix inversion for upper triangular matrices. For the definition of inverse relations see [37]. Indeed if $A = (a_{i,j}), B = (b_{i,j})$ are two upper infinite triangular matrices and $B = A^{-1}$, and $\{u_n(x)\}$ is a sequence of polynomials with u_n of degree n then

$$P_n(x) = \sum_{j=0}^{n} a_{n,j} A_j u_j(x) \iff A_m u_m(x) = \sum_{n=0}^{m} b_{m,n} P_n(x). \tag{2.160}$$

We now find that for any basis $\{v_n(x)\}$ of the space of polynomials we have

$$\sum_{m=0}^{\infty} A_m B_m u_m(x) v_m(y)$$

$$= \sum_{n=0}^{\infty} \left(\sum_{j=0}^{n} a_{n,j} A_j u_j(x) \right) \sum_{m=0}^{\infty} B_{m+n} v_{m+n}(y) b_{m+n,n}. \qquad (2.161)$$

Since we can interchange A and B we also have the dual expansion

$$\sum_{m=0}^{\infty} A_m B_m u_m(x) v_m(y)$$

$$= \sum_{n=0}^{\infty} \left(\sum_{j=0}^{n} b_{n,j} A_j u_j(x) \right) \sum_{m=0}^{\infty} B_{m+n} v_{m+n}(y) a_{m+n,n}. \qquad (2.162)$$

Formulas of the type in (2.159) have a long history. Fields and Wimp [13] expanded hypergeometric functions into Jacobi type polynomials. In [42], Verma generalized their expansion to expansion with arbitrary coefficients in Jacobi type polynomials. His formula is

$$\sum_{m=0}^{\infty} a_m b_m \frac{(zw)^m}{m!}$$

$$= \sum_{n=0}^{\infty} \frac{(-z)^n}{n!\,(\gamma+n)_n} \left(\sum_{r=0}^{\infty} \frac{b_{n+r}\, z^r}{r!\,(\gamma+2n+1)_r} \right) \left[\sum_{s=0}^{n} \frac{(-n)_s (n+\gamma)_s}{s!} a_s w^s \right].$$
$$\qquad (2.163)$$

Fields and Wimp as well as Verma noted a Laguerre type expansion where w is replaced by w/γ, b_n is replaced by $\gamma^n b_n$ and $\gamma \to \infty$. The expansion (2.159) extends all those expansions to expansions in Askey–Wilson type polynomials.

Remark 2.54 One may take Λ_n to be 0 unless $n \equiv a \pmod{b}$ for fixed integers $a, b, a > 0, b \geq 0$. This leads to hypergeometric expansions where the differences of consecutive parameters in a certain group is $1/b$.

2.11 Askey–Wilson expansions

Andrews [2] proved the terminating basic hypergeometric identity

$$\,_5\phi_4 \left(\begin{matrix} q^{-N}, \rho_1, \rho_2, b, c \\ \rho_1 \rho_2 q^{-N}/a, e, f, g \end{matrix} \bigg| q, q \right) = \frac{(aq/\rho_1, aq/\rho_2; q)_N}{(aq, aq/\rho_1\rho_2; q)_N}$$

$$\times \sum_{n=0}^{N} \frac{(q^{-N}, \rho_1, \rho_2, a; q)_n (1 - aq^{2n})}{(q, aq/\rho_1, aq/\rho_2, aq^{N+1}; q)_n (1-a)} \left(\frac{aq^{N+1}}{\rho_1 \rho_2} \right)^n u_n, \qquad (2.164)$$

where N is a non-negative integer, $qabc = efg$, and

$$u_n = {}_4\phi_3\left(\begin{matrix} q^{-n}, aq^n, b, c \\ e, \quad f, \quad g \end{matrix}\,\middle|\, q, q\right). \qquad (2.165)$$

Remark 2.55 The Andrews formula (2.164) is the case $p = 4$ in Corollary 2.48 with the parameter identification

$$a_1 = q^{-N}, \quad a_2 = \rho_1, \quad a_3 = \rho_2, \quad b_1 = \rho_1\rho_2 q^{-N}/a, \quad \zeta = q.$$

In this case the ${}_3\phi_2$ can be summed by the q-Pfaff–Saalschütz theorem, [16, (II.12)].

Remark 2.56 Another application of Corollary 2.48 is the case $p = 6$ with

$$a_1 = q^{-N}, \quad a_2 = c_1c_2c_3t_4q^{N-1}, \quad a_j = t_{j-2}t_4 \quad \text{for } 3 \le j \le 5,$$
$$b_k = t_4c_k \quad \text{for } 1 \le j \le 3.$$

The reader is encouraged to write down the resulting formula.

Upon setting $z = t_1$ in Corollary 2.48, we have the next corollary.

Corollary 2.57 *We have the following identity*

$${}_p\phi_{p-1}\left(\begin{matrix} a_1,\ldots,a_{p-1},t_4/t_1 \\ t_2t_4, t_3t_4, b_1,\ldots, b_{p-3} \end{matrix}\,\middle|\, q, \zeta\right)$$

$$= \sum_{k=0}^{\infty} \frac{(a_1,\ldots,a_{p-1}, t_1t_2, t_1t_3; q)_k}{(t_2t_4, t_3t_4, b_1, \cdots, b_{p-3}; q)_k}$$

$$\times \frac{(-t_4\zeta/t_1)^k q^{\binom{k}{2}}}{(q, t_1t_2t_3t_4q^{k-1}; q)_k} \,{}_{p-1}\phi_{p-2}\left(\begin{matrix} q^ka_1, \cdots, q^ka_{p-1} \\ q^kb_1, \cdots, q^kb_{p-3}, t_1t_2t_3t_4q^{2k} \end{matrix}\,\middle|\, q, \zeta\right).$$

We note that by equating coefficients of ζ^n on both sides of the equation in Corollary 2.57, it is equivalent to the sum of a terminating very well-poised ${}_6\phi_5$, [16, (II.21)].

We proceed to derive other generating functions for the Askey–Wilson polynomials. In this section we give two generating functions for Askey–Wilson polynomials: Theorem 2.58, which follows from Proposition 2.45, and Theorem 2.59, for which we provide an independent proof.

Theorem 2.58 *The Askey–Wilson polynomials have the generating func-*

tion

$$\frac{(be^{i\theta}, be^{-i\theta}; q)_\infty}{(bt_4, b/t_4; q)_\infty} = \sum_{k=0}^{\infty} \frac{(-b)^k q^{\binom{k}{2}}}{(q, bt_4, t_1t_2t_3t_4q^{k-1}; q)_k} p_k(x; \mathbf{t}|q)$$
$$\times {}_3\phi_2 \left(\begin{array}{c} t_1t_4q^k, t_2t_4q^k, t_3t_4q^k \\ bt_4q^k, t_1t_2t_3t_4q^{2k} \end{array} \bigg| q, \frac{b}{t_4} \right) \qquad (2.166)$$

and satisfy the relationship

$$\frac{(t_1z, t_1/z, t_1t_2t_3t_4; q)_\infty}{(t_1t_2, t_1t_3, t_1t_4; q)_\infty}$$
$$= \sum_{k=0}^{\infty} \frac{(-t_1)^k (t_1t_2t_3t_4/q; q)_k}{(q, t_1t_2, t_1t_3, t_1t_4; q)_k} q^{\binom{k}{2}} \frac{1 - t_1t_2t_3t_4q^{2k-1}}{1 - t_1t_2t_3t_4/q} p_k(x; \mathbf{t}|q). \qquad (2.167)$$

Proof To prove (2.166) we let $n \to \infty$ in Proposition 2.45. Taking the limit inside the sum is justified by Tannery's theorem, [9], the discrete analogue of the Lebesgue dominated convergence theorem. We omit the details. The identity (2.167) is the case $b = t_1$ of (2.166), because the ${}_3\phi_2$ becomes a ${}_2\phi_1$ and is summed by the q-Gauss theorem, [16, (II.8)]. $\qquad \square$

One may ask for a version of Theorem 2.58 in which the infinite products in z are in the denominator.

Theorem 2.59 *The Askey–Wilson polynomials have the generating function*

$$\frac{1}{(be^{i\theta}, be^{-i\theta}; q)_\infty} = \sum_{n=0}^{\infty} c_n(\mathbf{t}, b) p_n(x; \mathbf{t}|q), \qquad (2.168)$$

where

$$c_n(\mathbf{t}, b) = \frac{b^n (t_2t_3t_4bq^n; q)_\infty}{(q, t_1t_2t_3t_4q^{n-1}; q)_n \prod_{j=2}^{4}(t_jb; q)_\infty}$$
$$\times {}_3\phi_2 \left(\begin{array}{c} q^n t_2t_3, q^n t_2t_4, q^n t_3t_4 \\ q^{2n}t_1t_2t_3t_4, q^n t_2t_3t_4b \end{array} \bigg| q, t_1b \right). \qquad (2.169)$$

Of course Theorem 2.59 is a special case of Theorem 2.46 but we will give a proof of the special case since we did not provide a prove of the more general result.

Proof of Theorem 2.59 We use two facts to prove Theorem 2.59. The first fact is the orthogonality relation for Askey–Wilson polynomials (2.123), where the coefficient of $\delta_{m,n}$ is $h_n(\mathbf{t})A(\mathbf{t})$,

$$
\left.
\begin{aligned}
h_n(\mathbf{t}) &= \frac{(q;q)_n \prod_{1\leq j<k\leq 4}(t_jt_k;q)_n (t_1t_2t_3t_4q^{n-1};q)_n}{(t_1t_2t_3t_4;q)_{2n}}, \\
A(\mathbf{t}) &= \frac{2\pi(t_1t_2t_3t_4;q)_\infty}{(q;q)_\infty \prod_{1\leq j<k\leq 4}(t_jt_k;q)_\infty},
\end{aligned}
\right\} \tag{2.170}
$$

and we have assumed that $\max\{|t_1|,|t_2|,|t_3|,|t_4|\}<1$. The second fact is

$$
\begin{aligned}
&\frac{(q;q)_\infty}{2\pi}\int_0^\pi \frac{w(\cos\theta;x_1,x_2,x_3,x_4)}{(x_5e^{i\theta},x_5e^{-i\theta};q)_\infty}\sin\theta\,d\theta \\
&= \frac{(x_1x_2x_3x_4,x_2x_3x_4x_5,x_1x_5;q)_\infty}{\prod_{1\leq r<s\leq 5}(x_rx_s;q)_\infty} \\
&\times {}_3\phi_2\left(\begin{matrix} x_2x_3,x_2x_4,x_3x_4 \\ x_1x_2x_3x_4,\ x_2x_3x_4x_5 \end{matrix}\;\middle|\; q,x_1x_5\right).
\end{aligned} \tag{2.171}
$$

The integral (2.171) is a special case of the Nassrallah–Rahman integral (2.140).

For symmetry we replace b by t_5. We shall find the coefficient $c_n(\mathbf{t},t_5)$ of $p_n(x;\mathbf{t}|q)$ using orthogonality, setting

$$
\sum_{n=0}^\infty c_n(\mathbf{t},t_5)p_n(x;\mathbf{t}|q) = \frac{1}{(t_5e^{i\theta},t_5e^{-i\theta})_\infty}.
$$

Such a formula exists because the right-hand side is $\in L^2[-1,1,w(x;\mathbf{t}|q)]$. Moreover

$$
c_n(\mathbf{t},t_5)h_n(\mathbf{t})A(\mathbf{t}) = \int_{-1}^1 \frac{w(x;\mathbf{t}|q)}{(t_5e^{i\theta},t_5e^{-i\theta};q)_\infty}\,p_n(x;\mathbf{t}|q)dx.
$$

Therefore, using (2.122) we see that

$$
\begin{aligned}
c_n(\mathbf{t},t_5)h_n(\mathbf{t})A(\mathbf{t}) &= \frac{(t_1t_2,t_1t_3,t_1t_4;q)_n}{t_1^n}\sum_{k=0}^n \frac{(q^{-n},t_1t_2t_3t_4q^{n-1};q)_k}{(q,t_1t_2,t_1t_3,t_1t_4;q)_k}q^k \\
&\times \int_0^\pi \frac{w(\cos\theta;t_1q^k,t_2,t_3,t_4|q)}{(t_5e^{i\theta},t_5e^{-i\theta};q)_\infty}\sin\theta d\theta.
\end{aligned}
$$

The integral is now evaluated by (2.171) and we obtain

$$
\begin{aligned}
\frac{(q;q)_\infty}{2\pi}c_n(\mathbf{t},t_5)h_n(\mathbf{t})A(\mathbf{t}) &= \frac{(t_1t_2,t_1t_3,t_1t_4;q)_n}{t_1^n}\sum_{k=0}^n \frac{(q^{-n},t_1t_2t_3t_4q^{n-1};q)_k}{(q,t_1t_2,t_1t_3,t_1t_4;q)_k}q^k \\
&\times \frac{(q^kt_1t_2t_3t_4,t_2t_3t_4t_5,q^kt_1t_5;q)_\infty}{\prod_{j=2}^5(q^kt_1t_j;q)_\infty \prod_{2\leq r<s\leq 5}(t_rt_s;q)_\infty}\,{}_3\phi_2\left(\begin{matrix} t_2t_3,t_2t_4,t_3t_4 \\ q^kt_1t_2t_3t_4,\ t_2t_3t_4t_5 \end{matrix}\;\middle|\; q,q^kt_1t_5\right).
\end{aligned}
$$

Write the $_3\phi_2$ as a sum over s and interchange the k and s sums to see that

$$
\frac{(q;q)_\infty}{2\pi} c_n(\mathbf{t},t_5) h_n(\mathbf{t}) A(\mathbf{t}) = \frac{(t_1t_2,t_1t_3,t_1t_4;q)_n (t_1t_2t_3t_4,t_2t_3t_4t_5;q)_\infty}{t_1^n \prod_{j=2}^{4}(t_1t_j;q)_\infty \prod_{2<r<s\le5}(t_rt_s;q)_\infty}
$$
$$
\times \sum_{s=0}^{\infty} \frac{(t_2t_3,t_2t_4,t_3t_4;q)_s}{(q,t_1t_2t_3t_4,t_2t_3t_4t_5;q)_s} (t_1t_5)^s \sum_{k=0}^{n} \frac{(q^{-n},t_1t_2t_3t_4q^{n-1};q)_k}{(q,q^s t_1t_2t_3t_4;q)_k} q^{k(s+1)}.
$$

The k sum is an evaluable terminating $_2\phi_1$, and we obtain

$$
\frac{(q;q)_\infty}{2\pi} c_n(\mathbf{t},t_5) h_n(\mathbf{t}) A(\mathbf{t}) = \frac{(t_1t_2,t_1t_3,t_1t_4;q)_n (t_1t_2t_3t_4,t_2t_3t_4t_5;q)_\infty}{t_1^n \prod_{j=2}^{4}(t_1t_j;q)_\infty \prod_{2\le r<s\le5}(t_rt_s;q)_\infty}
$$
$$
\times \sum_{s=0}^{\infty} \frac{(t_2t_3,t_2t_4,t_3t_4;q)_s}{(q,t_1t_2t_3t_4,t_2t_3t_4t_5;q)_s} (t_1t_5)^s \frac{(q^{s+1-n};q)_n}{(q^s t_1t_2t_3t_4;q)_n}.
$$

Thus $s \ge n$, so shift s by n. Therefore the left-hand side in the above equation is the statement of the theorem. $\qquad\square$

An attractive special case of Theorem 2.59 is a corollary due to Kim and Stanton [29].

Corollary 2.60 *A generating function of the continuous dual q-Hahn polynomials $p_n(x;t_1,t_2,t_3,0|q)$ is*

$$
\sum_{k=0}^{\infty} \frac{p_k(x;t_1,t_2,t_3,0|q)}{(q,bt_1t_2t_3;q)_k} b^k = \frac{(bt_1,bt_2,bt_3;q)_\infty}{(be^{i\theta},be^{-i\theta},bt_1t_2t_3;q)_\infty}.
$$

Proof Take $t_2 = 0$ in Theorem 2.59 and relabel the t_j. The $_3\phi_2$ becomes a $_1\phi_0$ which we sum by the q-binomial theorem. $\qquad\square$

We note that the case $t_3 = 0$ is the generating function (2.148) of the Al-Salam–Chihara polynomials, while the case $t_1 = t_2 = t_3 = 0$ is the generating function for the q-Hermite polynomials.

We now recast some of the results proved so far in integral form using the orthogonality relation (2.123).

Proposition 2.45 becomes

$$
\int_0^\pi \frac{(be^{i\theta},be^{-i\theta};q)_n(e^{2i\theta},e^{-2i\theta};q)_\infty}{\prod_{j=1}^{4}(t_je^{i\theta},t_je^{-i\theta};q)_\infty} p_k(\cos\theta;\mathbf{t}|q)\, d\theta
$$
$$
= \frac{(-b)^k q^{\binom{k}{2}}(q;q)_n (b/t_4,bt_4q^k;q)_{n-k}}{(q;q)_{n-k}} \frac{2\pi(t_1t_2t_3t_4q^{2k};q)_\infty}{(q;q)_\infty \prod_{1\le r<s\le4}(t_rt_sq^k;q)_\infty}
$$
$$
\times {}_4\phi_3\left(\begin{matrix} q^{k-n},t_1t_4q^k,t_2t_4q^k,t_3t_4q^k \\ bt_4q^k,t_1t_2t_3t_4q^{2k},q^{1+k-n}t_4/b \end{matrix} \Big| q,q \right). \tag{2.172}
$$

In view of (2.166), the limiting case $n \to \infty$ of (2.172) is

$$\int_0^\pi \frac{(be^{i\theta}, be^{-i\theta}, e^{2i\theta}, e^{-2i\theta}; q)_\infty}{\prod_{j=1}^4 (t_j e^{i\theta}, t_j e^{-i\theta}; q)_\infty} p_k(\cos\theta; t|q)\, d\theta$$

$$= (-b)^k q^{\binom{k}{2}} (b/t_4, bt_4 q^k; q)_\infty \frac{2\pi (t_1 t_2 t_3 t_4 q^{2k}; q)_\infty}{(q;q)_\infty \prod_{1 \leq r < s \leq 4} (t_r t_s q^k; q)_\infty}$$

$$\times {}_3\phi_2 \left(\begin{matrix} t_1 t_4 q^k,\ t_2 t_4 q^k,\ t_3 t_4 q^k \\ bt_4 q^k, t_1 t_2 t_3 t_4 q^{2k} \end{matrix} \ \middle|\ q, \frac{b}{t_4} \right). \tag{2.173}$$

When $b = t_1$, for example, the ${}_3\phi_2$ in (2.173) sums. The result is known because it is the constant term in the expansion of $p_k(x; t_1, t_2, t_3, t_4|q)$ in $p_k(x; 0, t_2, t_3, t_4|q)$, see Theorem 2.41.

It is important to note that the left-hand side of (2.173) is symmetric under permutations of $\{t_1, t_2, t_3, t_4\}$. But the symmetry of the right-hand side under interchanging t_1 and t_4 gives the transformation

$$(b/t_4, bt_4 q^k; q)_\infty\, {}_3\phi_2 \left(\begin{matrix} t_1 t_4 q^k,\ t_2 t_4 q^k,\ t_3 t_4 q^k \\ bt_4 q^k, t_1 t_2 t_3 t_4 q^{2k} \end{matrix} \ \middle|\ q, \frac{b}{t_4} \right)$$

$$= (b/t_1, bt_1 q^k; q)_\infty\, {}_3\phi_2 \left(\begin{matrix} t_1 t_4 q^k,\ t_1 t_2 q^k,\ t_1 t_3 q^k \\ bt_1 q^k, t_1 t_2 t_3 t_4 q^{2k} \end{matrix} \ \middle|\ q, \frac{b}{t_1} \right). \tag{2.174}$$

Remark 2.61 By multiplying (2.172) by $\Lambda_n \zeta^n$ and adding for $n \geq 0$ we obtain a general expansion formula with arbitrary coefficients. This can be specialized to derive a general formula involving integrals of power series with arbitrary coefficients times a product of an Askey–Wilson polynomial and its weight function. This can further specialized by taking Λ_n to be of hypergeometric or basic hypergeometric form.

We mention without proof the following theorem which, among other things, evaluates the moments of the Askey–Wilson weight function.

Theorem 2.62 *We have the integral evaluation*

$$
\int_0^\pi \frac{(e^{2i\theta}, e^{-2i\theta}; q)_\infty (be^{i\theta}, be^{-i\theta}; p)_n}{\prod_{j=1}^4 (a_j e^{i\theta}, a_j e^{-i\theta}; q)_\infty} \, d\theta
$$

$$
= \frac{2\pi (a_1 a_2 a_3 a_4; q)_\infty}{(q; q)_\infty \prod_{1 \le j < k \le 4}(a_j a_k; q)_\infty} \frac{(a_1 a_2, q a_1 a_3, q a_1 a_4; q)_n}{(q, q a_1^2, a_1 a_2 a_3 a_4; q)_n}
$$

$$
\times \sum_{k=0}^n \frac{1 - a_1^2 q^{2k}}{1 - a_1^2} \frac{(a_1^2, q^{-n}; q)_k}{(q, a_1^2 q^{n+1}; q)_k} (a_1 b q^k, b q^{-k}/a_1; p)_n
$$

$$
\times q^{k(n+1)} \frac{(1 - a_1 a_3)(1 - a_1 a_4)}{(1 - a_1 a_3 q^k)(1 - a_1 a_4 q^k)}
$$

$$
\times \, {}_4\phi_3 \left(\begin{matrix} q^{k-n}, q, a_3 a_4, a_1 q^{k+1}/a_2 \\ a_1 a_3 q^{k+1}, a_1 a_4 q^{k+1}, q^{1-n}/a_1 a_2 \end{matrix} \,\middle|\, q, q \right). \tag{2.175}
$$

Theorem 2.62 is due to Ismail and Rahman in [20]. Note that if $p = 1$ then, with $x = \cos\theta, y = (b + 1/b)/2$, we have

$$
(be^{i\theta}, be^{-i\theta}; p)_n \big|_{p=1} = (1 - 2bx + b^2)^n = (-2b)^n (x - y)^n. \tag{2.176}
$$

In particular when $y = 0$, Theorem 2.62 gives the moments of the Askey–Wilson weight function.

Exercises

Exercise 2.63 Show that for any positive integer n, $p_n(z, \mathbf{t}|q) = 0$ if

$$
z = -\gamma, \quad t_1 = \gamma, \quad t_2 = \gamma^3, \quad t_3 = \gamma^5, \quad t_4 = 0, \quad \gamma = e^{2\pi i/6}.
$$

Exercise 2.64 Let ω be a primitive cubic root of unity. Prove that

$$
\sum_{k=0}^n \begin{bmatrix} n \\ k \end{bmatrix}_q \frac{1}{(c^3; q)_k} p_k(-1/2; c, \omega c, \omega^2 c, 0|q) = \begin{cases} 0 & \text{if } 3 \nmid n, \\ \frac{(q, q^2; q^3)_{n/3}}{(c^3 q, c^3 q^2; q^3)_{n/3}} & \text{if } 3 \mid n. \end{cases}
$$

2.12 A q-exponential function

Ismail and Zhang [28] introduced the q-exponential function

$$
\mathscr{E}_q(\cos\theta; t) :=
$$

$$
\frac{(t^2; q^2)_\infty}{(qt^2; q^2)_\infty} \sum_{n=0}^\infty (-i e^{i\theta} q^{(1-n)/2}, -i e^{-i\theta} q^{(1-n)/2}; q)_n \frac{(-it)^n}{(q; q)_n} q^{n^2/4}. \tag{2.177}
$$

Define $u_n(x, y)$ by

$$
u_n(\cos\theta, \cos\phi) = e^{-in\phi} (-e^{i(\phi+\theta)} q^{(1-n)/2}, -e^{i(\phi-\theta)} q^{(1-n)/2}; q)_n. \tag{2.178}
$$

It is easy to see that $u_n(x, y) \to 2^n(x+y)^n$ as $q \to 1$. Hence

$$\lim_{q \to 1} \mathscr{E}_q(x; (1-q)t/2) = \exp(tx).$$

This shows that $\mathscr{E}_q(x; t)$ is a q-analogue of e^{tx}. The factor in front of the sum in (2.177) is to normalize $\mathscr{E}_q$ by $\mathscr{E}_q(0; t) = 1$.

Following Rainville [36] we say that a polynomial sequence $\{p_n(x)\}$ belongs to a linear operator T which reduces the degree of a polynomial by 1 if $T p_n(x) = p_{n-1}(x)$. One can repeat the same arguments used by Rainville and prove the following theorem.

Theorem 2.65 *Two polynomial sequences $\{p_n(x)\}$ and $\{q_n(x)\}$ belong to the same operator T if and only if there is a sequence of constants $\{a_n\}$ with $a_0 \neq 0$ such that*

$$p_n(x) = \sum_{k=0}^{n} a_k q_{n-k}(x). \tag{2.179}$$

This is equivalent to

$$\sum_{n=0}^{\infty} p_n(x) t^n = \left[\sum_{n=0}^{\infty} a_n t^n\right]\left[\sum_{k=0}^{\infty} q_k(x) t^k\right] = h(t) \sum_{k=0}^{\infty} q_k(x) t^k, \tag{2.180}$$

and $h(0) \neq 0$.

We now apply Theorem 2.65 to the operator $T = \mathscr{D}_q$ with different p_n. Let

$$s_n(x; a) = (aq^{-n/2} e^{i\theta}, aq^{-n/2} e^{-i\theta}; q)_n. \tag{2.181}$$

It is easy to see that

$$\mathscr{D}_q s_n(x; a) = \frac{-2aq^{-n/2}}{1-q}(1-q^n) s_{n-1}(x; a). \tag{2.182}$$

Therefore $\{s_n(x; a)(1-q)^n(-2a)^{-n} q^{n(n+1)/4}/(q; q)_n\}$ belongs to $\mathscr{D}_q$. Formulas (2.182), (2.35), and (2.36) show that $[(q-1)/2]^n \phi_n(x) q^{-n/4}/(q; q)_n$, and $[(1-q)/2]^n \rho_n(x) q^{n(n-1)/4}/(q; q)_n$, also belong to $\mathscr{D}_q$. This proves the following theorem.

Theorem 2.66 *We have the expansions*

$$\left.\begin{aligned}
\mathscr{E}_q(\cos\theta; \alpha) &= \frac{(-t; q^{1/2})_\infty}{(qt^2; q^2)_\infty} {}_2\phi_1\left(\begin{array}{c} q^{1/4} e^{i\theta}, q^{1/4} e^{-i\theta} \\ -q^{1/2} \end{array}\middle| q^{1/2}, -t\right), \\
\mathscr{E}_q(x; t) &= g(t) \sum_{n=0}^{\infty} \frac{(aq^{-n/2} e^{i\theta}, aq^{-n/2} e^{-i\theta}; q)_n}{(q; q)_n}\left(-\frac{t}{a}\right)^n,
\end{aligned}\right\} \tag{2.183}$$

where g(t) is given by

$$\frac{\mathcal{E}_q(x_0;t)}{g(t)} = \sum_{n=0}^{\infty} \frac{(aq^{-n/2}c, aq^{-n/2}/c; q)_n}{(q;q)_n} \left(-\frac{t}{a}\right)^n, \tag{2.184}$$

and $x_0 - (c + 1/c)/2$, for $c \neq 0$.

We next define a function $\mathcal{E}_q(x, y; t)$ by

$$\mathcal{E}_q(\cos\theta, \cos\phi; t) := \frac{(t^2; q^2)_\infty}{(qt^2; q^2)_\infty} \sum_{n=0}^{\infty} \frac{(te^{-i\phi})^n}{(q;q)_n} q^{n^2/4} \tag{2.185}$$
$$\times (-e^{i(\phi+\theta)} q^{(1-n)/2}, -e^{i(\phi-\theta)} q^{(1-n)/2}; q)_n.$$

This function was introduced in [28].

We note that the symmetry $\mathcal{E}_q(x, y; t) = \mathcal{E}_q(y, x; t)$ follows from the definition (2.185). It follows from (2.177) and (2.182) that

$$\mathcal{D}_q \mathcal{E}_q(x; t) = \frac{2tq^{1/4}}{1-q} \mathcal{E}_q(x; t), \quad \text{and} \quad \mathcal{D}_q \mathcal{E}_q(x, y; t) = \frac{2tq^{1/4}}{1-q} \mathcal{E}_q(x, y; t),$$
$$\tag{2.186}$$

and $\mathcal{D}_q$ acts on x. Therefore $\mathcal{E}_q(x, y; t) = g(y, t) \mathcal{E}_q(x; t)$. On the other hand the definition (2.185) shows that $\mathcal{E}_q(0, y; t) = \mathcal{E}_q(y; t)$. This establishes the addition theorem

$$\mathcal{E}_q(x, y; t) = \mathcal{E}_q(x; t) \mathcal{E}_q(y; t). \tag{2.187}$$

Further properties of $\mathcal{E}_q(x; t)$ were developed in [21, 28] and many are recorded in [19].

There are several formulas expanding $\mathcal{E}_q(x; t)$ in a series of orthogonal polynomials. One sample from [26] is

$$\frac{(q^2t^4; q^4)_\infty}{(-t; q)_\infty} \mathcal{E}_{q^2}(x; t) = \sum_{k=0}^{\infty} \frac{t^k q^{k^2/2}}{(q, -q, t_2 t_3 t_4 q^{k-1/2}; q)_k} p_k(x; q^{1/2}, t_2, t_3, t_4 | q)$$
$$\times {}_3\phi_2 \left(\begin{matrix} q^{k+1/2} t_2, q^{k+1/2} t_3, q^{k+1/2} t_4 \\ -q^{k+1}, t_2 t_3 t_4 q^{2k+1/2} \end{matrix} \middle| q, -t \right).$$

Acknowledgements The author is very grateful to Howard Cohl and Kasso Okoudjou for organizing the summer school and for taking care of all the logistics. Thanks also to the students who attended the lectures and for their feedback. Special thanks to Howard Cohl, Michael Schlosser, Plamen Simeonov, and Dennis Stanton, for their careful reading of the manuscript and for pointing out many misprints in the early version of these notes.

References

[1] W.A. Al-Salam and T.S. Chihara, Convolutions of orthogonal polynomials, *SIAM J. Math. Anal.* **7** (1976), 16–28.

[2] G.E. Andrews, q-orthogonal polynomials, Rogers–Ramanujan identities, and mock theta functions, *Tr. Mat. Inst. Steklova* **276** (2012), *Teoriya Chisel, Algebra i Analiz*, 27–38; translation in *Proc. Steklov Inst. Math.* **276** (2012), 21–32.

[3] G.E. Andrews, *q-series: Their Development and Application in Analysis, Number Theory, Combinatorics, Physics, and Computer Algebra*, CBMS Regional Conference Series, Number 66, American Mathematical Society, Providence, RI, 1986.

[4] G.E. Andrews, R.A. Askey, and R. Roy, *Special Functions*, Cambridge University Press, Cambridge, 1999.

[5] M.H. Annaby and Z.S Mansour, q-Taylor and interpolation series for Jackson q-difference operators. *J. Math. Anal. Appl.* **344** (2008), 472–483.

[6] R.A. Askey and M.E.H. Ismail, Recurrence relations, continued fractions and orthogonal polynomials, *Memoirs Amer. Math. Soc.* **300** (1984), 112 pages.

[7] R.A. Askey and J.A. Wilson, Some basic hypergeometric orthogonal polynomials that generalize Jacobi polynomials, *Memoirs Amer. Math. Soc.* **54**, Number 319, (1985).

[8] R.P. Boas, *Entire Functions*, Academic Press, New York, 1954

[9] T.J. Bromwich, *An Introduction to the Theory of Infinite Series*, Revised edition, Macmillan, London, 1926.

[10] B.M. Brown, W.D. Evans and M.E.H. Ismail, The Askey–Wilson polynomials and q-Sturm–Liouville problems, *Math. Proc. Cambridge Phil. Soc.* **119** (1996), 1–16.

[11] S. Cooper, The Askey–Wilson operator and the $_6\phi_5$ summation formula, *South East Asian J. Math. Math. Sci.* **1** (2002), 71–82.

[12] J.L. Fields and M.E.H. Ismail, Polynomial expansions, *Math. Comp.* **29** (1975), 894–902.

[13] J. Fields and J. Wimp, Expansions of hypergeometric functions in hypergeometric functions, *Math. Comp.* **15** (1961), 390–395.

[14] I. Gessel and D. Stanton, Applications of q-Lagrange inversion to basic hypergeometric series, *Trans. Amer. Math. Soc.* **277** (1983), 173–201.

[15] G. Gasper, Elementary derivations of summation and transformation formulas for q-series. In *Special Functions, q-Series and Related Topics*, M.E.H. Ismail, D.R. Masson, and M. Rahman, eds., Fields Institute Communications, American Mathematical Society, Providence, RI, 1997.

[16] G. Gasper and M. Rahman, *Basic Hypergeometric Series*, second edition Cambridge University Press, Cambridge, 2004.

[17] M.E.H. Ismail, A simple proof of Ramanujan's $_1\psi_1$ sum, *Proc. Amer. Math. Soc.* **63** (1977), 185–186.

[18] M.E.H. Ismail, The Askey–Wilson operator and summation theorems. In *Mathematical Analysis, Wavelets, and Signal Processing*, M.E.H. Ismail,

M.Z. Nashed, A. Zayed and A. Ghaleb, eds., Contemporary Mathematics, **190**, American Mathematical Society, Providence, RI, 1995, pp. 171–178.

[19] M.E.H. Ismail, *Classical and Quantum Orthogonal Polynomials in one Variable*, Cambridge University Press, Cambridge, 2005.

[20] M.E.H. Ismail and M. Rahman, Connection relations and expansions, *Pac. J. Math.* **252** (2011), 427–446.

[21] M.E.H. Ismail and M. Rahman and R. Zhang, Diagonalization of certain integral operators II, *J. Comp. Appl. Math.* **68** (1996), 163–196.

[22] M.E.H. Ismail and P. Simeonov, Formulas and identities involving the Askey–Wilson operator, *Adv. Appl. Math.* **76** (2016), 1–29.

[23] M.E.H. Ismail and D. Stanton, On the Askey–Wilson and Rogers polynomials, *Canadian J. Math.* **40** (1988), 1025–1045.

[24] M.E.H. Ismail and D. Stanton, Applications of q-Taylor theorems, *J. Comp. Appl. Math.* **153** (2003), 259–272.

[25] M.E.H. Ismail and D. Stanton, q-Taylor theorems, polynomial expansions, and interpolation of entire functions, *J. Approx. Theory* **123** (2003), 125–146.

[26] M.E.H. Ismail, and D. Stanton, Expansions in the Askey–Wilson polynomials, *J. Math. Anal. Appl.* **424** (2015), 664–674.

[27] M.E.H. Ismail and J. Wilson, Asymptotic and generating relations for the q-Jacobi and $_4\phi_3$ polynomials, *J. Approx. Theory* **36** (1982), 43–54.

[28] M.E.H. Ismail and R. Zhang, Diagonalization of certain integral operators, Advances in Math. **109** (1994), 1–33.

[29] J.S. Kim and D. Stanton, Bootstrapping and the Askey–Wilson polynomials, *J. Math. Anal. Appl.* **421** (2015), 501–520..

[30] K. Knopp, *Theory of Functions, Parts I-III*, Dover, New York, 1945.

[31] R. Koekoek and R. Swarttouw, The Askey scheme of hypergeometric orthogonal polynomials and its q-analogues. Reports of the Faculty of Technical Mathematics and Informatics no. 98-17, Delft University of Technology, Delft, 1998.

[32] Z.G. Liu, q-Hermite polynomials and a q-beta integral, *Northeast. Math. J.* **13**, (1997), 361–366.

[33] Z.G. Liu, Two q-difference equations and q-operator identities. *J. Differerence Equ. Appl.* **16** (2010), 1293–1307.

[34] Z.G. Liu, On the q-partial differential equations and q-series. In *The Legacy of Srinivasa Ramanujan*. Ramanujan Math. Soc. Lect. Notes Ser., vol. 20, Ramanujan Math. Soc., Mysore (2013), pp. 213–250.

[35] A.F. Nikiforov and S.K. Suslov, *Classical Orthogonal Polynomials of a Discrete Variable* Springer, New York, 1991.

[36] E.D. Rainville, *Special Functions*, Macmillan, New York, 1960.

[37] J. Riordan, Inverse relations and combinatorial identities, *Amer. Math. Monthly* **71** (1964), 485-498.

[38] L. Schendel, Zur Theorie der Functionen, *Journal für die reine und angewandte Mathematik* **84** (1877), 80–84. (Available online at https://eudml.org/doc/148346).

[39] M.J. Schlosser, A Taylor expansion theorem for an elliptic extension of the Askey–Wilson operator. In *Special Functions and Orthogonal Polynomials*, Contemporary Mathematics, **471**, Amer. Math. Soc., Providence, RI, 2008, 175–186.

[40] Michael J. Schlosser and M. Yoo, Elliptic hypergeometric summations by Taylor series expansion and interpolation, *Symmetry, Integrability and Geometry: Methods and Applications SIGMA* **12** (2016), 21 pages.

[41] G. Szegő, *Orthogonal Polynomials*, fourth edition, American Mathematical Society, Providence, 1975.

[42] A. Verma, Some transformations of series with arbitrary terms, *Ist. Lombardo Accad. Sci. Lett. Rend. A* **106** (1972), 342–353.

[43] E.T. Whittaker and G.N. Watson, *A Course of Modern Analysis*, fourth edition, Cambridge University Press, Cambridge, 1927.

3

Applications of Spectral Theory to Special Functions

Erik Koelink

Abstract: Many special functions are eigenfunctions to explicit operators, such as difference and differential operators, which is in particular true for the special functions occurring in the Askey scheme, its q-analogue and extensions. The study of the spectral properties of such operators leads to explicit information for the corresponding special functions. We discuss several instances of this application, involving orthogonal polynomials and their matrix-valued analogues.

Preamble

According to Paul Túran, special functions should be renamed useful functions. Their usefulness is shown throughout the ages in various applications in astronomy, physics, etc., as well is in mathematics itself, such as number theory, combinatorics, representation theory, and so on. Special functions are very often related to solutions of differential or difference equations. As an example, the Bessel functions, Jacobi polynomials and many other cases are eigenfunctions to specific second order differential operators. On the one hand this relations gives information on the special functions, whereas on the other hand it gives the opportunity to study the corresponding differential operator in more detail.

Spectral theory is a part of functional analysis, vastly generalising the theory of eigenvalues and eigenvectors to linear operators on infinite-dimensional Hilbert spaces. We restrict to symmetric operators, and in particular self-adjoint operators on suitable Hilbert spaces. For such operators there is a general spectral decomposition, known as the spectral theorem. The corresponding spectral decomposition can be interpreted as orthogonality or as generalized orthogonality in the sense of an integral transform. For the differential operator for the Bessel functions this leads to the Hankel transform, and for the Jacobi polynomials this leads to the orthogonality relations. This has been been generalised to various situations, and it is an important tool in research.

The spectral theorem, being the essential tool, is described in some detail. In particular how to obtain the spectral measure from the resolvent operator using the Perron-Stieltjes inversion formula. For specific classes of operators we can describe the resolvent in more detail. As a first set of examples we consider the three-term recurrence for orthogonal polynomials, both scalar and matrix valued. We apply this set-up to various operators which have a tridiagonalisation with respect to a suitable basis. A prototypical example is the Schrödinger operator on the real line with the Morse potential, which models the disassociation of diatomic molecules. In terms of special functions, one of the results from this prototypical example is that Laguerre polynomials are mapped to Meixner–Pollaczek polynomials by the Whittaker transform. This can be generalised to other settings as well, including other transforms and other sets of polynomials as well as to higher order recurrence and matrix-valued orthogonal polynomials.

For the Askey scheme and its q-analogue the entries always satisfy a bispectral property, implying that there are at least two relevant operators for these family of polynomials. Occasionally, the 'dual' (in a suitable sense) operator can be viewed as a symmetric or self-adjoint operator on a suitable Hilbert space and this then gives additional information on the family of polynomials from the (q-)Askey scheme at hand. In particular, we show that the family of N-extremal measures of the continuous q^{-1}-Hermite polynomials can be obtained in this way. This proof is of a completely different nature than the original proof by Ismail and Masson.

In the final part we apply these methods to non-polynomial special functions, which can be viewed as a non-polynomial extension of the (q-)Askey scheme. We give various examples, and we show how this can be connected to some of the examples discussed in the beginning of the course.

We use standard notation for hypergeometric series, basic hypergeometric series (also known as q-hypergeometric series) and special functions following standard references, such as e.g., Andrews, Askey and Roy [5], Gasper and Rahman [29], Ismail [47], Koekoek and Swarttouw [54, 55], Szegő [93], Temme [94]. There is an abundance of references, and apart from the references in the books in the bibliography, the review paper by Damanik, Pushnitski and Simon [19] contains many references. The appendix discusses the spectral theorem, and references are given there. All measures discussed are Borel measures on the real line, and we denote the

σ-algebra of Borel sets on $\mathbb{R}$ by $\mathcal{B}$. Furthermore, $\mathbb{N} = \{0,1,2,\ldots\}$. All the results in these notes have appeared in the literature.

3.1 Introduction

Spectral decompositions of self-adjoint operators on Hilbert spaces can at least be traced back to the work of Fredholm on the solutions of integral equations. The study of Sturm–Liouville differential operators was a great impetus for the development of spectral analysis, see e.g., [22]. For some explicit Sturm–Liouville type differential operators there is a link to well-known special functions, such as e.g., Jacobi polynomials, which shows the close connection between special functions and spectral theory. At the moment, this is for instance an important ingredient in the study of so-called exceptional orthogonal polynomials, see e.g., [24].

Spectral theory is, loosely speaking, essentially a study of the eigenvalues, or spectral data, of a suitable operator, and to determine such an operator completely in terms of its eigenvalues. For a self-adjoint matrix this means that we look for its eigenvalues, which are real in this case, and the corresponding eigenspaces, which are orthogonal in this case. So we can write the self-adjoint matrix as a sum of multiplication and projection operators, and this is the most basic form of the spectral theorem for self-adjoint operators. We recall the spectral theorem in its most general form in Appendix 3.7.

The application to differential operators, and also to various developments in physics, such as quantum mechanics, is still very important. Via this application, there have been many developments for special functions. One of the classical applications is to study the second order differential operator

$$D^{\alpha,\beta} = (1-x^2)\frac{d^2}{dx^2} + (\beta - \alpha - (\alpha+\beta+2)x)\frac{d}{dx}$$

on the weighted $L^2(w^{\alpha,\beta})$ space for the weight $w^{\alpha,\beta}(x) = C(1-x)^{\alpha}(1+x)^{\beta}$ on $[-1,1]$ for a suitable normalisation constant C. Then $D^{\alpha,\beta}$ can be understood as an unbounded self-adjoint operator with compact resolvent. The spectral measure is then given by projections on the orthonormal Jacobi polynomials, which are eigenfunctions of $D^{\alpha,\beta}$. Similarly, the differential operator can also be studied on $[1,\infty)$ with respect to a suitable weight, and then its spectral decomposition leads to the Jacobi-function transform, see e.g., [23, Ch. XIII], [69] and references therein.

Another classical application of spectral analysis is a proof of Favard's theorem, see Corollary 3.22, stating that polynomials satisfying a suitable three-term recurrence relation, are orthogonal polynomials. This follows from studying a so-called Jacobi operator on the Hilbert space $\ell^2(\mathbb{N})$ of square summable sequences. The spectral analysis of such a Jacobi operator is closely related to the moment problem, and this link can be found at several places in the literature such as e.g., [20, 23, 57, 87, 88, 89]. The Haussdorf moment problem, i.e., on a finite interval, played an important role in the development of functional analysis, notably the development of functionals and related theorems, see [79, §I.3].

One particular application is to have other explicit operators, e.g., differential operators or difference operators, realised as Jacobi operators and next use this connection to obtain results for these explicit operators. In Section 3.6 we give a couple of examples, including the original (as far as we are aware) motivating example of the Schrödinger operator with the Morse potential due to the chemist Broad, see references in Section 3.6.1.

As is well known, the Askey scheme of hypergeometric orthogonal polynomials consists of those polynomials that are also eigenfunctions to a second-order differential or difference operator. This also holds for the q-analogue of the Askey scheme. This was initially observed by Askey in [9, Appendix] and the first Askey scheme was drawn by hand by Labelle in [74].

See Figures 3.1 and 3.2, taken from Koekoek, Lesky, Swarttouw [54] for the current state of affairs. Naturally, many of these operators, such as the differential operator for the Jacobi polynomials, have been studied in detail. This is in particular valid for the operators occurring in the Askey scheme. For indeterminate moment problems in the q-Askey scheme, the subject of the related operators, especially the difference operators, is studied in [15]. On the other hand, it is natural to extend the $(q-)$Askey scheme to include also integral transforms with kernels in terms of (basic) hypergeometric series, such as the Hankel, Jacobi, Wilson transform, and its q-analogues and to study these transforms and their properties from a spectral analytic point of view using the associated operators. We refer to the schemes [65, Figures 1.1, 1.2] remarking that in the meantime that the first of these figures has been vastly extended by Groenevelt [32] to include the Wilson function transform, and various transformations that can be obtained as limiting cases. In the terminology of Grünbaum and coworkers, all the instances of the $(q-)$Askey scheme are examples of the bispectral property. This means that the polynomials are eigenfunctions to a three-term recurrence operator (acting in the degree) and at the same time are eigenfunctions of a suitable second order differential or difference opera-

ASKEY SCHEME

OF

HYPERGEOMETRIC

ORTHOGONAL POLYNOMIALS

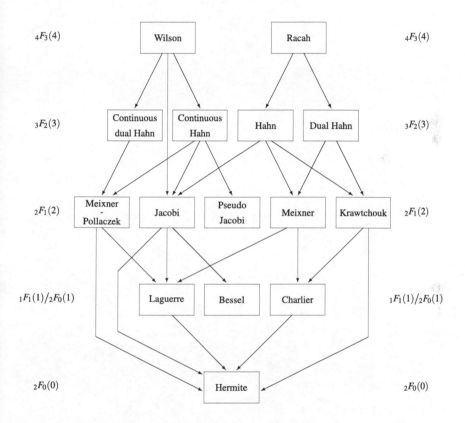

Figure 3.1 The Askey scheme as in [54].

tor in the variable. In particular, all these instances give rise to bispectral families of special functions.

Motivated by one of the second order q-difference operators arising in the q-analogue of the Askey scheme, we discuss the spectral analysis of three-term recurrence operators on $\ell^2(\mathbb{Z})$ in Section 3.2. We apply the spectral theorem to a particular example and we obtain a set of orthogonality measures for the continuous q^{-1}-Hermite polynomials. Here we follow the convention $0 < q < 1$, so that $q^{-1} > 1$. These measures turn out to be N-

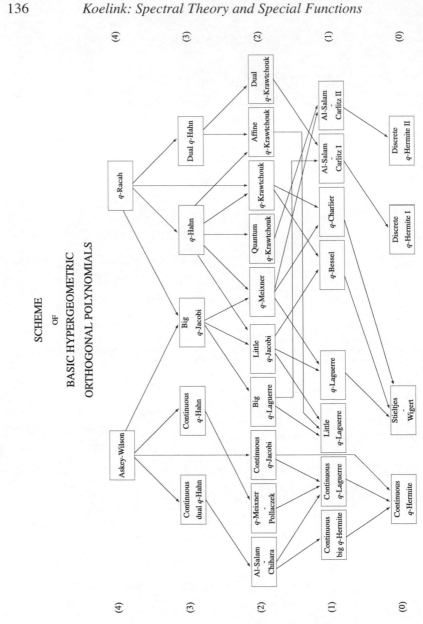

Figure 3.2 The *q*-Askey scheme as in [54].

extremal, where N stands for Nevanlinna. This result is originally obtained by Ismail and Masson [51], and this proof is due to Christiansen and the author [17] as a special case of results for the symmetric Al-Salam–Chihara polynomials for $q > 1$. This is partly based on [57, §4]. Similar ideas have been used in e.g., [16, 40], to study other moment problems and related orthogonal polynomials.

In Section 3.3 we briefly recall the relation between three-term recurrence operators and orthogonal polynomials. This is a well-known subject in the literature, and there are several books and review papers on this subject, e.g., [2, 14, 23, 76], [87, Ch. 16], [88, 89, 92]. We base our presentation on [57], and extend this approach to the case of matrix-valued orthogonal polynomials and block Jacobi operators. The spectral approach is essentially due to M.G. Kreĭn [72, 73], whose great mathematical legacy is discussed in [1]. We discuss briefly a rather general example of arbitrary size. In Section 3.5 we discuss some of the assumptions made in Section 3.4. Here we also make use of the previous lecture series by Berg, [12], and Durán and López-Rodríguez, [26], but also [6, 31, 72, 73].

In Section 3.6 we show how realisations of explicit operators, such as differential operators, as recurrence operators can be used to study the spectral theory. This gives rise to relations between the spectral decomposition of such an operator and the related orthogonal polynomials. In the physics literature such a method is known as the J-matrix method, and there is a vast literature of physics applications, see e.g., references to the works of Al-Haidari, Bahlouli, Bender, Dunne, Yamani and others in [48]. The first example of Section 3.6 is the study of the Schrödinger operator with the Morse potential, originally due to Broad [13], see also [21]. The second example of Section 3.6 is in the same vein, and due to Ismail and the author [48]. This case has recently been generalised by Genest et al. [30] to include more parameters and to cover the full family of Wilson polynomials. Moroever, in [30] a link to the Bannai-Ito algebra is established. The last example of Section 3.6 leads to a more general family of matrix-valued orthogonal polynomials for operators which have a realisation as a 5-term recurrence operator. We then discuss an example of such a case, extending the second example of Section 3.6. We apply this approach to an explicit second order differential operator. The same realisation of suitable operators as tridiagonal operators has useful implications in e.g., representation theory, see e.g., [18, 33, 34, 37, 39, 42, 43, 58, 67, 78, 80] for the case of representation theory of the Lie algebra $\mathfrak{su}(1,1)$ and its quantum group analogue. Using explicit realisations of representations, these results have given very explicit bilinear generating functions, see e.g., [35, 68].

All the general results as well as the explicit examples have appeared in the literature before. There are many other references available in the literature, and apart from the books – and the references mentioned there

– mentioned in the bibliography, one can especially consult the references in [19], where a list of more than 200 references can be found. Many results in (scalar) orthogonal polynomials have an analogous statement for matrix-valued orthogonal polynomials. Several researchers have worked on these generalisations, including Berg, Cantero, Castro, Durán, Geronimo, Grünbaum, de la Iglesia, Lopéz-Rodríguez, Marcellán, Pacharoni, Tirao, Van Assche, etc. and we point again to [19] for references.

Let us note that, in these notes, the emphasis is on explicit operators related to explicit sets of special functions, so that information on these special functions is obtained from the spectral analysis. On the other hand, there are also many results on the spectral analysis of more general classes of operators. For this subject one can consult Simon's book [90] and the extensive list of references given there.

It may happen that a differential or difference operator with suitable eigenfunctions in terms of well-known special functions cannot be suitably realised as a three-term recurrence operator on a Hilbert space such as $\ell^2(\mathbb{Z})$ or $\ell^2(\mathbb{N})$. It can then be very useful to look for a larger Hilbert space, and an extension of the operator to the larger Hilbert space. This is different from the extension of a Hilbert space in order to find self-adjoint extensions. Then one needs to find a way of obtaining the extended Hilbert space and the extension of the operator. This is usually governed by the interpretation of these operators and special functions in a different context, such as, for example, representation theory. Examples are in [32, 36, 63, 66, 80] for instance. This leads to extensions of the Askey and q-Askey scheme of Figures 3.1 and 3.2 with non-polynomial function transforms arising as the spectral decomposition of suitable differential and difference operators on Hilbert spaces of functions, see e.g., Figures 1.1 and 1.2 in [65]. Figure 1.2 of [65] is still valid as an extension of the q-Askey scheme, but Figure 1.1 of [65] now has Groenevelt's Wilson function transforms [32] at the top level.

Acknowledgements Thanks to René Swarttouw and Roelof Koekoek for their version of the Askey scheme in Figures 3.1 and 3.2. I thank the organisers of the summer school, in particular Howard Cohl, Mourad Ismail and Kasso Okoudjou, for the opportunity to give the lectures at *Orthogonal Polynomials and Special Functions Summer School OPSF-S6*, July 2016, University of Maryland. I also thank all the participants of OPSF-S6 for their feedback. I thank Wolter Groenevelt and Luud Slagter for their input. The referees have pointed out many errors and oversights, and I thank them for their help in improving these lecture notes.

3.2 Three-term recurrences in $\ell^2(\mathbb{Z})$

In this section we discuss three-term recurrence relations on the Hilbert space $\ell^2(\mathbb{Z})$. We apply to this one particular example, which is motivated by a second order difference operator arising in the q-Askey scheme.

We consider sequence spaces and the associated Hilbert spaces as in Example 3.75. For the Hilbert space $\ell^2(\mathbb{Z})$ with orthonormal basis $\{e_l\}_{l\in\mathbb{Z}}$ we consider for complex sequences $\{a_l\}_{l\in\mathbb{Z}}$, $\{b_l\}_{l\in\mathbb{Z}}$, $\{c_l\}_{l\in\mathbb{Z}}$ the operator

$$Le_l = a_l e_{l+1} + b_l e_l + c_l e_{l-1}, \qquad l \in \mathbb{Z}, \tag{3.1}$$

with dense domain $\mathscr{D}$ the subspace of finite linear combinations of the basis vectors.

Lemma 3.1 *L extends to a bounded operator on $\ell^2(\mathbb{Z})$ if and only if the sequences $\{a_l\}_{l\in\mathbb{Z}}$, $\{b_l\}_{l\in\mathbb{Z}}$, $\{c_l\}_{l\in\mathbb{Z}}$ are bounded.*

In Exercise 3.10 you are requested to prove Lemma 3.1.

In the case when L is bounded, we see that L acting on $v = \sum_{k\in\mathbb{Z}} v_k e_k \in \ell^2(\mathbb{Z})$ is given by

$$Lv = \sum_{k\in\mathbb{Z}} \left(a_{k-1} v_{k-1} + b_k v_k + c_{k+1} v_{k+1} \right) e_k. \tag{3.2}$$

When L is not bounded, we have to interpret this in a suitable fashion, by, e.g., initially allowing only for $v \in \ell^2(\mathbb{Z})$ with only finitely many non-zero coefficients, i.e., for $v \in \mathscr{D}$. In general, we view L as an operator acting on the sequence space of sequences labeled by $\mathbb{Z}$, and we are in particular interested in the case of square-summable sequences.

Lemma 3.2 *For $v = \sum_{k\in\mathbb{Z}} v_k e_k \in \ell^2(\mathbb{Z})$ define*

$$L^* v = \sum_{k=-\infty}^{\infty} \left(\overline{a_k} v_{k+1} + \overline{b_k} v_k + \overline{c_k} v_{k-1} \right) e_k,$$

which, in general, is not an element of $\ell^2(\mathbb{Z})$. Define

$$\mathscr{D}^* = \{v \in \ell^2(\mathbb{Z}) \mid L^* v \in \ell^2(\mathbb{Z})\}.$$

The adjoint of $(L, \mathscr{D})$ is $(L^*, \mathscr{D}^*)$.

In Exercise 3.11 you are requested to prove Lemma 3.2. As for L in (3.2), we apply L^* to arbitrary sequences. Note that $\mathscr{D} \subset \mathscr{D}^*$, so that $(L, \mathscr{D})$ is symmetric in case $L^*|_{\mathscr{D}} = L$, which is the case for $\overline{a_k} = c_{k+1}$ and $\overline{b_k} = b_k$ for all $k \in \mathbb{Z}$.

From now on we assume that $\overline{a_k} = c_{k+1}$ and $\overline{b_k} = b_k$ for all $k \in \mathbb{Z}$, and moreover, that $a_k > 0$ for all $k \in \mathbb{Z}$. This last assumption is not essential, since changing each of the basis elements by a phase factor shows that we can assume this in case $a_k \neq 0$ for all $k \in \mathbb{Z}$. Note that in case $a_{k_0} = 0$ for some k_0 we have L-invariant subspaces, and we can consider L on such an invariant subspace. In particular, the dimension of the space of formal solutions to $L^* f = zf$ is two.

Example 3.3 The first example is related to explicit orthogonal polynomials, namely the symmetric Al-Salam–Chihara polynomials in base q^{-1}, see [29, 54, 55]. We are in particular interested in the limit case of the continuous q^{-1} Hermite polynomials introduced by Askey in [7]. These polynomials correspond to an indeterminate moment problem, see Section 3.3, and have been studied in detail by Ismail and Masson, [51], who have determined the explicit expression of the N-extremal measures, where N stands for Nevanlinna. The N-extremal measures are the measures for which the polynomials are dense in the corresponding weighted L^2-space.

The details of Example 3.3 are taken from [17], in which the case of the general symmetric Al-Salam–Chihara polynomials is studied and, in the notation of [17], this example corresponds to $\beta \downarrow 0$. The polynomials, after rescaling, are eigenfunctions to a second order q-difference equation for functions supported on a set labeled by $\mathbb{Z}$. After rewriting, we find the following three-term recurrence operator:

$$Le_l = a_l e_{l+1} + b_l e_l + a_{l-1} e_{l-1},$$

$$a_l = \frac{\alpha^2 q^{2l+\frac{1}{2}}}{1 + \alpha^2 q^{2l+1}} \frac{1}{\sqrt{(1 + \alpha^2 q^{2l})(1 + \alpha^2 q^{2l+2})}},$$

$$b_l = \frac{\alpha^2 (1+q) q^{2l-1}}{(1 + \alpha^2 q^{2l+1})(1 + \alpha^2 q^{2l-1})},$$

where $\alpha \in (q, 1]$. We emphasise that the polynomials being eigenfunctions to L follows from the second order q-difference operator for the continuous q^{-1}-Hermite polynomials [7], [55, (3.26.5)], and not from the three-term recurrence relation for the orthogonal polynomials. Recall that $0 < q < 1$. It follows immediately from the explicit expressions that

$$a_l = \begin{cases} \alpha^2 q^{2l+\frac{1}{2}} + \mathcal{O}(q^{4l}), & l \to \infty, \\ \alpha^{-2} q^{-2l-\frac{3}{2}} + \mathcal{O}(q^{-4l}), & l \to -\infty, \end{cases}$$

$$b_l = \begin{cases} \alpha^2 (1+q) q^{2l-1} + \mathcal{O}(q^{4l}), & l \to \infty, \\ \alpha^{-2} (1+q) q^{-2l-1} + \mathcal{O}(q^{-4l}), & l \to -\infty. \end{cases}$$

The exponential decay of the coefficients a_l and b_l in this case for $l \to \pm\infty$, show that we can approximate L by the finite rank operators $P_n L$, where P_n is the projection on the finite-dimensional subspace spanned by the basis vectors $\{e_{-n}, e_{-n+1}, \ldots, e_{n-1}, e_n\}$. The approximation holds true in operator norm, $\|L - P_n L\| = \mathcal{O}(q^n)$, so that L is a compact operator. So the operator L has a discrete spectrum accumulating at zero, and each of the eigenspaces for the non-zero eigenvalues is finite-dimensional.

Next we consider the formal eigenspaces for $z \in \mathbb{C}$ of L^*;

$$S_z^+ = \{f = \sum_{k \in \mathbb{Z}} f_k e_k \mid L^* f = zf, \ \sum_{k > 0} |f_k|^2 < \infty\},$$
$$S_z^- = \{f = \sum_{k \in \mathbb{Z}} f_k e_k \mid L^* f = zf, \ \sum_{k < 0} |f_k|^2 < \infty\}. \tag{3.3}$$

So $\dim S_z^\pm \leq 2$. Note that $S_z^\pm$ consist of those eigenvectors that are square summable at $\pm\infty$, which we call the free solutions at $\pm\infty$.

For any two sequences $\{v\}_{l \in \mathbb{Z}}, \{f\}_{l \in \mathbb{Z}}$, we define the Wronskian or Casorati determinant by

$$[v, f]_l = a_l (v_{l+1} f_l - f_{l+1} v_l),$$

which is a sequence. However, for eigenvectors of L^*, the Wronskian or Casorati determinant is a constant sequence.

Lemma 3.4 *Let v and f be formal solutions to $L^* u = zu$. Then*

$$[v, f] = [v, f]_l = a_l (v_{l+1} f_l - f_{l+1} v_l)$$

is independent of $l \in \mathbb{Z}$.

In particular, Lemma 3.4 can be applied to the solutions in $S_z^\pm$. Note that the Casorati determinant $[v, f] \neq 0$ for non-trivial solutions unless v and f span a one-dimensional subspace of solutions.

Proof Since v and f are formal solutions, we have for all $l \in \mathbb{Z}$

$$a_l v_{l+1} + b_l v_l + a_{l-1} v_{l-1} = z v_l,$$
$$a_l f_{l+1} + b_l f_l + a_{l-1} f_{l-1} = z f_l,$$

since we assume the self-adjoint case. Multiplying the first equation by f_l and the second by v_l and subtracting gives

$$a_l (v_{l+1} f_l - f_{l+1} v_l) + a_{l-1} (v_{l-1} f_l - f_{l-1} v_l) = 0$$

which means that $[v, f]_l$ is indeed independent of $l \in \mathbb{Z}$. $\qquad \square$

Theorem 3.5 *Assume that* $\dim S_z^{\pm} = 1$ *for all* $z \in \mathbb{C} \setminus \mathbb{R}$ *and that* $S_z^+ \cap S_z^- = \{0\}$ *for all* $z \in \mathbb{C} \setminus \mathbb{R}$. *Then* $(L^*, \mathscr{D}^*)$ *is self-adjoint. The resolvent operator is given by* (3.4), (3.5).

In Section 3.3 we show that $\dim S_z^{\pm} \geq 1$ for all $z \in \mathbb{C} \setminus \mathbb{R}$. Note that $S_z^+ \cap S_z^-$ gives the deficiency space at $z \in \mathbb{C} \setminus \mathbb{R}$ of $(L^*, \mathscr{D}^*)$, which has constant dimension on the upper and lower half plane; see Appendix 3.7.5. Since L has real coefficients, it commutes with complex conjugation; i.e., for $f = \sum_l f_l e_l$ define the vector $\bar{f} = \sum_l \overline{f_l} e_l$: then $L^* \bar{f} = \overline{L^* f}$, and we see that the deficiency spaces N_z and $N_{\bar{z}}$ have the same dimension. So we can replace the assumption $S_z^+ \cap S_z^- = \{0\}$ for all $z \in \mathbb{C} \setminus \mathbb{R}$ in Theorem 3.5 by $S_z^+ \cap S_z^- = \{0\}$ for some $z \in \mathbb{C} \setminus \mathbb{R}$.

Proof Since the deficiency index $n_z = \dim(S_z^+ \cap S_z^-) = 0$, we see that $(L^*, \mathscr{D}^*)$ has deficiency indices $(0, 0)$, so that by Proposition 3.79 it is self-adjoint.

Now take non-zero $\phi_z \in S_z^+$, $\Phi_z \in S_z^-$, which are unique up to a scalar by assumption. Moreover, the Wronskian $[\phi_z, \Phi_z] \neq 0$, since ϕ_z and Φ_z are not multiples of each other. We define the Green kernel for $z \in \mathbb{C} \setminus \mathbb{R}$ by

$$G_{k,l}(z) = \frac{1}{[\phi_z, \Phi_z]} \begin{cases} (\Phi_z)_k (\phi_z)_l, & k \leq l, \\ (\Phi_z)_l (\phi_z)_k, & k > l. \end{cases} \tag{3.4}$$

So $\{G_{k,l}(z)\}_{k=-\infty}^{\infty}, \{G_{k,l}(z)\}_{l=-\infty}^{\infty} \in \ell^2(\mathbb{Z})$ and $\ell^2(\mathbb{Z}) \ni v \mapsto G(z)v$ given by

$$G(z)v = \sum_{k \in \mathbb{Z}} (G(z)v)_k e_k, \qquad (G(z)v)_k = \sum_{l=-\infty}^{\infty} v_l G_{k,l}(z) = \langle v, \overline{G_{k,\cdot}(z)} \rangle \tag{3.5}$$

is well-defined. Note that $v \in \mathscr{D}$ implies

$$|(G(z)v)_k| \leq \underbrace{\sum_{l=-\infty}^{\infty} |v_l G_{k,l}(z)|}_{\text{finite}} \leq \underbrace{\left(\sum_{l=-\infty}^{\infty} |v_l|^2\right)^{1/2}}_{\text{finite}} \underbrace{\left(\sum_{l=-\infty}^{\infty} |G_{k,l}(z)|^2\right)^{1/2}}_{\text{finite}}$$

$$= \|v\| \underbrace{\left(\sum_{l=-\infty}^{\infty} |G_{k,l}(z)|^2\right)^{1/2}}_{\text{finite}}$$

and

$$\|G(z)v\|^2 = \sum_{k \in \mathbb{Z}} |(G(z)v)_k|^2 \leq \|v\|^2 \underbrace{\sum_{k \in \mathbb{Z}} \sum_{l=-\infty}^{\infty} |G_{k,l}(z)|^2}_{\text{finite}}$$

$$= \|v\|^2 \underbrace{\sum_{l=-\infty}^{\infty} \sum_{k \in \mathbb{Z}} |G_{k,l}(z)|^2}_{\text{finite}} < \infty$$

since $\sum_{k\in\mathbb{Z}} |G_{k,l}(z)|^2 < \infty$ by the definition (3.4) and $\phi_z \in S_z^+$, $\Phi_z \in S_z^-$. So $G(z)v \in \ell^2(\mathbb{Z})$.

We first check $(L^* - z)G(z)v = v$ for v in the dense subspace $\mathscr{D}$. We do so by calculating the kth entry of $[\phi_z, \Phi_z](L^* - z)G(z)v$ as a sum over $l \in \mathbb{Z}$, which we split up in a sum until $k - 1$, from $k + 1$ and a single term. Explicitly,

$$
\begin{aligned}
[\phi_z, \Phi_z] & \big((L^* - z)G(z)v \big)_k \\
&= [\phi_z, \Phi_z] \Big(a_k (G(z)v)_{k+1} + (b_k - z)(G(z)v)_k + a_{k-1}(G(z)v)_{k-1} \Big) \\
&= \sum_{l=-\infty}^{k-1} v_l \big(a_k(\phi_z)_{k+1} + (b_k - z)(\phi_z)_k + a_{k-1}(\phi_z)_{k-1} \big)(\Phi_z)_l \\
&\quad + \sum_{l=k+1}^{\infty} v_l \big(a_k(\Phi_z)_{k+1} + (b_k - z)(\Phi_z)_k + a_{k-1}(\Phi_z)_{k-1} \big)(\phi_z)_l \\
&\quad + v_k \big(a_k(\Phi_z)_k(\phi_z)_{k+1} + (b_k - z)(\Phi_z)_k(\phi_z)_k + a_{k-1}(\Phi_z)_{k-1}(\phi_z)_k \big) \\
&= v_k a_k \big((\Phi_z)_k(\phi_z)_{k+1} - (\Phi_z)_{k+1}(\phi_z)_k \big) = v_k [\phi_z, \Phi_z].
\end{aligned}
$$

The first term vanishes, since ϕ_z is a formal eigenfunction to L. Similarly, the second sum vanishes, since Φ_z is an eigenfunction to L. Finally, use $(b_k - z)(\Phi_z)_k(\phi_z)_k + a_{k-1}(\Phi_z)_{k-1}(\phi_z)_k = -a_k(\Phi_z)_{k+1}(\phi_z)_k$ and recognise the Casorati determinant.

By assumption, ϕ_z and Φ_z are not linearly dependent, so that the Casorati determinant $[\phi_z, \Phi_z] \neq 0$. Dividing both sides by the Casorati determinant gives the result. Note that this also shows that $G(z)v \in \mathscr{D}^*$. So we see see that $(L^* - z)G(z)$ is the identity on the dense subspace $\mathscr{D}$, and since L^* is selfadjoint, we have that $R(z) = (L^* - z)^{-1}$ is a bounded operator which is equal to $G(z)$. $\qquad\square$

Note that the determination of the spectral measure is governed by the structure of the function $z \mapsto [\phi_z, \Phi_z]$, which is analytic in the upper and lower half plane. In particular, if it extends to a function on $\mathbb{C}$ with poles at the real axis, we see that the spectral measure is discrete. This happens in case of Example 3.3.

Example 3.6 We continue Example 3.3, and we describe the solution space in some detail. Define the constant

$$
C_l(\alpha) = \frac{\alpha^{2l} q^{l^2 - \frac{1}{2}l} \sqrt{1 + \alpha^2 q^{2l}}}{(-\alpha^2 q; q)_{2l}} = \begin{cases} \mathscr{O}(\alpha^{2l} q^{l^2 - \frac{1}{2}l}), & l \to \infty, \\ \mathscr{O}(q^{-l^2 - \frac{1}{2}l}), & l \to -\infty, \end{cases}
$$

and for $z \in \mathbb{C} \setminus \mathbb{R}$ the functions

$$(\phi_z)_l = C_l(\alpha) z^{-l} {}_0\varphi_1 \left(\begin{matrix} - \\ -\alpha^2 q^{l+1} \end{matrix} ; q, -\frac{\alpha^2 q^{2l+1}}{z} \right),$$

$$(\Phi_z)_l = \frac{1}{C_l(\alpha)} z^l {}_0\varphi_1 \left(\begin{matrix} - \\ -\alpha^{-2} q^{1-l} \end{matrix} ; q, -\frac{q^{1-2l}}{\alpha^2 z} \right).$$

Then the corresponding elements $\phi_z \in S_z^+$ and $\Phi_z \in S_z^-$. The ℓ^2-behaviour follows easily from the asymptotic behaviour of the constant $C_l(\alpha)$. The fact that these functions actually are a solution for the three-term recurrence relation follows from contiguous relations for basic hypergeometric series, and we do not give the details, see [17] and Exercise 3.12. Next we calculate $[\phi_z, \Phi_z] = -z(1/z; q)_\infty$ using a limiting argument, see Exercise 3.12 as well.

Now that in the situation of Theorem 3.5 we have explicitly determined the resolvent operator $R(z) = (L^* - z)^{-1}$ we can apply the Stieltjes–Perron inversion formula of Theorem 3.78. For this we need

$$\left\langle (L^* - z)^{-1} v, w \right\rangle = \sum_{k \leq j} \frac{(\phi_z)_j (\Phi_z)_k}{[\phi_z, \Phi_z]} (v_k \overline{w}_j + v_j \overline{w}_k)(1 - \tfrac{1}{2}\delta_{j,k}) \qquad (3.6)$$

for $v, w \in \ell^2(\mathbb{Z})$, which follows by plugging in the expression of the Green kernel for the resolvent as in Theorem 3.5 and its proof, see (3.4), (3.5). So the outcome of the Stieltjes–Perron inversion formula of Theorem 3.78 depends on the behaviour of the extension of the function, initially defined on $\mathbb{C} \setminus \mathbb{R}$,

$$z \mapsto \frac{(\phi_z)_j (\Phi_z)_k}{[\phi_z, \Phi_z]} \qquad (3.7)$$

when approaching the real axis from above and below.

We assume that the function in (3.7) is analytic in the upper and lower half plane, which can be proved in general, see e.g., [72, 87]. Assume now that it has an extension to a function exhibiting a pole at $x_0 \in \mathbb{R}$. Then Theorem 3.78 shows that the spectral measure has a mass point at x_0 and

$$\left\langle E(\{x_0\}) v, w \right\rangle = -\frac{1}{2\pi i} \oint_{(x_0)} \left\langle (L^* - s)^{-1} v, w \right\rangle ds, \quad v, w \in \ell^2(\mathbb{Z}).$$

Moreover, assuming that the pole x_0 corresponds to a zero of the Casorati determinant or Wronskian $[\phi_z, \Phi_z]$, we find

$$\frac{1}{2\pi i} \oint_{(x_0)} \frac{(\phi_s)_j (\Phi_s)_k}{[\phi_s, \Phi_s]} ds = (\phi_{x_0})_j (\Phi_{x_0})_k \operatorname*{Res}_{z=x_0} \frac{1}{[\phi_z, \Phi_z]}.$$

When ϕ_{x_0} is a multiple of Φ_{x_0}, the Casorati determinant vanishes, so assume $\Phi_{x_0} = A(x_0)\phi_{x_0}$ and that $\phi_{x_0} \in \ell^2(\mathbb{Z})$, so that

$$\langle E(\{x_0\})v,w\rangle$$
$$- -\Lambda(x_0)\sum_{k\leq j}(\phi_{x_0})_j(\phi_{x_0})_k(v_k\overline{w}_j + v_j\overline{w}_k)(1 - \tfrac{1}{2}\delta_{j,k})\operatorname*{Res}_{z=x_0}\frac{1}{[\phi_z,\Phi_z]}$$
$$= -A(x_0)\operatorname*{Res}_{z=x_0}\frac{1}{[\phi_z,\Phi_z]}\langle v,\phi_{x_0}\rangle\langle\phi_{x_0},w\rangle$$

assuming that $\phi_{x_0} = \sum_{l\in\mathbb{Z}}(\phi_{x_0})_l e_l$ has real-valued coefficients $(\phi_{x_0})_l$ for real x_0. See Exercise 3.14 for the general case.

Example 3.7 We continue Example 3.3, 3.6. Since $[\phi_z,\Phi_z] = -z(1/z;q)_\infty$ for $z \neq 0$, we see that we can take $x_0 = q^n$ for $n \in \mathbb{N}$ which is a simple zero of the Casorati determinant. Now the residue calculation can be done explicitly;

$$\operatorname*{Res}_{z=q^n}\frac{1}{[\phi_z,\Phi_z]} = \lim_{z\to q^n}\frac{z-q^n}{[\phi_z,\Phi_z]} = \lim_{z\to q^n}\frac{z-q^n}{-z(1/z;q)_\infty}$$
$$= \lim_{z\to q^n}\frac{z-q^n}{-z(1/z;q)_n(1 - q^n/z)(q^{n+1}/z;q)_\infty}$$
$$= \frac{-1}{(q^{-n};q)_n(q;q)_\infty} = \frac{(-1)^{n+1}q^{-\frac{1}{2}n(n+1)}}{(q;q)_n(q;q)_\infty}.$$

Moreover, since the Casorati determinant vanishes, the two solutions of interest are proportional;

$$(-1)^n\alpha^{2n+2}(\Phi_{q^n})_l = \frac{(-\alpha^2 q;q)_\infty}{(-1/\alpha^2;q)_\infty}(\phi_{q^n})_l, \quad \text{for all } l \in \mathbb{Z},$$

which can be proved by manipulations of basic hypergeometric series, and we refer to [17] for the details. In particular, $\phi_{q^n} \in \ell^2(\mathbb{Z})$ for $n \in \mathbb{N}$ and $L^*\phi_{q^n} = q^n\phi_{q^n}$. So the spectral measure in this case has a discrete mass point at q^n, $n \in \mathbb{N}$, satisfying

$$\langle E(\{q^n\})v,w\rangle$$
$$= -(-1)^n\alpha^{-(2n+2)}\frac{(-\alpha^2 q;q)_\infty}{(-1/\alpha^2;q)_\infty}\frac{(-1)^{n+1}q^{-\frac{1}{2}n(n+1)}}{(q;q)_n(q;q)_\infty}\langle v,\phi_{x_0}\rangle\langle\phi_{x_0},w\rangle$$
$$= \frac{(-\alpha^2 q;q)_\infty}{(-1/\alpha^2,q;q)_\infty}\frac{\alpha^{-(2n+2)}q^{-\frac{1}{2}n(n+1)}}{(q;q)_n}\langle v,\phi_{x_0}\rangle\langle\phi_{x_0},w\rangle.$$

It follows that the eigenspace is one-dimensional and spanned by the ϕ_{q^n}, since $E(\{q^n\})$ is a rank one projection onto the space spanned the eigen-vector ϕ_{q^n}. Plugging in $v = w = \phi_{q^n}$ then gives

$$\|\phi_{q^n}\|^2 = \langle E(\{q^n\})\phi_{q^n}, \phi_{q^n}\rangle$$

$$= \frac{(-\alpha^2 q;q)_\infty}{(-1/\alpha^2,q;q)_\infty} \frac{\alpha^{-(2n+2)}q^{-\frac{1}{2}n(n+1)}}{(q;q)_n}\|\phi_{q^n}\|^4 \implies$$

$$\|\phi_{q^n}\|^2 = \frac{(-1/\alpha^2,q;q)_\infty}{(-\alpha^2 q;q)_\infty}\alpha^{(2n+2)}q^{\frac{1}{2}n(n+1)}(q;q)_n.$$

Since $\{0\}$ is not a discrete mass point, see Exercise 3.13, we see that the spectrum of L is $q^{\mathbb{N}} \cup \{0\}$ and that we have an orthogonal basis of eigen-vectors $\{\phi_{q^n}\}_{n\in\mathbb{N}}$ for $\ell^2(\mathbb{Z})$.

It turns out that we can rewrite the orthogonality of the eigenvectors $\{\phi_{q^n}\}_{n\in\mathbb{N}}$ in terms of orthogonality relations for orthogonal polynomials, namely for the continuous q^{-1}-Hermite polynomials. This is not a coincidence, since we started out with the second order q-difference operator having these polynomials as eigenfunctions. Of course, this can be done since the continuous q^{-1}-Hermite polynomials are in the q-Askey scheme. Writing down the orthogonality relations explicitly gives

$$\sum_{l=-\infty}^{\infty} \alpha^{4l}q^{2l^2-l}(1+\alpha^2 q^{2l})h_n(x_l(\alpha)|q)h_m(x_l(\alpha)|q)$$

$$= \delta_{n,m}q^{-n(n+1)/2}(q;q)_n(-\alpha^2,-q/\alpha^2,q;q)_\infty, \qquad (3.8)$$

where the polynomials are generated by the monic three-term recurrence relation

$$xh_n(x|q) = h_{n+1}(x|q) + q^{-n}(1-q^n)h_{n-1}(x|q), \quad h_{-1}(x|q) = 0, h_0(x|q) = 1,$$

and the mass points are $x_l(\alpha) = \frac{1}{2}((\alpha q^l)^{-1} - \alpha q^l)$. By the completeness of the basis of eigenvectors $\{\phi_{q^n}\}_{n\in\mathbb{N}}$ it follows that the polynomials are dense in the weighted L^2-space of the corresponding discrete measures in (3.8). Since $\alpha \in (q, 1]$, for each $\xi \in \mathbb{R}$ there is a measure of the type in (3.8) with positive mass in ξ. It follows from the general theory of moment problems [2, 14] that (3.8) gives all N-extremal measures for the continuous q^{-1}-Hermite polynomials. The same result (and more) on the N-extremal measures has been obtained previously by Ismail and Masson [51] by calculating explicitly the functions in the Nevanlinna parametrisation.

Example 3.8   The example discussed in Examples 3.3, 3.6, 3.7 is relatively easy, since L is bounded, and even compact. Another well studied three-term recurrence operator on $\ell^2(\mathbb{Z})$ is the following unbounded operator

$$2Le_k = a_k\, e_{k+1} + b_k\, e_k + a_{k-1}\, e_{k-1},$$

$$a_k = \sqrt{\left(1 - \frac{y^{k+1}}{z}\right)\left(1 - \frac{cq^{k+1}}{d^2 z}\right)}, \qquad b_k = \frac{q^k(c+q)}{dz},$$

assuming $z < 0, 0 < c < 1, d \in \mathbb{R}\setminus\{0\}$. The operator L is essentially self-adjoint for $0 < c \leq q^2$, and the spectral decomposition has an absolutely continuous part and a discrete part, with infinite number of points. This can be proved in the same way as in this section, where basic hypergeometric series play an important role in finding the (free) solutions to the eigenvalue equation $L^*f = zf$. The corresponding spectral decomposition leads to an integral transform known as the little q-Jacobi function transform, see [65]. The quantum group-theoretic interpretation goes back to Kakehi [52], see also [64, App. A]. This result, including a suitable self-adjoint extension for the case $c = q$ and its spectral decomposition, can be found in [42, App. B, C]. In [65] it is described how the little q-Jacobi function transform can be viewed as a non-polynomial addition to the q-Askey scheme.

Remark 3.9 The solution space of the three-term recurrence is two-dimensional, so that the dimension of $\dim S_z^{\pm}$ is determined by summability conditions at $\pm\infty$. When one of $\dim S_z^{\pm}$ is bigger than 1, we have higher deficiency indices. If one of $S_z^{\pm}$ is one-dimensional, and the other is 2-dimensional, and we have deficiency indices $(1,1)$. If both spaces are two-dimensional, the deficiency indices are $(2,2)$. This is an observation essentially due to Masson and Repka [78]. For an example of such a three-term recurrence relation with deficiency indices $(1,1)$, see [56].

3.2.1 Exercises

Exercise 3.10 Prove Lemma 3.1.

Exercise 3.11 Prove Lemma 3.2.

(a) Recall the definition of the domain of the adjoint operator of $(L, \mathscr{D})$ from Section 3.7.5, so we have to find all $w \in \ell^2(\mathbb{Z})$ for which $\mathscr{D} \ni v \mapsto \langle Lv, w \rangle$ is continuous. This is the same as requiring the existence of a constant C so that $|\langle Lv, w \rangle| \leq C\|v\|$ for all $v \in \mathscr{D}$. Write for $v \in \mathscr{D}$

$$\langle Lv, w \rangle = \sum_{\substack{k \in \mathbb{Z} \\ \text{finite}}} v_k \left(\overline{a_k} w_{k+1} + \overline{b_k} w_k + \overline{c_k} w_{k-1} \right)$$

and use Cauchy–Schwarz to prove that $\mathscr{D}^*$ is contained in the domain of the adjoint of $(L, \mathscr{D})$.

(b) Show conversely that any element in the domain of the adjoint is an element of $\mathscr{D}^*$. (Hint: Use the identity in (a) and take a special choice for $v \in \mathscr{D}$ which converges to an element of $\mathscr{D}^*$.)

(c) Finish the proof of Lemma 3.2.

Exercise 3.12   Prove that in Example 3.6 the spaces $S_z^{\pm}$ are indeed spanned by the elements given.

(a) Show that $\sum_{l \in \mathbb{Z}} (\phi_z)_l e_l$ is a formal eigenvector of L. (Hint: This is not directly deducible from the expression as $_0\varphi_1$, first transform to a $_2\varphi_1$, see [29], and use contiguous relations for $_2\varphi_1$. See [17] for details.)

(b) Next show that $\sum_{l>0} |(\phi_z)_l|^2 < \infty$. (Hint: use the asymptotic behaviour of $C_l(\alpha)$ as $l \to \infty$.)

(c) Conclude that $\phi_z \in S_z^+$.

(d) Let $V : \ell^2(\mathbb{Z}) \to \ell^2(\mathbb{Z})$ be the unitary involution $e_l \mapsto e_{-l}$. Write $L(\alpha) = L$ for the operator L, as in Example 3.3, to stress the dependence on α. Show that $L(1/\alpha) = VL(\alpha)V^*$. Conclude that $\Phi_z \in S_z^-$.

(e) Calculate the Casorati determinant or Wronskian $[\phi_z, \Phi_z]$ by taking the limit $l \to \infty$ in Lemma 3.4 using the asymptotic behaviour of a_l as in Example 3.3 and

$$\lim_{x \downarrow 0} {_0\varphi_1} \left(\begin{matrix} - \\ 1/x \end{matrix} ; q, \frac{z}{x} \right) = (z; q)_{\infty}.$$

Show that $[\phi_z, \Phi_z] = -z(1/z; q)_{\infty}$ for $z \neq 0$, by taking the limit $l \to \infty$ in the Casorati determinant or Wronskian using Lemma 3.4.

Exercise 3.13   Show that in Example 3.6 there is no eigenvector, i.e., in ℓ^2, for the eigenvalue 0. (Hint: show that both $(-1)^l q^{-\frac{1}{2}l} \sqrt{1 + \alpha^2 q^{2l}}$ and $(-1)^l q^{-\frac{3}{2}l}(1 - q^l)(1 + \alpha^2 q^l)\sqrt{1 + \alpha^2 q^{2l}}$ give two linearly independent solutions for the recurrence for $z = 0$, and that there is no linear combination which is square summable.)

Exercise 3.14   Show that in general we can take $\overline{\phi_{\bar{z}}} \in S^+(z)$, next put

$$G_{k,l}(z) = \frac{1}{[\phi_{\bar{z}}, \Phi_z]} \begin{cases} (\Phi_z)_k (\overline{\phi_{\bar{z}}})_l, & k \leq l, \\ (\Phi_z)_l (\overline{\phi_{\bar{z}}})_k, & k > l, \end{cases}$$

and show that the resolvent $R(z)$ can be obtained as in the proof of Theorem 3.5.

Exercise 3.15 Rewrite the operator L as a three-term recurrence relation labeled by $\mathbb{N}$ by considering $\mathbb{C}^2$-vectors

$$u_k = \begin{pmatrix} e_k \\ e_{-k-1} \end{pmatrix}, \quad k \in \mathbb{N}$$

and define

$$\mathscr{L} u_k = \begin{pmatrix} Le_k \\ Le_{-k-1} \end{pmatrix},$$

and write $\mathscr{L}$ as a three-term recurrence in terms of u_k with 2×2 matrices acting naturally on $\ell^2(\mathbb{N}) \hat{\otimes} \mathbb{C}^2 \cong \ell^2(\mathbb{Z})$. Determine the matrices in the three-term recurrence explicitly in terms of the coefficients of L in (3.1). See also Section 3.5.3.

3.3 Three-term recurrence relations and orthogonal polynomials

In this section we consider three-term recursion relations labeled by $l \in \mathbb{N}$, and we relate such operators to orthogonal polynomials and the moment problem.

3.3.1 Orthogonal polynomials

Assume μ is a positive Borel measure on the real line $\mathbb{R}$ with infinite support such that all moments

$$m_k = \int_{\mathbb{R}} x^k \, d\mu(x) < \infty$$

exist. We assume the normalisation of μ by $m_0 = \mu(\mathbb{R}) = 1$, so that we have a probability measure.

Note that all polynomials are contained in the Hilbert space $L^2(\mu)$. Then we can apply the Gram–Schmidt procedure to $\{1, x, x^2, x^3, \ldots\}$ to obtain a sequence of polynomials $p_n(x)$ of degree n so that

$$\int_{\mathbb{R}} p_m(x) \overline{p_n(x)} \, d\mu(x) = \delta_{m,n}. \tag{3.9}$$

These polynomials form a family of orthogonal polynomials. We normalise the leading coefficient of p_n to be positive, which can also be viewed as part of the Gram–Schmidt procedure. Observe also that, since all moment m_k are real, the polynomials have real coefficients, so we do not require complex conjugation in (3.9).

Theorem 3.16 (Three-term recurrence relation)   *Let $\{p_k\}_{k=0}^{\infty}$ be the orthonormal polynomials in $L^2(\mu)$; then there exist sequences $\{a_k\}_{k=0}^{\infty}$, $\{b_k\}_{k=0}^{\infty}$, with $a_k > 0$ and $b_k \in \mathbb{R}$, such that*

$$x p_k(x) = a_k p_{k+1}(x) + b_k p_k(x) + a_{k-1} p_{k-1}(x), \qquad k \geq 1,$$
$$x p_0(x) = a_0 p_1(x) + b_0 p_0(x).$$

If μ is compactly supported, then the sequences $\{a_k\}_{k=0}^{\infty}$, $\{b_k\}_{k=0}^{\infty}$ are bounded.

We leave the proof of Theorem 3.16 as Exercise 3.24, where a_n and b_n are expressed as integrals.

Conversely, given arbitrary coefficient sequences $\{a_n\}_{n \in \mathbb{N}}$ and $\{b_n\}_{n \in \mathbb{N}}$ with $a_n > 0, b_n \in \mathbb{R}$ for all $n \in \mathbb{N}$, we see that the recursion of Theorem 3.16 determines the polynomials $p_n(x)$ with the initial condition $p_0(x) = 1$. In order to study these polynomials, one can study the Jacobi operator

$$J e_k = \begin{cases} a_k e_{k+1} + b_k e_k + a_{k-1} e_{k-1}, & k \geq 1, \\ a_0 e_1 + b_0 e_0, & k = 0, \end{cases} \tag{3.10}$$

as an operator on the Hilbert space $\ell^2(\mathbb{N})$ with orthonormal basis $\{e_k\}_{k \in \mathbb{N}}$. Note that we can study such a Jacobi operator without assuming the situation of Theorem 3.16, i.e., arising from a Borel measure with finite moments. So we generate polynomials $\{p_n\}_{n \in \mathbb{N}}$ from the three-term recurrence relation of Theorem 3.16, but now with the coefficients from the Jacobi operator. Note that once $p_0(z)$ is fixed, the polynomials are determined. We assume that $p_0(z) = 1$. See Section 3.3.2 for more information.

Initially, J is defined on the dense linear subspace $\mathscr{D}$ of finite linear combinations of the orthonormal basis $\{e_k\}_{k \in \mathbb{N}}$. It follows from (3.10) and Theorem 3.16 that, at least formally, we have found eigenvectors for J;

$$J\left(\sum_{k=0}^{\infty} p_k(z) e_k \right) = z \sum_{k=0}^{\infty} p_k(z) e_k. \tag{3.11}$$

However, we haven't defined J on arbitrary vectors and in general

$$\sum_{k=0}^{\infty} p_k(z) e_k \notin \ell^2(\mathbb{N}),$$

but (3.11) indicates that there is a relation between the spectrum of J and the orthonormal polynomials. By looking at a partial sum of (3.11), the left-hand side is well defined.

Lemma 3.17 *For $M \in \mathbb{N}$,*

$$J \left(\sum_{k=0}^{M} p_k(z) e_k \right) = z \sum_{k=0}^{M} p_k(z) e_k + a_M p_M(z) e_{M+1} - a_M p_{M+1}(z) e_M.$$

Truncating J to a $(M+1) \times (M+1)$-matrix, which we denote by J_M, we see that –using $\{e_0, \ldots, e_M\}$ as the standard basis–

$$J_M \left(\sum_{k=0}^{M} p_k(z) e_k \right) = z \sum_{k=0}^{M} p_k(z) e_k - a_M p_{M+1}(z) e_M.$$

Since J_M is a self-adjoint matrix, and since its eigenspaces are one-dimensional, we obtain the following corollary.

Corollary 3.18 *For $M \in \mathbb{N}$, the zeroes of p_{M+1} are real and simple.*

We now study the orthonormal polynomials of Theorem 3.16 by studying the Jacobi operator $(J, \mathscr{D})$.

Lemma 3.19 *The adjoint $(J^*, \mathscr{D}^*)$ is given by*

$$\mathscr{D}^* = \{ v = \sum_{k=0}^{\infty} v_k e_k \in \ell^2(\mathbb{N}) \mid \sum_{k=0}^{\infty} (a_k v_k + b_k v_k + a_{k-1} v_{k-1}) e_k \in \ell^2(\mathbb{N}) \}$$

and $J^* v = \sum_{k=0}^{\infty} (a_k v_k + b_k v_k + a_{k-1} v_{k-1}) e_k$ for $v \in \mathscr{D}^*$ of this form.

The proof of Lemma 3.19 is completely analogous to that of Lemma 3.2, see Exercise 3.25.

In order to study the Jacobi operator, we find another solution to the corresponding eigenvalue equation for J. Since the formal eigenspace of J is 1-dimensional, we can only find a solution of the equation $\langle Jv, e_k \rangle = x \langle v, e_k \rangle$ for $k \geq 1$. Let $r_k(x)$ be the sequence of polynomials generated by the three-term recurrence of Theorem 3.16 for $k \geq 1$ with initial conditions $r_0(x) = 0$ and $r_1(x) = a_0^{-1}$. Obviously, r_k is a polynomial of degree $k-1$. The polynomials $\{r_k\}_{k=0}^{\infty}$ are known as the associated polynomials or polynomials of the second kind. If we assume that the Jacobi operator (3.10) comes from the three-term recurrence relation for orthogonal polynomials as in Theorem 3.16, we can describe the polynomials r_k explicitly in terms of the measure μ. This is done in Lemma 3.20.

Lemma 3.20 *Let*

$$w(z) = \int_{\mathbb{R}} \frac{1}{x - z} d\mu(x)$$

be the Stieltjes transform of the measure μ, which is well-defined for $z \in$

$\mathbb{C} \setminus \mathbb{R}$. *We have that*

$$r_k(x) = \int_{\mathbb{R}} \frac{p_k(x) - p_k(y)}{x - y} \, d\mu(y)$$

and for $z \in \mathbb{C} \setminus \mathbb{R}$

$$\sum_{k=0}^{\infty} |w(z)p_k(z) + r_k(z)|^2 \leq \int_{\mathbb{R}} \frac{1}{|x - z|^2} \, d\mu(x) \leq \frac{1}{|\Im(z)|^2} < \infty.$$

Proof We leave the explicit expression of r_k as Exercise 3.26. In the Hilbert space $L^2(\mu)$ we consider the expansion of the function $x \mapsto \frac{1}{x-z}$ for $z \in \mathbb{C} \setminus \mathbb{R}$, which is an element of $L^2(\mu)$ by the estimate $|\frac{1}{x-z}| \leq \frac{1}{|\Im(z)|}$ and μ being a probability measure. We calculate the inner product of $x \mapsto \frac{1}{x-z}$ with an orthonormal polynomial p_k;

$$\int_{\mathbb{R}} \frac{p_k(x)}{x - z} \, d\mu(x)$$

$$= \int_{\mathbb{R}} \frac{p_k(x) - p_k(z)}{x - z} \, d\mu(x) + p_k(z) \int_{\mathbb{R}} \frac{1}{x - z} \, d\mu(x) = r_k(z) + w(z)p_k(z).$$

By the Bessel inequality for the orthonormal sequence $\{p_k\}_{k \in \mathbb{N}}$ in $L^2(\mu)$ the result follows. $\qquad \square$

As a corollary to Lemma 3.20 we get that

$$\lim_{k \to \infty} \frac{r_k(z)}{p_k(z)} = -w(z) = \int_{\mathbb{R}} \frac{1}{z - x} \, d\mu(x), \qquad z \in \mathbb{C} \setminus \mathbb{R}. \qquad (3.12)$$

which is known as Markov's theorem, see [11] for an overview.

In particular, we see that the vector

$$f(z) = \sum_{k=0}^{\infty} \left(r_k(z) + w(z)p_k(z) \right) e_k \in \ell^2(\mathbb{N})$$

for $z \in \mathbb{C} \setminus \mathbb{R}$, and satisfying $\langle J^* f(z), e_k \rangle = z \langle f(z), e_k \rangle$ for $k \geq 1$. We view $f(z)$ as the free solution in this case. So it is a square summable solution for the three-term recurrence relation for $k \gg 0$. From here we can define the Green function and calculate the resolvent explicitly. Under the assumption that $\sum_{n \in \mathbb{N}} |p_n(z)|^2$ diverges for $z \in \mathbb{C} \setminus \mathbb{R}$ this can be obtained from Section 3.4.3 by specialising to $N = 1$.

3.3.2 Jacobi operators

The converse problem, namely finding the orthogonality measure μ for the polynomials $\{p_k\}_{k \in \mathbb{N}}$ generated by a three-term recurrence relation as of

Theorem 3.16, can be solved by studying the Jacobi operator of (3.10). The operator $(J, \mathscr{D})$, with adjoint $(J^*, \mathscr{D}^*)$ as in Lemma 3.19, can be studied from a spectral point of view.

Proposition 3.21 *The deficiency indices (n_+, n_-) of $(J, \mathscr{D})$ are $(0,0)$ or $(1,1)$. In the case $(n_+, n_-) = (0,0)$ the operator $(J, \mathscr{D})$ is essentially self-adjoint. Let E be the spectral decomposition of $(J^*, \mathscr{D}^*)$ in case $(n_+, n_-) = (0,0)$ and of a self-adjoint extension $(J_\theta, D(J_\theta))$, $(J, \mathscr{D}) \subset (J_\theta, D(J_\theta)) \subset (J^*, \mathscr{D}^*)$ in case $(n_+, n_-) = (1,1)$. Then an orthogonality measure for the polynomials is given by $\mu(B) = \langle E(B)e_0, e_0 \rangle$, $B \in \mathcal{B}$.*

Proof The deficiency indices are equal, since J^* commutes with conjugation. Since the eigenvalue equation $J^* v = z v$ is completely determined by the initial value $\langle v, e_0 \rangle$, the deficiency space is at most 1-dimensional. Note that $J^* v = z v$ gives $\langle v, e_n \rangle = p_n(z)\langle v, e_0 \rangle$, so that the defect indices are $(1,1)$ if and only if $\sum_{n=0}^{\infty} |p_n(z)|^2 < \infty$.

Also, e_0 is a cyclic vector of $\ell^2(\mathbb{N})$ for J, i.e., $\ell^2(\mathbb{N})$ equals the closure of the space of $J^k e_0$, $k \in \mathbb{N}$ and even $e_k = p_k(J)e_0$, which follows by induction on k. Since J^* or J_θ extend J, we have

$$\delta_{k,l} = \langle e_k, e_l \rangle = \langle p_k(J)e_0, p_l(J)e_0 \rangle = \langle p_l(J)p_k(J)e_0, e_0 \rangle$$
$$= \int_{\mathbb{R}} p_l(\lambda)p_k(\lambda)\, dE_{e_0, e_0}(\lambda) = \int_{\mathbb{R}} p_l(\lambda)p_k(\lambda)\, d\mu(\lambda)$$

using the spectral theorem for self-adjoint operators in Appendix 3.7. □

Corollary 3.22 (Favard's theorem) *Let the polynomials p_n of degree n be generated by the recursion $p_0(z) = 1$, $p_1(z) = a_1^{-1}(z - b_0)$ and*

$$z p_n(z) = a_n p_{n+1}(z) + b_n p_n(z) + a_{n-1} p_{n-1}(z)$$

for sequences $\{a_n\}_{n \in \mathbb{N}}$, $\{b_n\}_{n \in \mathbb{N}}$ with $a_n > 0$ and $b_n \in \mathbb{R}$ for all n. Then there exists a Borel measure on $\mathbb{R}$ with finite moments so that

$$\int_{\mathbb{R}} p_n(x)p_m(x)\, d\mu(x) = \delta_{m,n}.$$

Remark 3.23 According to Proposition 3.79 the labeling of the self-adjoint extension of Proposition 3.21 is given by $U(n_+) = U(1)$, so we can think of $\theta \in [0, 2\pi)$ as parametrising the self-adjoint extensions of $(J, \mathscr{D})$ in Proposition 3.21. It can then be proved that the corresponding orthogonality measures for different self-adjoint extensions lead to different Borel measures for the orthogonal polynomials, see e.g., [23, Ch. XII.8], [57, Thm. (3.4.5)], [87, Ch. 16],

Note that in particular, we see that the condition $\dim S_z^\pm \geq 1$, mentioned immediately after Theorem 3.5, follows by considering the two Jacobi operators associated to L by considering $k \to \infty$ and $k \to -\infty$. In fact, a theorem by Masson and Repka [78], states the deficiency indices of the operator L of Section 3.2 can be obtained by adding the deficiency indices of the two Jacobi operators corresponding to $k \to \infty$ and $k \to -\infty$.

3.3.3 Moment problems

The moment problem is the following:

(1) Given a sequence $\{m_0, m_1, m_2, \ldots\}$, does there exist a positive Borel measure μ on $\mathbb{R}$ such that $m_k = \int x^k \, d\mu(x)$?
(2) If the answer to (1) is yes, is the measure obtained unique?

We exclude the case of finite discrete orthogonal polynomials, so we assume $\mathrm{supp}(\mu)$ is not a finite set. This is equivalent to the Hankel matrix $(m_{i+j})_{0 \leq i,j \leq N}$ being regular for all $N \in \mathbb{N}$. We do not discuss the conditions for existence of such a measure. The Haussdorf moment problem (1920) requires $\mathrm{supp}(\mu) \subset [0,1]$. The Stieltjes moment problem (1894) requires $\mathrm{supp}(\mu) \subset [0,\infty)$. The Hamburger moment problem (1922) does not require a condition on the support of the measure. See Akhiezer [2], Buchwalter and Cassier [14], Dunford and Schwartz [23, Ch. XII.8], Schmüdgen [87, Ch. 16], Shohat and Tamarkin [88], Simon [89], Stieltjes [91], and Stone [92] for more information.

The fact that the measure is not determined by its moments was first noticed by Stieltjes in his famous memoir, [91], published posthumously. See Kjeldsen [53] for an overview of the early history of the moment problem. Stieltjes's example is discussed in Exercise 3.27.

So we see that the moment problem is determinate – i.e., the answer to (2) is yes – if and only if the corresponding Jacobi operator is essentially self-adjoint.

3.3.4 Exercises

Exercise 3.24   Prove Theorem 3.16.

(a) Prove that there is a three-term recurrence relation. (Hint: Expand $x p_n(x)$ in the polynomials, and use that multiplying by x is (a possibly unbounded) symmetric operator on the space of polynomials in $L^2(\mu)$, since μ is a real Borel measure.)

(b) Establish $a_n = \int_{\mathbb{R}} x p_n(x) p_{n+1}(x) \, d\mu(x)$ and $b_n = \int_{\mathbb{R}} x (p_n(x))^2 \, d\mu(x)$.

(c) Show that if μ has bounded support that the coefficients a_n and b_n are bounded. (Hint: if supp $(\mu) \subset [-M, M]$ then one can estimate x in the integrals by M, and next use the Cauchy–Schwarz inequality in $L^2(\mu)$.)

Exercise 3.25 Prove Lemma 3.19. (Hint: consider the proof as in Exercise 3.11.)

Exercise 3.26 Prove the explicit expression for r_k of Lemma 3.20. (Hint: write

$$x(p_k(x) - p_k(y)) + (x - y)p_k(y) =$$
$$a_k(p_{k+1}(x) - p_{k+1}(y)) + b_k(p_k(x) - p_k(y)) + a_{k-1}(p_{k-1}(x) - p_{k-1}(y))$$

using the three-term recurrence relation. Divide by $x - y$ and integrate with respect to μ. Then the second term on the left-hand side vanishes for $k \geq 1$. Check the initial values as well.)

Exercise 3.27 (a) Show that for $\gamma > 0$

$$\int_0^\infty x^n e^{-\gamma^2 \ln^2 x} \sin(2\pi\gamma^2 \ln x) \, dx = 0, \qquad \text{for all } n \in \mathbb{N}.$$

(Hint: switch to $y = \gamma \ln(x) - \frac{1}{2\gamma}(n+1)$.)

(b) Conclude that the moments $\int_0^\infty x^n e^{-\gamma^2 \ln^2 x}(1 + r\sin(2\pi\gamma^2 \ln x)) \, dx$ are independent of r, and this is a positive measure for $r \in \mathbb{R}$ with $|r| \leq 1$.

Exercise 3.28 Prove the Christoffel–Darboux formula for the orthonormal polynomials using the three-term recurrence relation:

$$(x - y) \sum_{k=0}^{n-1} p_k(x) p_k(y) = a_{n-1}(p_n(x) p_{n-1}(y) - p_{n-1}(x) p_n(y))$$

and derive the limiting case

$$\sum_{k=0}^{n-1} p_k(x)^2 = a_{n-1}(p_n'(x) p_{n-1}(x) - p_{n-1}'(x) p_n(x)).$$

3.4 Matrix-valued orthogonal polynomials

In this section we study matrix-valued orthogonal polynomials using a spectral analytic description of the corresponding Jacobi operator. We follow [6, 19, 31], and references given there, in particular those in [19].

3.4.1 Matrix-valued measures and related polynomials

We consider $\mathbb{C}^N$ as a finite-dimensional inner product space with standard orthonormal basis $\{e_i\}_{i=1}^N$. By $M_N(\mathbb{C})$ we denote the matrix algebra of linear maps $T\colon \mathbb{C}^N \to \mathbb{C}^N$. Let $E_{i,j} \in M_N(\mathbb{C})$ be the rank one operators $E_{i,j}v = \langle v, e_j\rangle e_i$, so that $E_{i,j}e_k = \delta_{k,j}e_i$. So $E_{i,j}$ is the $N \times N$-matrix with all zeroes, except one 1 at the (i,j)th entry. Note that in particular $\mathbb{C}^N$ is a (finite-dimensional) Hilbert space, see Example 3.75, so that $M_N(\mathbb{C})$ carries a norm and with this norm $M_N(\mathbb{C})$ is a C*-algebra, see Section 3.7.2.

A linear map $T\colon \mathbb{C}^N \to \mathbb{C}^N$ is positive, or positive-definite, if $\langle Tv, v\rangle > 0$ for all $v \in \mathbb{C}^N \setminus \{0\}$; we write this as $T > 0$. The map T is positive semi-definite if $\langle Tv, v\rangle \geq 0$ for all $v \in \mathbb{C}^N$, written as $T \geq 0$. The space of positive linear semi-definite maps, or positive semi-definite matrices (after fixing a basis), is denoted by $P_N(\mathbb{C})$. $P_N(\mathbb{C})$ is a closed cone in $M_N(\mathbb{C})$. Its interior $P_N^o(\mathbb{C})$ is the open cone of positive matrices. Note that each positive linear map is Hermitean, see [46, §7.1]. Then we set $T > S$ if $T - S > 0$ and $T \geq S$ if $T - S \geq 0$, see Section 3.7.2.

Definition 3.29   A matrix-valued measure (or matrix measure) is a σ-additive map $\mu\colon \mathcal{B} \to P_N(\mathbb{C})$ where $\mathcal{B}$ is the Borel σ-algebra on $\mathbb{R}$.

Recall that σ-additivity means that for any sequence $\{E_1, E_2, \ldots\}$ of pairwise disjoint Borel sets, we have

$$\mu\left(\bigcup_{k=1}^{\infty} E_k\right) = \sum_{k=1}^{\infty} \mu(E_k)$$

where the right-hand side is unconditionally convergent in $M_N(\mathbb{C})$.

Note $\mu_{v,w}\colon \mathcal{B} \to \mathbb{C}$, $\mu_{v,w}(B) = \langle \mu(B)w, v\rangle$ is a complex-valued Borel measure on $\mathbb{R}$, and in particular $\mu_{v,v}\colon \mathcal{B} \to \mathbb{R}$ is a positive Borel measure on $\mathbb{R}$. Let $\tau_\mu = \sum_{i=1}^N \mu_{i,i}$ be the positive Borel measure on $\mathbb{R}$ corresponding to the trace of μ, i.e., $\tau_\mu(B) = \mathrm{Tr}(\mu(B))$ for all $B \in \mathcal{B}$. Here we use the notation $\mu_{i,j} = \mu_{e_i, e_j}$, but note that the trace measure τ_μ is independent of the choice of basis for $\mathbb{C}^N$. The following result is [12, Thm. 1.12], see also [19, §1.2], [72, §3], [83].

Theorem 3.30   *For a matrix measure μ there exist functions $W_{i,j} \in L^1(\tau_\mu)$ such that*

$$\mu_{i,j}(B) = \int_B W_{i,j}(x)\,d\tau_\mu(x), \quad \text{for all } B \in \mathcal{B}$$

and $W(x) = \left(W_{i,j}(x)\right)_{1 \leq i,j \leq N} \in P_N(\mathbb{C})$ for τ_μ-almost everywhere.

The proof is based on the fact that for a positive-definite matrix $A =$

$(a_{i,j})_{i,j=1}^N$ we have $|a_{i,j}| \leq \sqrt{a_{i,i}a_{j,j}} \leq \frac{1}{2}(a_{i,i}+a_{j,j}) \leq \mathrm{Tr}\,(A)$ and the Radon–Nikodym theorem, see [12] for details. The first inequality follows from considering a positive-definite 2×2-submatrix, and the second by the arithmetic–geometric mean inequality. Note that this inequality also implies that $W(x) \leq I$ τ_μ-almost everywhere (a.e.), see also [83, Lemma 2.3]. The measure τ_μ is regular, see [84, Thm. 2.18], [96, Satz I.2.14].

Assumption 3.31 From now on we assume for Section 3.4 that μ is a matrix measure for which τ_μ has infinite support and for which all moments exist; i.e., $(x \mapsto x^k W_{i,j}(x)) \in L^1(\tau_\mu)$ for all $1 \leq i, j \leq N$ and all $k \in \mathbb{N}$. Moreover, we assume that, in the notation of Theorem 3.30, the matrix W is positive-definite τ_μ-a.e., i.e., $W(x) > 0$ τ_μ-a.e.

Note that we do not assume that the weight is irreducible in a suitable sense, but we discuss the reducibility issue briefly in Section 3.5.4.

By

$$M_k = \int_{\mathbb{R}} x^k \, d\mu(x) \in M_N(\mathbb{C}), \qquad (M_k)_{i,j} = \int_{\mathbb{R}} x^k W_{i,j}(x) \, d\tau_\mu,$$

we denote the corresponding moments in $M_N(\mathbb{C})$. Note that the even moments are positive-definite, i.e., $M_{2k} \in P_N^o(\mathbb{C})$.

Given a weight function as Assumption 3.31, we can associate matrix-valued orthogonal polynomials P_n so that

$$\int_{\mathbb{R}} P_n(x) W(x) P_m^*(x) \, d\tau_\mu(x) = \delta_{m,n} I, \qquad (3.13)$$

where $P_m^*(z) = (P_m(\bar{z}))^*$ for $z \in \mathbb{C}$, so that the $P_m^*(z) = \sum_{k=0}^m A_k^* z^k$ if $P_m(z) = \sum_{k=0}^m A_k z^k$, where $A_k \in M_N(\mathbb{C})$ are the coefficients of the polynomial P_m. Moreover, for all $m \in \mathbb{N}$, the leading coefficient A_m of P_m is regular, see e.g., [19, 12]. See Exercise 3.47.

Note that we do not normalise the first $M_0 = \int_{\mathbb{R}} d\mu(x)$ as the identity matrix $I \in M_N(\mathbb{C})$. So we normalise $P_0(z) = M_0^{-1/2}$, which can be done since M_0 is a positive-definite matrix, hence having a square root and an inverse having a square root as well.

Consider the space of $M_N(\mathbb{C})$-valued functions F so that

$$\int_{\mathbb{R}} F(x) W(x) F^*(x) \, d\tau_\mu(x)$$

exists entry wise in $M_N(\mathbb{C})$. Here, as before, $F^*(z) = (F(\bar{z}))^*$. So this means that integrals

$$\int_{\mathbb{R}} \sum_{i,j=1}^N F_{k,i}(x) W_{i,j}(x) F_{j,l}^*(x) \, d\tau_\mu(x)$$

exist for $1 \leq k, l \leq N$. In general, the sum and integral cannot be interchanged, see [83, Example, p. 292], but note that this can be done in case F is polynomial by Assumption 3.31. The Hilbert C*-module $L^2_C(\mu)$ is obtained by modding out by the space of functions for which the integral is zero (as the element in the cone of positive matrices in $M_N(\mathbb{C})$). Because of Assumption 3.31 these are the $M_N(\mathbb{C})$-valued functions which are zero τ_μ-a.e. If we do not assume W to be positive-definite τ_μ-a.e., we have to mod out by a larger space in general, see Section 3.5.1.

Then $L^2_C(\mu)$ is a left $M_N(\mathbb{C})$-module and the $M_N(\mathbb{C})$-valued inner product on $L^2_C(\mu)$ is defined by

$$\langle F, G \rangle = \int_{\mathbb{R}} F(x) W(x) G^*(x) \, d\tau_\mu(x) \in M_N(\mathbb{C})$$

and satisfying for $F, G, H \in L^2(\mu), A, B \in M_N(\mathbb{C})$,

$$\langle AF + BG, H \rangle = A\langle F, H \rangle + B\langle G, H \rangle, \qquad \langle F, G \rangle = \left(\langle G, F \rangle \right)^*,$$
$$\langle F, F \rangle \geq 0, \quad \langle F, F \rangle = 0 \iff F = 0,$$

so that we have a Hilbert C*-module. The completeness is proved in [83, Thm. 3.9], using the fact that the Hilbert–Schmidt norm on $M_N(\mathbb{C})$ is equivalent to the operator norm. So in particular, the polynomials P_n give an orthonormal collection for the Hilbert C*-module $L^2_C(\mu)$.

With μ we also associate the Hilbert space $L^2_v(\mu)$, which is the space of $\mathbb{C}^N$-valued functions f so that

$$\int_{\mathbb{R}} f^*(x) W(x) f(x) \, d\tau_\mu(x) = \int_{\mathbb{R}} \sum_{i,j=1}^N f_i^*(x) W_{i,j}(x) f_j(x) \, d\tau_\mu(x) < \infty,$$

where $f(z)$ is a column vector and $f^*(z) = \left(f(\bar{z}) \right)^*$ is a row vector. Then the inner product in $L^2_v(\mu)$ is given by

$$\langle f, g \rangle = \int_{\mathbb{R}} g^*(x) W(x) f(x) \, d\tau_\mu(x).$$

Again, we assume we have modded out by $\mathbb{C}^N$-valued functions f with $\langle f, f \rangle = 0$, which in this case are functions $f: \mathbb{R} \to \mathbb{C}^N$ which are zero τ_μ-a.e. by Assumption 3.31. The space $L^2_v(\mu)$ is studied in detail, and in greater generality, in [23, XIII.5.6-11].

If $F \in L^2_C(\mu)$ then $z \mapsto F^*(z)v$ is an element of $L^2_v(\mu)$. For $f_i \in L^2_v(\mu)$, the $M_N(\mathbb{C})$-valued function F having f_i^* as its ith column is in $L^2_C(\mu)$.

Theorem 3.32   *There exist sequence of matrices $\{A_n\}_{n \in \mathbb{N}}$, $\{B_n\}_{n \in \mathbb{N}}$ so* *that* $\det(A_n) \neq 0$ *for all* $n \in \mathbb{N}$ *and* $B_n^* = B_n$ *for all* $n \in \mathbb{N}$ *and*

$$zP_n(z) = \begin{cases} A_n P_{n+1}(z) + B_n P_n(z) + A_{n-1}^* P_{n-1}(z), & n \geq 1 \\ A_0 P_1(z) + B_0 P_0(z), & n = 0. \end{cases}$$

We leave the proof of Theorem 3.32 as Exercise 3.49.

Remark 3.33 (i) For a sequence of unitary matrices $\{U_n\}_{n\in\mathbb{N}}$, the polynomials $\tilde{P}_n(z) = U_n P_n(z)$ are also orthonormal polynomials with respect to the same matrix-valued measure μ and with matrices A_n, B_n replaced by $\tilde{A}_n = U_n A_n U_{n+1}^*$, $\tilde{B}_n = U_n B_n U_n^*$. Conversely, if $\{\tilde{P}_n\}_{n\in\mathbb{N}}$ is a family of orthonormal polynomials, then there exist unitary matrices $\{U_n\}_{n\in\mathbb{N}}$ such that polynomials $\tilde{P}_n(z) = U_n P_n(z)$.

(ii) By (i), there is always a choice in fixing the matrix A_n. One possible choice is to take A_n upper (or lower) triangular. Another normalisation is to consider monic matrix-valued polynomials $\{R_n\}_{n\in\mathbb{N}}$ instead. The three-term recurrence for $n \geq 1$ becomes

$$zR_n(z) = R_{n+1}(z) + \left(\mathrm{lc}\,(P_n)^{-1} B_n \mathrm{lc}\,(P_n)\right) R_n(z)$$
$$+ \left(\mathrm{lc}\,(P_{n-1})^{-1} A_{n-1} A_{n-1}^* \mathrm{lc}\,(P_{n-1})\right) R_{n-1}(z)$$

since $\mathrm{lc}\,(P_n) = A_n \mathrm{lc}\,(P_{n+1})$, where $\mathrm{lc}\,(P_n) \in M_N(\mathbb{C})$ denotes the leading coefficient of the polynomial P_n, which is a regular matrix.

Example 3.34 This example gives an explicit example of a matrix-valued measure and corresponding three-term recurrence relation for arbitrary size, which can be considered as a matrix-valued analogue of the Gegenbauer or ultraspherical polynomials, see e.g., [8, 47, 54, 55, 93, 94] for the scalar case. It is one of few examples for arbitrary size where most, if not all, of the important properties are explicitly known. The case $v = 1$ was originally obtained using group theory and analytic methods, see [59, 60], motivated by [44, 71], and later analytically extended in v, see [61]. A q-analogue for the case $v = 1$, viewed as a matrix-valued analogue of a subclass of continuous q-ultraspherical polynomials can be found in [3]. For 2×2-matrix-valued cases, Pacharoni and Zurrián [81] have also derived analogues of the Gegenbauer polynomials, and there is some overlap with the irreducible subcases specialised to the 2×2-cases of this general example. This family of matrix-valued orthogonal polynomials is studied in [61], and we refer to this paper for details.

In this example $N = 2\ell + 1$, where $\ell \in \frac{1}{2}\mathbb{N}$, and we use the numbering from 0 to 2ℓ for the indices. We use the standard notation for Gegenbauer polynomials, see e.g., [47, §4.5]. For $v > 0$, $W^{(v)}(x)$ has the following LDU-decomposition

$$W^{(v)}(x) = L^{(v)}(x)T^{(v)}(x)L^{(v)}(x)^t, \qquad x \in (-1,1), \tag{3.14}$$

where $L^{(v)} \colon [-1,1] \to M_{2\ell+1}(\mathbb{C})$ is the unipotent lower triangular matrix-valued polynomial

$$\left(L^{(v)}(x)\right)_{m,k} = \begin{cases} 0 & \text{if } m < k \\ \dfrac{m!}{k!(2v+2k)_{m-k}} C_{m-k}^{(v+k)}(x) & \text{if } m \geq k, \end{cases}$$

and $T^{(v)} \colon (-1,1) \to M_{2\ell+1}(\mathbb{C})$ is the diagonal matrix-valued function

$$\left(T^{(v)}(x)\right)_{k,k} = t_k^{(v)} \left(1 - x^2\right)^{k+v-1/2},$$

$$t_k^{(v)} = \frac{k!\,(v)_k}{(v+1/2)_k} \frac{(2v+2\ell)_k\,(2\ell+v)}{(2\ell-k+1)_k\,(2v+k-1)_k}.$$

From this expression it immediately follows that $W^{(v)}$ is positive-definite on $(-1,1)$, since for $v > 0$ all the constants are positive. The definition (3.14) is not used as the definition in [61], but it has the advantage that it proves that $W^{(v)}$ is positive-definite immediately.

So we can consider the corresponding monic matrix-valued orthogonal polynomials for which we have the orthogonality relations, see [61, Thm. 3.1],

$$\int_{-1}^{1} P_n^{(v)}(x)\,W^{(v)}(x)\left(P_m^{(v)}\right)^*(x)\,dx = \delta_{n,m} H_n^{(v)},$$

$$\left(H_n^{(v)}\right)_{k,l} = \delta_{k,l}\sqrt{\pi}$$

$$\times \frac{\Gamma(v+\frac{1}{2})}{\Gamma(v+1)}\frac{v(2\ell+v+n)}{v+n}$$

$$\times \frac{n!\,(\ell+\frac{1}{2}+v)_n(2\ell+v)_n(\ell+v)_n}{(2\ell+v+1)_n(v+k)_n(2\ell+2v+n)_n(2\ell+v-k)_n}$$

$$\times \frac{k!\,(2\ell-k)!\,(n+v+1)_{2\ell}}{(2\ell)!\,(n+v+1)_k(n+v+1)_{2\ell-k}}$$

where Γ denotes the standard Γ-function, $\Gamma(z) = \int_0^\infty t^{z-1}e^{-t}dt$, see e.g., [5, 47, 94]. The three-term recurrence relation for the monic matrix-valued orthogonal polynomials is

$$x P_n^{(v)}(x) = P_{n+1}^{(v)}(x) + B_n^{(v)} P_n^{(v)}(x) + C_n^{(v)} P_{n-1}^{(v)}(x)$$

where the matrices $B_n^{(v)}, C_n^{(v)}$ are given by

$$B_n^{(v)} = \sum_{j=1}^{2\ell} \frac{j(j+v-1)}{2(j+n+v-1)(j+n+v)} E_{j,j-1}$$

$$+ \sum_{j=0}^{2\ell-1} \frac{(2\ell-j)(2\ell-j+v-1)}{2(2\ell-j+n+v-1)(2\ell+n-j+v)} E_{j,j+1},$$

$$C_n^{(v)} =$$

$$\sum_{j=0}^{2\ell} \frac{n(n+v-1)(2\ell+n+v)(2\ell+n+2v-1)}{4(2\ell+n+v-j-1)(2\ell+n+v-j)(j+n+v-1)(j+n+v)} E_{j,j}.$$

The proofs of the orthogonality relations and the three-term recurrence relations involve shift operators, where the lowering operator is essentially the derivative and the raising operator is a suitable adjoint (in the context of a Hilbert C*-module) of the derivative. The explicit value for C_n follows easily from the quadratic norm, and the calculation of B_n requires the use of these shift operators.

Putting $P_n(x) = \left(H_n^{(v)}\right)^{-1/2} P_n^{(v)}(x)$ as the corresponding orthonormal polynomials, we find the three-term recurrence relation of Theorem 3.32 with $A_n = \left(H_n^{(v)}\right)^{-1/2} \left(H_{n+1}^{(v)}\right)^{1/2}$, so that $A_{n-1}^* = \left(H_n^{(v)}\right)^{-1/2} C_n^{(v)} \left(H_{n-1}^{(v)}\right)^{1/2}$, and $B_n = \left(H_n^{(v)}\right)^{-1/2} B_n^{(v)} \left(H_n^{(v)}\right)^{1/2}$. Finally, note that we have not written the weight measure in terms of the corresponding tracial weight. Note that

$$\mathrm{Tr}\left(W^{(v)}(x)\right) = \sum_{p=0}^{2\ell} \left(\sum_{k=p}^{2\ell} \left(L_{k,p}^{(v)}(x)\right)^2\right) T_{p,p}^{(v)}(x),$$

so that by (3.14), the trace measure τ_μ is absolutely continuous with respect to the standard Gegenbauer weight $(1-x^2)^{v-1/2}dx$ on $[-1,1]$. Now a result by Rosenberg [83, p. 294] states that the abstractly defined, i.e., using the trace measure τ_μ, spaces $L_C^2(\mu)$ and $L_v^2(\mu)$ are indeed the same as the corresponding spaces using the weight $W^{(v)}$ on $[-1,1]$. Finally, note that in the limit $n \to \infty$ the recurrence relation reduces to a diagonal recurrence, in which the matrices are multiples of the identity. So this example fits into the approach of Aptekarev and Nikishin [6], Geronimo [31], Durán [25].

Starting with the matrix-valued measure and choosing a corresponding set of matrix-valued orthonormal polynomials $\{P_n\}_{n\in\mathbb{N}}$, we can associate the corresponding matrix-valued polynomials of the second kind

$$Q_n(z) = \int_{\mathbb{R}} \frac{P_n(z) - P_n(x)}{z - x} d\mu(x) = \int_{\mathbb{R}} \frac{P_n(z) - P_n(x)}{z - x} W(x) d\tau_\mu(x) \quad (3.15)$$

so that $Q_0(z) = 0$ (as a matrix in $M_N(\mathbb{C})$) and, since $P_1(z) = A_0^{-1}(xM_0^{-1/2} - B_0)$, we have $Q_1(z) = A_0^{-1}M_0^{-1/2}M_0 = A_0^{-1}M_0^{1/2}$. Note that, in the context of Remark 3.33, we have $\tilde{Q}_n(z) = U_n Q_n(z)$.

In the case $\ell = 0$ or $N = 1$ of Example 3.34 the associated polynomials can be expressed in terms of the Gegenbauer polynomials $C_{n-1}^{(\nu+1)}$. This breaks down in the general case of the matrix-valued Gegenbauer polynomials in Example 3.34.

Lemma 3.35 *With the notation of Theorem* 3.32 *and* (3.15) *we have for* $n \geq 1$

$$zQ_n(z) = A_n Q_{n+1}(z) + B_n Q_n(z) + A_{n-1}^* Q_{n-1}(z).$$

See Exercise 3.50 for the proof of Lemma 3.35.

There are many relations between the two solutions, however an easy analogue of Lemma 3.4 is not available, since the non-commutativity of $M_N(\mathbb{C})$ has to be taken into account. For our purposes we need the matrix-valued analogue of the Liouville–Ostrogradsky result in order to describe the Green kernel for the corresponding Jacobi operator (see [47, §22.2] for the classical (scalar) case and [12, Theorem 5.2] for the matrix case).

Lemma 3.36 *Let* $z \in \mathbb{C}$. *For* $k \geq 1$ *we have*

$$Q_k(z)P_{k-1}^*(z) - P_k(z)Q_{k-1}^*(z) = A_{k-1}^{-1}$$

and for $k \geq 0$ *we have* $Q_k(z)P_k^*(z) = P_k(z)Q_k^*(z)$.

We follow [12, §5] for its proof.

Proof We proceed by joint induction on k. The case $k = 1$ is

$$Q_1(z)P_0^*(z) - P_1(z)Q_0^*(z) = A_0^{-1}M_0^{1/2}(M_0^{-1/2})^* - 0 = A_0^{-1}.$$

The case $k = 0$ of the second statement is trivial, since $Q_0(z) = 0$. For $k = 1$, we see that both sides equal

$$A_0^{-1}(zM_0^{-1/2} - B_0)(A_0^{-1})^*$$

since M_0 and B_0 are self-adjoint.

Now assume that both statements have been proved for $k \leq n$. Use Theorem 3.32 multiplied from the right by $Q_n^*(z)$ and Lemma 3.35 multiplied from the right by $P_n^*(z)$ and subtract to get

$$A_n\left(P_{n+1}(z)Q_n^*(z) - Q_{n+1}(z)P_n^*(z)\right) + (B_n - z)\left(P_n(z)Q_n^*(z) - Q_n(z)P_n^*(z)\right)$$
$$+ A_{n-1}^*\left(P_{n-1}(z)Q_n^*(z) - Q_{n-1}(z)P_n^*(z)\right) = 0.$$

By the induction hypothesis the middle term vanishes, and the last term is $A_{n-1}^*(A_{n-1}^{-1})^* = I$ by taking adjoints. Hence,

$$A_n\left(P_{n+1}(z)Q_n^*(z) - Q_{n+1}(z)P_n^*(z)\right) = -I$$

which is the first statement for $k = n+1$.

To prove the second statement for $k = n+1$, write

$$zP_n(z)Q_{n+1}^*(z)$$
$$= A_n P_{n+1}(z)Q_{n+1}^*(z) + B_n P_n(z)Q_{n+1}^*(z) + A_{n-1}^* P_{n-1}(z)Q_{n+1}^*(z) \quad\Longrightarrow$$
$$A_n P_{n+1}(z)Q_{n+1}^*(z) = (z - B_n)P_n(z)Q_{n+1}^*(z) -$$
$$A_{n-1}^* P_{n-1}(z)\left(Q_n^*(z)(z - B_n) - Q_{n-1}^*(z)A_{n-1}\right)(A_n^*)^{-1}$$

since $Q_{n+1}^*(z) = \left(Q_n^*(z)(z - B_n) - Q_{n-1}^*(z)A_{n-1}\right)(A_n^*)^{-1}$ by taking adjoints in Lemma 3.35 using the regularity of A_k and B_k being self-adjoint. Since this argument only uses the recursion for $k \geq 1$ we can interchange the roles of the polynomials P_k and Q_k. Subtracting the two identities then gives

$$A_n\left(P_{n+1}(z)Q_{n+1}^*(z) - Q_{n+1}(z)P_{n+1}^*(z)\right)$$
$$= (z - B_n)\left(P_n(z)Q_{n+1}^*(z) - Q_n(z)P_{n+1}^*(z)\right)$$
$$- A_{n-1}^*\left(P_{n-1}(z)Q_n^*(z) - Q_{n-1}(z)P_n^*(z)\right)(z - B_n)(A_n^*)^{-1}$$
$$- A_{n-1}^*\left(P_{n-1}(z)Q_{n-1}^*(z) - Q_{n-1}(z)P_{n-1}^*(z)\right)A_{n-1}(A_n^*)^{-1}.$$

Applying the induction hypothesis for the second statement, the last term vanishes. Since we assume the first statement for $k \leq n$, and we have already proved the first statement for $k = n+1$, we find

$$A_n\left(P_{n+1}(z)Q_{n+1}^*(z) - Q_{n+1}(z)P_{n+1}^*(z)\right)$$
$$= (z - B_n)(A_n^{-1})^* - A_{n-1}^*(A_{n-1}^{-1})^*(z - B_n)(A_n^*)^{-1}.$$

Since the right-hand side is zero and A_n is invertible, the second statement follows for $k = n+1$. So we have established the induction step, and the lemma follows. $\qquad\square$

3.4.2 *The corresponding Jacobi operator*

We now consider the Hilbert space $\ell^2(\mathbb{N}) \hat\otimes \mathbb{C}^N$, which we denote by $\ell^2(\mathbb{C}^N)$, as the Hilbert space tensor product of the Hilbert spaces $\ell^2(\mathbb{N})$ equipped with the standard orthonormal basis $\{e_n\}_{n\in\mathbb{N}}$ and $\mathbb{C}^N$ with the standard orthonormal basis $\{e_n\}_{n=1}^N$, see Example 3.75(iv). In explicit examples, such as Example 3.34, it is convenient to have a slightly different labeling. Then we can denote

$$V = \sum_{n=0}^{\infty} e_n \otimes v_n \in \ell^2(\mathbb{C}^N) = \ell^2(\mathbb{N}) \hat{\otimes} \mathbb{C}^N$$

where $v_n \in \mathbb{C}^N$. The inner product in the Hilbert space $\ell^2(\mathbb{C}^N) = \ell^2(\mathbb{N}) \hat{\otimes} \mathbb{C}^N$ is then

$$\langle V, W \rangle = \sum_{n=0}^{\infty} \langle v_n, w_n \rangle$$

where $W = \sum_{n=0}^{\infty} e_n \otimes w_n \in \ell^2(\mathbb{C}^N)$. We denote the inner products in $\ell^2(\mathbb{C}^N)$ and $\mathbb{C}^N$ by the same symbol $\langle \cdot, \cdot \rangle$, where the context dictates which inner product to take. This space can also be thought of sequences $(v_0, v_1, \dots)$ with $v_n \in \mathbb{C}^N$ which are square summable $\sum_{n=0}^{\infty} \|v_n\|^2 < \infty$. The case $N = 1$ gives back the Hilbert space $\ell^2(\mathbb{N})$ of square summable sequences.

Given the sequences $\{A_n\}_{n \in \mathbb{N}}$ and $\{B_n\}_{n \in \mathbb{N}}$ in $M_N(\mathbb{C})$ with all matrices A_n regular and all matrices B_n self-adjoint, we define the Jacobi operator J with domain $\mathscr{D}$ by

$$
\begin{aligned}
JV &= e_0 \otimes (A_0 v_1 + B_0 v_0) + \sum_{n=1}^{\infty} e_n \otimes \left(A_n v_{n+1} + B_n v_n + A_{n-1}^* v_{n-1} \right), \\
\mathscr{D} &= \{ V = \sum_{\substack{n=0 \\ \text{finite}}}^{\infty} e_n \otimes v_n \} \subset \ell^2(\mathbb{C}^N),
\end{aligned}
$$

$$(3.16)$$

so that $(J, \mathscr{D})$ is a symmetric operator

$$\langle JV, W \rangle = \langle V, JW \rangle, \qquad \text{for all } V, W \in \mathscr{D}.$$

Note that

$$
J(e_k \otimes v) = \begin{cases} e_{k+1} \otimes A_k^* v + e_k \otimes B_k v + e_{k-1} \otimes A_{k-1} v, & k \geq 1, \\ e_1 \otimes A_1^* v + e_0 \otimes B_0 v, & k = 0, \end{cases}
$$

so that

$$e_{k+1} \otimes v = J(e_k \otimes (A_k^*)^{-1} v) - e_k \otimes B_k (A_k^*)^{-1} v - e_{k-1} \otimes A_{k-1}(A_k^*)^{-1} v$$

for $k \geq 1$ and

$$e_1 \otimes v = J(e_0 \otimes (A_1^*)^{-1} v) - e_0 \otimes B_0 (A_1^*)^{-1} v.$$

Using induction with respect to $k \in \mathbb{N}$ we immediately obtain Lemma 3.37.

Lemma 3.37 *The closure of the linear span of $J^p v$ where $v \in \mathbb{C}^N$ and $p \in \mathbb{N}$ is equal to $\ell^2(\mathbb{C}^N)$.*

It is clear from (3.16) and Theorem 3.32 that we can consider $\sum_{n=0}^{\infty} e_n \otimes P_n(z)v$ formally as eigenvectors for J, and we first take a look at the truncated version.

Lemma 3.38 *Let $V = \sum_{n=0}^{M} e_n \otimes P_n(z)v \in \mathscr{D}$, $M \geq 1$, for some $v \in \mathbb{C}^N$, then*

$$JV = zV - e_M \otimes A_M P_{M+1}(z)v + e_{M+1} \otimes A_M^* P_M(z)v.$$

Let $\mathscr{P}_M \colon \ell^2(\mathbb{C}^N) \to \ell^2(\mathbb{C}^N)$ be the projection onto the span of $e_n \otimes v$, $0 \leq n \leq M$ and $v \in \mathbb{C}^N$, we see that V is an eigenvector of the truncated $\mathscr{P}_M J \mathscr{P}_M$ matrix for the eigenvalue z if and only if $\det(P_{N+1}(z)) = 0$ and $v \in \mathrm{Ker}\,(P_{N+1}(z))$. In particular, the zeroes of $\det(P_{N+1}(z))$ are real.

Proof The expression for JV follows from (3.16). Taking the truncated version kills the last term. Then the eigenvectors of the truncated Jacobi operator can only occur if $A_M P_{M+1}(z)v = 0 \in \mathbb{C}^N$, since A_M invertible. This gives the statement, and since the truncated Jacobi operator is self-adjoint, we find that the zeroes of $\det(P_{N+1}(z))$ are real. □

In the case $\{\|A_n\|\}_{n \in \mathbb{N}}$ and $\{\|B_n\|\}_{n \in \mathbb{N}}$ are bounded sequences, then J is a bounded operator. In that case J extends to a bounded self-adjoint operator on $\ell^2(\mathbb{C}^N)$. If this is not so, then we can determine its adjoint by the same action on its maximal domain, which is the content of Proposition 3.39.

Proposition 3.39 *The adjoint of $(J, \mathscr{D})$ is given by $(J^*, \mathscr{D}^*)$ with*

$$\mathscr{D}^* = \{W = \sum_{n=0}^{\infty} e_n \otimes w_n \in \ell^2(\mathbb{C}^N) \mid$$

$$\|A_0 w_1 + B_0 w_0\|^2 + \sum_{n=1}^{\infty} \|A_{n-1}^* w_{n-1} + B_n w_n + A_n w_{n+1}\|^2 < \infty\},$$

$$J^* W = e_0 \otimes (A_0 w_1 + B_0 w_0) + \sum_{n=1}^{\infty} e_n \otimes (A_{n-1}^* w_{n-1} + B_n w_n + A_n w_{n+1}).$$

Proof Recall the definition of the adjoint operator for an unbounded operator, see Section 3.7.5. Take $W \in \ell^2(\mathbb{C}^N)$ and consider for $V = \sum_{n=0}^{\infty} e_n \otimes$

$v_n \in \mathscr{D}$, so the sum for V is finite,

$$\langle JV, W \rangle = \langle A_0 v_1 + B_0 v_0, w_0 \rangle + \sum_{n=1}^{\infty} \langle A_n v_{n+1} + B_n v_n + A_{n-1}^* v_{n-1}, w_n \rangle$$

$$= \sum_{n=1}^{\infty} \langle v_{n+1}, A_n^* w_n \rangle + \sum_{n=1}^{\infty} \langle v_n, B_n w_n \rangle$$

$$+ \sum_{n=1}^{\infty} \langle v_{n-1}, A_{n-1} w_n \rangle + \langle v_1, A_0^* w_0 \rangle + \langle v_0, B_0 w_0 \rangle$$

$$= \sum_{n=2}^{\infty} \langle v_n, A_{n-1}^* w_{n-1} \rangle + \sum_{n=1}^{\infty} \langle v_n, B_n w_n \rangle$$

$$+ \sum_{n=0}^{\infty} \langle v_n, A_n w_{n+1} \rangle + \langle v_1, A_0^* w_0 \rangle + \langle v_0, B_0 w_0 \rangle$$

$$= \sum_{n=1}^{\infty} \langle v_n, A_{n-1}^* w_{n-1} + B_n w_n + A_n w_{n+1} \rangle + \langle v_0, A_0 w_1 + B_0 w_0 \rangle$$

since B_n is self-adjoint for all $n \in \mathbb{N}$ and all sums are finite since $V \in \mathscr{D}$. First assume that $W \in \mathscr{D}^*$, then by the above calculation we have

$$|\langle JV, W \rangle| \le \|V\| \, \|J^* W\| \le C \|V\|, \quad \text{for all } V \in \mathscr{D}$$

so that $\mathscr{D}^*$ is contained in the domain of the adjoint of $(J, \mathscr{D})$.

Conversely, for W in the domain of the adjoint of $(J, \mathscr{D})$, we have by definition that for all $V \in \mathscr{D}$

$$|\langle JV, W \rangle| \le C \|V\| \tag{3.17}$$

for some constant C. Take

$$V = e_0 \otimes (A_0 w_1 + B_0 w_0) + \sum_{k=1}^{M} e_n \otimes \left(A_{n-1}^* w_{n-1} + B_n w_n + A_n w_{n+1} \right)$$

in (3.17) and using the above calculation we find

$$\left(\|A_0 w_1 + B_0 w_0\|^2 + \sum_{n=1}^{M} \|A_{n-1}^* w_{n-1} + B_n w_n + A_n w_{n+1}\|^2 \right)^{1/2} \le C.$$

Since C is independent of M, by taking $M \to \infty$ we see $W \in \mathscr{D}^*$. The expression for the action of the adjoint of $(J, \mathscr{D})$ follows from the above calculation. Hence, the lemma follows. $\qquad \square$

3.4.3 The resolvent operator

Define the Stieltjes transform of the matrix-valued measure by

$$S(z) = \int_{\mathbb{R}} \frac{1}{x - z} d\mu(x) = \int_{\mathbb{R}} \frac{1}{\lambda - z} W(x) d\tau_\mu(x), \qquad z \in \mathbb{C} \setminus \mathbb{R},$$

and note that $S^*(z) = (S(\bar{z}))^* = S(z)$, since the measure τ_μ is positive and $W(x)$ is positive-definite τ_μ-a.e. So $S \colon \mathbb{C} \setminus \mathbb{R} \to M_N(\mathbb{C})$. Note that S is holomorphic in the upper and lower half plane, meaning that each of its matrix entries is holomorphic. The Stieltjes transform encodes the moments as in the classical case, see [6].

Define for $z \in \mathbb{C} \setminus \mathbb{R}$ and $k \in \mathbb{N}$

$$F_k(z) = Q_k(z) + P_k(z)S(z),$$

then, by Theorem 3.32 and Lemma 3.35,

$$zF_n(z) = A_n F_{n+1}(z) + B_n F_n(z) + A_{n-1}^* F_{n-1}(z), \qquad n \geq 1. \qquad (3.18)$$

Moreover, by Lemma 3.36,

$$
\begin{aligned}
A_{k-1} &\left(F_k(z) P_{k-1}^*(z) - P_k(z) F_{k-1}^*(z) \right) \\
&= A_{k-1} \left(Q_k(z) P_{k-1}^*(z) + P_k(z) S(z) P_{k-1}^*(z) - P_k(z) Q_{k-1}^*(z) \right. \\
&\qquad \left. - P_k(z) S^*(z) P_{k-1}^*(z) \right) \\
&= A_{k-1} \left(Q_k(z) P_{k-1}^*(z) - P_k(z) Q_{k-1}^*(z) \right) = I
\end{aligned}
$$

since $S(z) = S^*(z)$.

Lemma 3.40 *For $v \in \mathbb{C}^N$, $\sum_{n=0}^{\infty} e_n \otimes F_n(z)v \in \ell^2(\mathbb{C}^N)$.*

Proof Start by rewriting

$$
\begin{aligned}
F_k(z) &= Q_k(z) + P_k(z) S(z) \\
&= \int_{\mathbb{R}} \frac{P_k(z) - P_k(x)}{z - x} d\mu(x) + \int_{\mathbb{R}} \frac{1}{z - x} P_k(z) d\mu(x) \\
&= \int_{\mathbb{R}} \frac{-P_k(x)}{z - x} d\mu(x) = \int_{\mathbb{R}} P_k(x) W(x) F^*(x) d\tau_\mu(x)
\end{aligned}
$$

where $F(x) = (x - \bar{z})^{-1} I$. Note that $F \in L_C^2(\mu)$ for $z \in \mathbb{C} \setminus \mathbb{R}$, so that by the

Bessel inequality for Hilbert C^*-modules, see Appendix 3.7.2,

$$\sum_{k=0}^{\infty} \left(F_k(z)\right)^* F_k(z) \leq \langle F, F \rangle = \int_{\mathbb{R}} F(x) W(x) F^*(x) d\tau_\mu(x) \quad \Longrightarrow$$

$$\sum_{n=0}^{\infty} \|F_n(z)v\|^2 \leq \int_{\mathbb{R}} v^* F(x) W(x) F^*(x) v \, d\tau_\mu(x) \leq \frac{v^* M_0 v}{|\Im(z)|^2} < \infty. \qquad \square$$

Since the series in Lemma 3.40 converges, we see that

$$S(z) = -\lim_{k \to \infty} P_k(z)^{-1} Q_k(z) \quad \text{for } z \in \mathbb{C} \setminus \mathbb{R}. \tag{3.19}$$

Note that $P_k(z)$ is invertible by Lemma 3.38 for $z \in \mathbb{C} \setminus \mathbb{R}$. The convergence (3.19) is in operator norm, leading to entrywise convergence. This is a matrix-valued analogue of Markov's theorem (3.12), see also [6, §1.4].

Definition 3.41   Define for $z \in \mathbb{C}$ the vector space

$$S_z^+ = \left\{ V = \sum_{n=0}^{\infty} e_n \otimes v_n \in \ell^2(\mathbb{C}^N) \mid \exists M \in \mathbb{N} \forall n \geq M \right.$$

$$\left. z v_n = A_n v_{n+1} + B_n v_n + A_{n-1}^* v_{n-1} \right\}.$$

Since for linearly independent vectors in $\mathbb{C}^N$, the corresponding elements in Lemma 3.40 are linearly independent, we see that $\dim S_z^+ \geq N$ for $z \in \mathbb{C} \setminus \mathbb{R}$. Note that the condition $V = \sum_{n=0}^{\infty} e_n \otimes v_n \in S_z^+$ only involves the behaviour of v_n for $n \gg 0$, and we can recursively adapt $v_{M-1}, v_{M-2}, \ldots, v_0$ by requiring the recursion relation. Note that in general $z v_0 \neq A_0 v_1 + B_0 v_1$, as can be seen for the element $\sum_{n=0}^{\infty} e_n \otimes F_n(z)v$ of Lemma 3.35 from the explicit values for $P_0(z), P_1(z), Q_0(z), Q_1(z)$ in Section 3.4.1.

So S_z^+ is not the deficiency space for $(J^*, \mathscr{D}^*)$, since we do not require that it satisfies the recurrence for all $n \in \mathbb{N}$. Moreover, any solution for the recurrence relation for all $n \in \mathbb{N}$ is of the form $\sum_{n=0}^{\infty} e_n \otimes P_n(z)v$, so we find for $z \in \mathbb{C} \setminus \mathbb{R}$

$$N_z = \{ V \in \mathscr{D}^* \mid J^* V = zV \} = \left\{ \sum_{n=0}^{\infty} e_n \otimes P_n(z)v \mid v \in \mathbb{C}^N \right\} \cap S_z^+. \tag{3.20}$$

In particular, we see that deficiency indices $0 \leq n_\pm \leq N$. In the case $A_n, B_n \in M_N(\mathbb{R})$ for all $n \in \mathbb{N}$ we see that $n_+ = n_-$, since conjugation induces an isomorphism of N_z onto $N_{\bar{z}}$. Note that also $n_+ = n_-$ if we can find a sequence $\{U_n\}_{n \in \mathbb{N}}$ of unitary operators such that $U_n A_n U_{n+1}^*, U_n B_n U_n^* \in M_N(\mathbb{R})$ for all $n \in \mathbb{N}$, see Remark 3.33. Note that it is always possible to find a unitary U_n so that $U_n B_n U_n^* \in M_N(\mathbb{R})$, since B_n is self-adjoint. For $N = 1$ this can always be done, so that in this case the deficiency indices are always the same; $(n_+, n_-) = (0,0)$ or $(1,1)$.

Assumption 3.42 says $N_z = \{0\}$ for all $z \in \mathbb{C} \setminus \mathbb{R}$. Hence, $(J, \mathscr{D})$ is essentially self-adjoint and thus $(J^*, \mathscr{D}^*)$ is self-adjoint.

Assumption 3.42 For all $v \in \mathbb{C}^N$, the element $\sum_{n=0}^\infty e_n \otimes P_n(z)v \notin S_z^+$ for $z \in \mathbb{C} \setminus \mathbb{R}$.

The assumption means that $\sum_{n=0}^\infty \|P_n(z)v\|^2$ diverges for all $v \in \mathbb{C}^N$.

Theorem 3.43 *Define the operator $G_z \colon \mathscr{D} \to \ell^2(\mathbb{C}^N)$ for $z \in \mathbb{C} \setminus \mathbb{R}$ by*

$$G_z V = \sum_{n=0}^\infty e_n \otimes (G_z V)_n, \qquad (G_z V)_n = \sum_{k=0}^\infty (G_z)_{n,k} v_k,$$

$$M_N(\mathbb{C}) \ni (G_z)_{n,k} = \begin{cases} P_n(z) F_k^*(z), & n \le k, \\ F_n(z) P_k^*(z), & n > k. \end{cases}$$

Then G_z is the resolvent operator for J^*, i.e., $G_z = (J^* - z)^{-1}$, so

$$G_z \colon \ell^2(\mathbb{C}^N) \to \ell^2(\mathbb{C}^N)$$

extends to a bounded operator.

Proof First, we prove $G_z V \in \mathscr{D}^* \subset \ell^2(\mathbb{C}^N)$ for $V = \sum_{n=0}^\infty e_n \otimes v_n \in \mathscr{D}$. In order to do so we need to see that $(G_z V)_n$ is well-defined; the sum over k is actually finite and, by the Cauchy–Schwarz inequality,

$$\underset{\substack{k=0 \\ \text{finite}}}{\overset{\infty}{\sum}} \|(G_z)_{n,k} v_k\| \le \underset{\substack{k=0 \\ \text{finite}}}{\overset{\infty}{\sum}} \|(G_z)_{n,k}\| \, \|v_k\|$$

$$\le \Big(\underset{\substack{k=0 \\ \text{finite}}}{\overset{\infty}{\sum}} \|(G_z)_{n,k}\|^2 \Big)^{1/2} \Big(\underset{\substack{k=0 \\ \text{finite}}}{\overset{\infty}{\sum}} \|v_k\|^2 \Big)^{1/2}$$

$$= \|V\| \Big(\underset{\substack{k=0 \\ \text{finite}}}{\overset{\infty}{\sum}} \|(G_z)_{n,k}\|^2 \Big)^{1/2}.$$

So in order to show that $G_z V \in \ell^2(\mathbb{C}^N)$ we estimate

$$\sum_{n=0}^\infty \Big\| \underset{\substack{k=0 \\ \text{finite}}}{\overset{\infty}{\sum}} (G_z)_{n,k} v_k \Big\|^2 \le \sum_{n=0}^\infty \Big(\underset{\substack{k=0 \\ \text{finite}}}{\overset{\infty}{\sum}} \|(G_z)_{n,k} v_k\| \Big)^2 \le \|V\|^2 \sum_{n=0}^\infty \underset{\substack{k=0 \\ \text{finite}}}{\overset{\infty}{\sum}} \|(G_z)_{n,k}\|^2.$$

Next note that the double sum satisfies, using K for the maximum term occurring in the finite sum,

$$\underset{\text{finite}}{\sum_{k=0}^{\infty} \sum_{n=0}^{\infty}} \|(G_z)_{n,k}\|^2 \le \underset{\text{finite}}{\sum_{k=0}^{\infty} \sum_{n=0}^{K}} \|(G_z)_{n,k}\|^2 + \underset{\text{finite}}{\sum_{k=0}^{\infty}} \|P_k^*(z)\|^2 \sum_{n=K+1}^{\infty} \|F_n(z)\|^2,$$

and it converges by Lemma 3.35. Hence, $G_z V \in \ell^2(\mathbb{C}^{\mathbb{N}})$.

Next we consider

$$(J^* - z)G_z V = e_0 \otimes (A_0(G_z V)_1 + (B_0 - z)(G_z V)_0) +$$

$$\sum_{n=1}^{\infty} e_n \otimes \Big(A_n(G_z V)_{n+1} + (B_n - z)(G_z V)_n + A_{n-1}^*(G_z V)_{n-1} \Big)$$

and we want to show that

$$\begin{aligned} A_0(G_z V)_1 + (B_0 - z)(G_z V)_0 &= v_0, \\ A_n(G_z V)_{n+1} + (B_n - z)(G_z V)_n + A_{n-1}^*(G_z V)_{n-1} &= v_n, \end{aligned} \tag{3.21}$$

for $n \ge 1$. Note that (3.21) in particular implies that $G_z V \in \mathscr{D}^*$.

In order to establish (3.21) we use the definition of the operator G to find for $n \ge 1$

$$A_n(G_z V)_{n+1} + (B_n - z)(G_z V)_n + A_{n-1}^*(G_z V)_{n-1}$$

$$= \sum_{k=0}^{\infty} \Big(A_n(G_z)_{n+1,k} v_k + (B_n - z)(G_z)_{n,k} v_k + A_{n-1}^*(G_z)_{n-1,k} v_k \Big)$$

$$= \sum_{k=0}^{n-1} \Big(A_n F_{n+1}(z) + (B_n - z)F_n(z) + A_{n-1}^* F_{n-1}(z) \Big) P_k^*(z) v_k$$

$$+ A_n(G_z)_{n+1,n} v_n + (B_n - z)(G_z)_{n,n} v_n + A_{n-1}^*(G_z)_{n-1,n} v_n$$

$$\sum_{k=n+1}^{\infty} \Big(A_n P_{n+1}(z) + (B_n - z)P_n(z) + A_{n-1}^* P_{n-1}(z) \Big) F_k^*(z) v_k,$$

where we note that all sums are finite, since we take $V \in \mathscr{D}$. But also for $V \in \ell^2(\mathbb{C}^{\mathbb{N}})$ the series converges, because of Lemma 3.35.

Because of (3.18) and Theorem 3.32, the first and the last term vanish. For the middle term we use the definition for G to find

$$A_n(G_z)_{n+1,n} v_n + (B_n - z)(G_z)_{n,n} v_n + A_{n-1}^*(G_z)_{n-1,n} v_n$$

$$= \Big(A_n F_{n+1}(z) P_n^*(z) + (B_n - z)P_n(z)F_n^*(z) + A_{n-1}^* P_{n-1}(z)F_n^*(z) \Big) v_n$$

$$= \Big(A_n F_{n+1}(z) P_n^*(z) - A_n P_{n+1}(z)F_n^*(z) \Big) v_n$$

$$= A_n \Big(F_{n+1}(z) P_n^*(z) - P_{n+1}(z)F_n^*(z) \Big) v_n = v_n,$$

where we use Theorem 3.32 once more and Lemma 3.36. This proves (3.21) for $n \geq 1$. We leave the case $n = 0$ for Exercise 3.51.

So we find that $G_z \colon \mathcal{D} \to \mathcal{D}^*$ and $(J^* - z)G_z$ is the identity on $\mathcal{D}$. Since $z \in \mathbb{C} \setminus \mathbb{R}$, $z \in \rho(J^*)$ and $(J^* - z)^{-1} \in B(\ell^2(\mathbb{C}^N))$ which coincides with G_z on a dense subspace. So $G_z = (J^* - z)^{-1}$. $\qquad\square$

3.4.4 The spectral measure

We will stick with the Assumptions 3.31, 3.42.

Having Theorem 3.43 we calculate the matrix entries of the resolvent operator G_z for $V = \sum_{n=0}^{\infty} e_n \otimes v_n, W = \sum_{n=0}^{\infty} e_n \otimes w_n \in \mathcal{D}$ and $z \in \mathbb{C} \setminus \mathbb{R}$;

$$
\begin{aligned}
\langle G_z V, W \rangle &= \sum_{n=0}^{\infty} \langle (G_z V)_n, w_n \rangle = \sum_{k,n=0}^{\infty} \langle (G_z)_{n,k} v_k, w_n \rangle \\
&= \sum_{\substack{k,n=0 \\ n \leq k}}^{\infty} \langle P_n(z) F_k^*(z) v_k, w_n \rangle + \sum_{\substack{k,n=0 \\ n > k}}^{\infty} \langle F_n(z) P_k^*(z) v_k, w_n \rangle \\
&= \sum_{\substack{k,n=0 \\ n \leq k}}^{\infty} \langle \big(Q_k^*(z) + S(z) P_k^*(z) \big) v_k, \big(P_n(z) \big)^* w_n \rangle \\
&\quad + \sum_{\substack{k,n=0 \\ n > k}}^{\infty} \langle P_k^*(z) v_k, \big(Q_n(z) + P_n(z) S(z) \big)^* w_n \rangle \\
&= \sum_{\substack{k,n=0 \\ n \leq k}}^{\infty} \langle P_n(z) Q_k^*(z) v_k, w_n \rangle + \sum_{\substack{k,n=0 \\ n > k}}^{\infty} \langle Q_n(z) P_k^*(z) v_k, w_n \rangle \\
&\quad + \sum_{k,n=0}^{\infty} \langle P_n(z) S(z) P_k^*(z) v_k, w_n \rangle,
\end{aligned}
$$

where all sums are finite since $V, W \in \mathcal{D}$. The first two terms are polynomial, hence analytic, in z, and do not contribute to the spectral measure

$$
E_{V,W}((a,b)) = \lim_{\delta \downarrow 0} \lim_{\varepsilon \downarrow 0} \frac{1}{2\pi i} \int_{a+\delta}^{b-\delta} \langle G_{x+i\varepsilon} V, W \rangle - \langle G_{x-i\varepsilon} V, W \rangle \, dx. \quad (3.22)
$$

Lemma 3.44 *Let τ_μ be a positive Borel measure on $\mathbb{R}$, $W_{i,j} \in L^1(\tau_\mu)$ so that $x \mapsto x^k W_{i,j}(x) \in L^1(\tau_\mu)$ for all $k \in \mathbb{N}$. Define for $z \in \mathbb{C} \setminus \mathbb{R}$*

$$
g(z) = \int_{\mathbb{R}} \frac{p(s) W_{i,j}(s)}{s - z} \, d\tau_\mu(s),
$$

where p is a polynomial, then for $-\infty < a < b < \infty$

$$\lim_{\delta\downarrow 0}\lim_{\varepsilon\downarrow 0}\frac{1}{2\pi i}\int_{a+\delta}^{b-\delta} g(x+i\varepsilon) - g(x-i\varepsilon)\,dx = \int_{(a,b)} p(x)W_{i,j}(x)\,d\tau_\mu(x).$$

The proof of Lemma 3.44 is in Exercise 3.52.

From Lemma 3.44 we find

$$E_{V,W}((a,b)) = \sum_{k,n=0}^{\infty}\int_{(a,b)} w_n^* P_n(x)\,W(x)\,P_k^*(x)v_k\,d\tau_\mu(x)$$

$$= \sum_{k,n=0}^{\infty} w_n^* \left(\int_{(a,b)} P_n(x)\,W(x)\,P_k^*(x)\,d\tau_\mu(x)\right) v_k. \qquad (3.23)$$

By extending the integral to $\mathbb{R}$ we find

$$\langle V, W\rangle = \sum_{k,n=0}^{\infty} w_n^* \left(\int_{\mathbb{R}} P_n(x)\,W(x)\,P_k^*(x)\,d\tau_\mu(x)\right) v_k \qquad (3.24)$$

so that in particular we find the orthogonality relations for the polynomials

$$\int_{\mathbb{R}} P_n(x)\,W(x)\,P_k^*(x)\,d\tau_\mu(x) = \delta_{n,m}I. \qquad (3.25)$$

We can rephrase (3.24) as the following theorem.

Theorem 3.45 *Let $(J,\mathscr{D})$ be essentially self-adjoint, then the unitary map*

$$\mathscr{U}: \ell^2(\mathbb{C}^N) \to L_v^2(\mu), \quad V = \sum_{n=0}^{\infty} e_n \otimes v_n \mapsto \sum_{n=0}^{\infty} P_n^*(\cdot)v_n,$$

intertwines its closure $(J^*,\mathscr{D}^*)$ with multiplication, i.e., $\mathscr{U}J^* = M_z\mathscr{U}$, where $M_z\colon \mathscr{D}(M_z) \subset L_v^2(\mu) \to L_v^2(\mu)$, $f \mapsto (z \mapsto zf(z))$, where $\mathscr{D}(M_z)$ is its maximal domain.

Remark 3.46 (i) Note that Theorem 3.45 shows that the closure $(J,\mathscr{D})$ has spectrum equal to the support of τ_μ, and that each point in the spectrum has multiplicity N. According to general theory, see e.g., [96, §VII.1], we can split the (separable) Hilbert space into N invariant subspaces $\mathscr{H}_i$, $1 \le i \le N$, which are J^*-invariant and which can each be diagonalised with multiplicity 1. In this case we can take for $f \in L_v^2(\mu)$ the function $(P_i f)$, where P_i is the projection on the basis vector $e_i \in \mathbb{C}^N$. Note that because $P_i^* W(x)P_i \le W(x)$, we see that $P_i f \in L_v^2(\mu)$ and note that $(P_i f) \in L^2(w_{i,i}d\tau_\mu)$. And the inverse image of the elements $P_i f$ for $f \in L_v^2(\mu)$ under $\mathscr{U}$ gives the invariant subspaces $\mathscr{H}_i$. Note that in practice this might be hard to do, and for this reason it is usually more convenient to have an easier description, but with higher multiplicity.

(ii) We have not discussed reducibility of the weight matrix. If the weight can be block-diagonally decomposed, the same is valid for the corresponding J-matrix (up to suitable normalisation, e.g., in the monic version). For the development as sketched here, this is not required. We give some information on reducibility issues in Section 3.5.4.

We leave the analogue of Favard's theorem 3.22 in this case as Exercise 3.54.

3.4.5 Exercises

Exercise 3.47 Show that under Assumption 3.31 there exist orthonormal matrix-valued polynomials satisfying (3.13). Show that the polynomials are determined up to left multiplication by a unitary matrix, i.e., if $\tilde{P}_n$ forms another set of polynomials satisfying (3.13) then there exist unitary matrices $U_n, n \in \mathbb{N}$, with $\tilde{P}_n(z) = U_n P_n(z)$.

Exercise 3.48 In the context of Example 3.34 define the map $J: \mathbb{C}^{2\ell+1} \to \mathbb{C}^{2\ell+1}$ by $J: e_n \mapsto e_{2\ell-n}$ (recall labeling of the basis e_n with $n \in \{0, \ldots, 2\ell\}$). Check that J is a self-adjoint involution. Show that J commutes with all the matrices $B_n^{(v)}, C_n^{(v)}$ in the recurrence relation for the corresponding monic matrix-valued orthogonal polynomials, and with all squared norm matrices $H_n^{(v)}$.

Exercise 3.49 Prove Theorem 3.32, and show that

$$A_n = \int_{\mathbb{R}} x P_n(x) W(x) P_{n+1}^*(x) \, d\tau_\mu(x)$$

is invertible and

$$B_n = \int_{\mathbb{R}} x P_n(x) W(x) P_n^*(x) \, d\tau_\mu(x)$$

is self-adjoint.

Exercise 3.50 Prove Lemma 3.35 by generalising Exercise 3.26.

Exercise 3.51 Prove the case $n = 0$ of (3.21) in the proof of Theorem 3.43.

Exercise 3.52 In this exercise we prove Lemma 3.44.

(a) Show that for $\varepsilon > 0$

$$g(x+i\varepsilon) - g(x-i\varepsilon) = \int_{\mathbb{R}} \frac{2i\varepsilon}{(s-x)^2 + \varepsilon^2} p(s) W_{i,j}(s) \, d\tau_\mu(s).$$

(b) Show that for $-\infty < a < b < \infty$

$$\frac{1}{2\pi i} \int_a^b g(x+i\varepsilon) - g(x-i\varepsilon)\,dx =$$

$$\int_{\mathbb{R}} \frac{1}{\pi}\left(\arctan\left(\frac{b-s}{\varepsilon}\right) - \arctan\left(\frac{a-s}{\varepsilon}\right)\right) p(s) W_{i,j}(s)\,d\tau_\mu(s).$$

(c) Finish the proof of Lemma 3.44.

Exercise 3.53   Prove the Christoffel–Darboux formula for the matrix-valued orthonormal polynomials;

$$(x-y)\sum_{k=0}^{n-1} P_k^*(x)P_k(y) = P_n^*(x)A_{n-1}^*P_{n-1}(y) - P_{n-1}^*(x)A_{n-1}P_n(y)$$

and derive an expression for $\sum_{k=0}^{n-1} P_k^*(x)P_k(x)$ as in Exercise 3.28.

Exercise 3.54   Assume that we have matrix-valued polynomials that generate the recurrence as in Theorem 3.32. Moreover, assume that $\{\|A_n\|\}_{n\in\mathbb{N}}$, $\{\|B_n\|\}_{n\in\mathbb{N}}$ are bounded. Show that the corresponding Jacobi operator is a bounded self-adjoint operator. Apply the spectral theorem, and show that there exists a matrix-valued weight for which the matrix-valued polynomials are orthogonal.

Exercise 3.55   Show that $\sum_{n=0}^{\infty} \|A_n\|^{-1} = \infty$ implies Assumption 3.42.

3.5 More on matrix weights, matrix-valued orthogonal polynomials and Jacobi operators

In Section 3.4 we have made several assumptions, notably Assumption 3.31 and Assumption 3.42. In this section we discuss how to weaken the Assumption 3.31.

3.5.1 Matrix weights

Assumption 3.31 is related to the space $L_C^2(\mu)$ for a matrix-valued measure μ. We will keep the assumption that τ_μ has infinite support as the case that τ_μ has finite support reduces to the case that $L_C^2(\mu)$ will be finite-dimensional and we are in a situation of finite discrete matrix-valued orthogonal polynomials. The second assumption in Assumption 3.31 is that W is positive-definite τ_μ-a.e.

Definition 3.56   For a positive-definite matrix $W \in P_N(\mathbb{C})$ define the projection $P_W \in M_N(\mathbb{C})$ on the range of W.

Note that $P_W W = W P_W = W$ and $W(I - P_W) = 0 = (I - P_W)W$.

In the context of Theorem 3.30 we have a Borel measure τ_μ, so we need to consider measurability with respect to the Borel sets of $\mathbb{R}$.

Lemma 3.57 *Put $J(x) = P_{W(x)}$, then $J \colon \mathbb{R} \to M_N(\mathbb{C})$ is measurable.*

Proof The matrix-entries $W_{i,j}$ are measurable by Theorem 3.30, so W is measurable. Then $p(W) \colon \mathbb{R} \to M_N(\mathbb{C})$ for any polynomial p is measurable. Since we have observed that $0 \le W(x) \le I$ τ_μ-a.e., we can use a polynomial approximation (in sup-norm) of $\sqrt[n]{\cdot}$ on the interval $[0,1] \supset \sigma(W(x))$ τ_μ-a.e. Hence, $\sqrt[n]{W(x)}$ is measurable, and next observe that $J(x) = \lim_{n \to \infty} \sqrt[n]{W(x)}$ to conclude that J is measurable. $\qquad\square$

Corollary 3.58 *The functions $d(x) = \dim \operatorname{Ran}(J(x))$ and $\sqrt{W(x)}$ are measurable. So the set $D_d = \{x \in \mathbb{R} \mid \dim \operatorname{Ran}(W(x)) = d\}$ is measurable for all d.*

We now consider all measurable $F \colon \mathbb{R} \to M_N(\mathbb{C})$ such that

$$\int_{\mathbb{R}} F(x)W(x)F^*(x)\,d\tau_\mu(x) < \infty,$$

which we denote by $\mathcal{L}_C^2(\mu)$, and we mod out by

$$\mathcal{N}_C = \{F \in \mathcal{L}_C^2(\mu) \mid \langle F, F \rangle = 0\};$$

then the completion of $\mathcal{L}_C^2(\mu)/\mathcal{N}_C$ is the corresponding Hilbert C^*-module $L_C^2(\mu)$.

Lemma 3.59 *$\mathcal{N}_C$ is a left $M_N(\mathbb{C})$-module, and*

$$\mathcal{N}_C = \{F \in \mathcal{L}_C^2(\mu) \mid \operatorname{Ran}(J(x)) \subset \operatorname{Ker}(F(x)) \quad \tau_\mu\text{-a.e.}\}.$$

By taking orthocomplements, the condition can be rephrased as

$$\operatorname{Ran}(F^*(x)) \subset \operatorname{Ker}(J(x)),$$

and since $\operatorname{Ran}(J(x)) = \operatorname{Ran}(W(x))$ and $\operatorname{Ker}(J(x)) = \operatorname{Ker}(W(x))$ it can also be rephrased in terms of the range and kernel of W.

Proof $\mathcal{N}_C$ is a left $M_N(\mathbb{C})$-module by construction of the $M_N(\mathbb{C})$-valued inner product.

Observe, with $J \colon \mathbb{R} \to M_N(\mathbb{C})$ as in Lemma 3.57, that we can split a function $F \in \mathcal{L}_C^2(\mu)$ in the functions FJ and $F(I-J)$, both again in $\mathcal{L}_C^2(\mu)$, so that $F = FJ + F(I-J)$ and

$$\langle F, F \rangle = \langle FJ, FJ \rangle + \langle F(I-J), FJ \rangle + \langle FJ, F(I-J) \rangle + \langle F(I-J), F(I-J) \rangle$$

$$= \langle FJ, FJ \rangle = \int_{\mathbb{R}} (FJ)(x)W(x)(JF)^*(x)\,d\tau_\mu(x)$$

since $(I - J(x))W(x) = 0 = W(x)(I - J(x))$ τ_μ-a.e. It follows that for any $F \in \mathcal{L}^2_C(\mu)$ with $\text{Ran}(J(x)) \subset \text{Ker}(F(x))$ τ_μ-a.e. the function FJ is zero, and then $F \in \mathcal{N}_C$.

Conversely, if $F \in \mathcal{N}_C$ and therefore

$$0 = \text{Tr}(\langle F, F \rangle) = \int_\mathbb{R} \text{Tr}\left(F(x)W(x)F^*(x)\right) d\tau_\mu(x).$$

Since $\text{Tr}(A^*A) = \sum_{k,j=1}^N |a_{k,j}|^2$, all matrix-entries of $x \mapsto F(x)(W(x))^{1/2}$ are zero τ_μ-a.e. Hence $x \mapsto F(x)(W(x))^{1/2}$ is zero τ_μ-a.e. This gives $x \mapsto \langle W(x)F^*(x)v, F^*(x)v \rangle = 0$ for all $v \in \mathbb{C}^N$ and τ_μ-a.e. Hence, $\text{Ran}(F^*(x)) \subset \text{Ker}(J(x))$ τ_μ-a.e., and so $\text{Ran}(J(x)) \subset \text{Ker}(F(x))$ τ_μ-a.e. $\qquad\square$

Similarly, we define the space $\mathcal{L}^2_v(\mu)$ of measurable functions $f \colon \mathbb{R} \to \mathbb{C}^N$ so that

$$\int_\mathbb{R} f^*(x) W(x) f(x) d\tau_\mu(x) < \infty$$

where f is viewed as a column vector and f^* as a row vector. Then we mod out by $\mathcal{N}_v = \{f \in \mathcal{L}^2_v(\mu) \mid \langle f, f \rangle = 0\}$ and we complete in the metric induced from the inner product

$$\langle f, g \rangle = \int_\mathbb{R} g^*(x) W(x) f(x) d\tau_\mu(x) = \int_\mathbb{R} \langle W(x) f(x), g(x) \rangle d\tau_\mu(x).$$

The analogue of Lemma 3.59 for $L^2_v(\mu)$ is examined in [23, XIII.5.8].

Lemma 3.60 $\mathcal{N}_v = \{f \in \mathcal{L}^2_v(\mu) \mid f(x) \in \text{Ker}(J(x)) \quad \tau_\mu\text{–a.e.}\}.$

The proof of Lemma 3.60 is Exercise 3.64.

3.5.2 Matrix-valued orthogonal polynomials

In general, for a not-necessarily positive-definite matrix measure $d\mu = W d\tau_\mu$ with finite moments, we cannot perform a Gram–Schmidt procedure so we have to impose another condition. Note that it is guaranteed by Theorem 3.30 that W is positive semi-definite.

Assumption 3.61   From now on we assume for Section 3.5 that μ is a matrix measure for which τ_μ has infinite support and for which all moments exist, i.e., $(x \mapsto x^k W_{i,j}(x)) \in L^1(\tau_\mu)$ for all $1 \le i, j \le N$ and all $k \in \mathbb{N}$. Moreover, we assume that all even moments $M_{2k} = \int_\mathbb{R} x^{2k} d\mu(x) = \int_\mathbb{R} x^{2k} W(x) d\tau_\mu(x)$ are positive-definite, $M_{2k} \in P^o_N(\mathbb{C}^N)$, for all $k \in \mathbb{N}$.

Note that Lemma 3.59 shows that $x^k \notin \mathcal{N}_C$ (except for the trivial case), so that $M_{2k} \neq 0$. This, however, does not guarantee that M_{2k} is positive definite.

Theorem 3.62   *Under the Assumption 3.61 there exists a sequence of matrix-valued orthonormal polynomials $\{P_n\}_{n\in\mathbb{N}}$ with regular leading coefficients. There exist sequences of matrices $\{A_n\}_{n\in\mathbb{N}}$, $\{B_n\}_{n\in\mathbb{N}}$ so that $\det(A_n) \neq 0$ for all $n \in \mathbb{N}$ and $B_n^* = B_n$ for all $n \in \mathbb{N}$, so that*

$$zP_n(z) = \begin{cases} A_nP_{n+1}(z) + B_nP_n(z) + A_{n-1}^*P_{n-1}(z), & n \geq 1, \\ A_0P_1(z) + B_0P_0(z), & n = 0. \end{cases}$$

Proof   Instead of showing the existence of the orthonormal polynomials we show the existence of the monic matrix-valued orthogonal polynomials R_n so that $\langle R_n, R_n \rangle$ positive-definite for all $n \in \mathbb{N}$. Then $P_n = \langle R_n, R_n \rangle^{-1/2}R_n$ gives a sequence of matrix-valued orthonormal polynomials.

We start with $n = 0$, then $R_0(x) = I$, and $\langle R_0, R_0 \rangle = M_0 > 0$ by Assumption 3.61. We now assume that the monic matrix-valued orthogonal polynomials R_k so that $\langle R_k, R_k \rangle$ is positive-definite have been constructed for all $k < n$. We now prove the statement for $k = n$.

Put, since R_n is monic,

$$R_n(x) = x^nI + \sum_{m=0}^{n-1} C_{n,m}R_m(x), \qquad C_{n,m} \in M_N(\mathbb{C}).$$

The orthogonality requires $\langle R_n, R_m \rangle = 0$ for $m < n$. This gives the solution

$$C_{n,m} = -\langle x^n, R_m \rangle \langle R_m, R_m \rangle^{-1}, \qquad m < n,$$

which is well-defined by the induction hypothesis. It remains to show that $\langle R_n, R_n \rangle > 0$, i.e., $\langle R_n, R_n \rangle$ is positive-definite. Write $R_n(x) = x^nI + Q(x)$, so that

$$\langle R_n, R_n \rangle = \int_{\mathbb{R}} x^nW(x)x^n\,d\tau_\mu(x) + \langle x^n, Q \rangle + \langle Q, x^n \rangle + \langle Q, Q \rangle$$

so that the first term equals the positive-definite moment M_{2n} by Assumption 3.61. It suffices to show that the other three terms are positive semi-definite, so that the sum is positive-definite. This is clear for $\langle Q, Q \rangle$, and a calculation shows

$$\langle x^n, Q \rangle + \langle Q, x^n \rangle = 2 \sum_{m=0}^{n-1} \langle x^n, R_m \rangle \langle R_m, R_m \rangle \langle R_m, x^n \rangle$$

and, since $ABA^* \geq 0$ for $B > 0$, the induction hypothesis shows that these terms are also positive-definite. Hence $\langle R_n, R_n \rangle$ is positive-definite.

Establishing that the corresponding orthonormal polynomials satisfy a three-term recurrence relation is achieved as in Section 3.4, see Exercise 3.65.   □

We can now go through the proofs of Section 3.4 and see that we can obtain in the same way the spectral decomposition of the self-adjoint operator $(J^*, \mathscr{D}^*)$ in Theorem 3.45, where the Assumption 3.31 is replaced by Assumption 3.61 and the Assumption 3.42 is still in force.

Corollary 3.63   *The spectral decomposition of the self-adjoint extension* $(J^*, \mathscr{D}^*)$ *of* $(J, \mathscr{D})$ *of Theorem 3.45 remains valid. The multiplicity of the spectrum is given by the function* $d \colon \sigma(J^*) \to \mathbb{N}$ τ_μ-*a.e. where d is defined in Corollary 3.58.*

Corollary 3.63 means that the operator $(J^*, \mathscr{D}^*)$ is abstractly realised as a multiplication operator on a direct integral of Hilbert spaces $\int H_{d(x)} \, dv(x)$, where H_d is the Hilbert space of dimension d and v is a measure on the spectrum of $(J^*, \mathscr{D}^*)$, see e.g., [92, Ch. VII] for more information.

3.5.3 Link to case of $\ell^2(\mathbb{Z})$

In [10, §VII.3] Berezanskiĭ discussed how three-term recurrence operators on $\ell^2(\mathbb{Z})$ can be related to 2×2-matrix recurrence on $\mathbb{N}$, so that we are in the case $N = 2$ of Section 3.4. Let us briefly discuss a way of doing this, following [10, §VII.3]; see also Exercise 3.15.

We identify $\ell^2(\mathbb{Z})$ with $\ell^2(\mathbb{C}^2) = \ell^2(\mathbb{N}) \hat{\otimes} \mathbb{C}^2$ by

$$e_n \mapsto e_n \otimes \begin{pmatrix} 1 \\ 0 \end{pmatrix}, \qquad e_{-n-1} \mapsto e_n \otimes \begin{pmatrix} 0 \\ 1 \end{pmatrix}, \quad n \in \mathbb{N}, \qquad (3.26)$$

where $\{e_n\}_{n \in \mathbb{Z}}$ is the standard orthonormal basis of $\ell^2(\mathbb{Z})$ and $\{e_n\}_{n \in \mathbb{N}}$ the standard orthonormal basis of $\ell^2(\mathbb{N})$, as before. The identification (3.26) is highly non-canonical. By calculating $L(ae_n + be_{-n-1})$ using Section 3.2 we get the corresponding operator J acting on $\mathscr{D} \subset \ell^2(\mathbb{C}^2)$

$$\sum_{n=0}^{\infty} e_n \otimes v_n \mapsto e_0 \otimes (A_0 v_1 + B_0 v_0) + \sum_{n=1}^{\infty} e_n \otimes (A_n v_{n+1} + B_n v_n + A_{n-1}^* v_{n-1}),$$

$$A_n = \begin{pmatrix} a_n & 0 \\ 0 & a_{-n-2} \end{pmatrix}, \, n \in \mathbb{N}, \quad B_n = \begin{pmatrix} b_n & 0 \\ 0 & b_{-n-1} \end{pmatrix}, \, n \geq 1,$$

$$B_0 = \begin{pmatrix} b_0 & a_{-1} \\ a_{-1} & b_{-1} \end{pmatrix}.$$

In the notation of Section 3.2, let $S_z^{\pm}$ be spanned by $\phi_z^{\pm} = \sum_{n \in \mathbb{Z}} (\phi_z)_n f_n \in$

S_z^+ and $\Phi_z = \sum_{n \in \mathbb{Z}} (\Phi_z)_n f_n \in S_z^-$. Then under the correspondence of this section, the 2×2-matrix-valued function

$$F_n(z) = \begin{pmatrix} (\phi_z)_n & 0 \\ 0 & (\Phi_z)_{-n-1} \end{pmatrix} \in S_z^+,$$

$$zF_n(z) = A_n F_{n+1}(z) + B_n F_n(z) + A_{n-1}^* F_{n-1}(z), \qquad n \geq 1.$$

The example discussed in Examples 3.3, 3.6 and 3.7 shows that the multiplicity of each element in the spectrum is 1, so we see that the corresponding 2×2-matrix weight measure is purely discrete and that $d(\{q^n\}) = 1$ for each $n \in \mathbb{N}$.

3.5.4 Reducibility

Naturally, if we have positive Borel measures $\mu_p, 1 \leq p \leq N$, we can obtain a matrix-valued measure μ by putting

$$\mu(B) = T \begin{pmatrix} \mu_1(B) & 0 & \cdots & 0 \\ 0 & \mu_2(B) & \cdots & 0 \\ \vdots & & \ddots & \vdots \\ 0 & \cdots & 0 & \mu_N(B) \end{pmatrix} T^* \qquad (3.27)$$

for an invertible $T \in M_N(\mathbb{C})$. Denoting the scalar-valued orthonormal polynomials for the measure μ_i by $p_{i;n}$, then

$$P_n(x) = \begin{pmatrix} p_{1;n}(x) & 0 & \cdots & 0 \\ 0 & p_{2;n}(x) & \cdots & 0 \\ \vdots & & \ddots & \vdots \\ 0 & \cdots & 0 & p_{N;n}(x) \end{pmatrix} T^{-1}$$

are the corresponding matrix-valued orthogonal polynomials. Similarly, we can build up a matrix-valued measure of size $(N_1 + N_2) \times (N_1 + N_2)$ starting from a $N_1 \times N_1$-matrix measure and a $N_2 \times N_2$-matrix measure. In such cases the Jacobi operator J can be reduced as well.

We consider the real vector space

$$\mathcal{A} = \mathcal{A}(\mu) = \{T \in M_N(\mathbb{C}) \mid T\mu(B) = \mu(B)T^* \text{ for all } B \in \mathcal{B}\}, \qquad (3.28)$$

and the commutant algebra

$$A = A(\mu) = \{T \in M_N(\mathbb{C}) \mid T\mu(B) = \mu(B)T \text{ for all } B \in \mathcal{B}\}, \qquad (3.29)$$

which is a $*$-algebra, for any matrix-valued measure μ.

Then, by Tirao and Zurrián [95, Thm. 2.12], the weight splits into a sum of weights of smaller dimension if and only of $\mathbb{R}I \subsetneq \mathcal{A}$. On the other hand, the commutant algebra A is easier to study, and in [62, Thm. 2.3], it is proved that $\mathcal{A} \cap \mathcal{A}^* = A_h$, the Hermitean elements in the commutant algebra A, so that we immediately get that $\mathcal{A} = A_h$ if $\mathcal{A}$ is $*$-invariant. The $*$-invariance of $\mathcal{A}$ can then be studied using its relation to moments, quadratic norms, the monic polynomials, and the corresponding coefficients in the three-term recurrence relation, see [62, Lemma 3.1]. See also Exercise 3.66.

In particular, for the case of the matrix-valued Gegenbauer polynomials of Example 3.34, we have that $A = \mathbb{C}I \oplus \mathbb{C}J$, where $J: \mathbb{C}^{2\ell+1} \to \mathbb{C}^{2\ell+1}$, $e_n \mapsto e_{2\ell-n}$ is a self-adjoint involution, see [61, Prop. 2.6], and that $\mathcal{A}$ is $*$-invariant, see [62, Example 4.2]. See also Exercise 3.48. So in fact, we can decompose the weight in Example 3.34 into a direct sum of two weights obtained by projecting on the ± 1-eigenspaces of J, and then there is no further reduction possible.

3.5.5 Exercises

Exercise 3.64   Prove Lemma 3.60 following Lemma 3.59.

Exercise 3.65   Prove the statement on the three-term recurrence relation of Theorem 3.62.

Exercise 3.66   Consider the following 2×2-weight function on $[0, 1]$ with respect to the Lebesgue measure;

$$W(x) = \begin{pmatrix} x^2 + x & x \\ x & x \end{pmatrix}.$$

Show that $W(x)$ is positive-definite a.e. on $[0, 1]$. Show that the commutant algebra A is trivial, and that the vector space $\mathcal{A}$ is non-trivial.

3.6 The J-matrix method

The J-matrix method consists of realising an operator to be studied, e.g., a Schrödinger operator, as a recursion operator in a suitable basis. If this recursion is a three-term recursion then we can try to bring orthogonal polynomials in play. When the recursion is more generally a $2N + 1$-term recursion we can use a result of Durán and Van Assche [27], see also [12, §4], to write it as a three-term recursion for $N \times N$-matrix-valued polynomials. The J-matrix method is used for a number of physics models, see e.g., references in [48].

We start with the case of a linear operator L acting on a suitable function space; typically L is a differential operator, or a difference operator. We look for linearly independent functions $\{y_n\}_{n=0}^{\infty}$ such that L is tridiagonal with respect to these functions, i.e., there exist constants A_n, B_n, C_n $(n \in \mathbb{N})$ such that

$$Ly_n = \begin{cases} A_n y_{n+1} + B_n y_n + C_n y_{n-1}, & n \geq 1, \\ A_0 y_1 + B_0 y_0, & n = 0. \end{cases} \tag{3.30}$$

Note that we do not assume that the functions $\{y_n\}_{n\in\mathbb{N}}$ form an orthogonal or orthonormal basis. We combine both equations by assuming $C_0 = 0$. Note also that in case some $A_n = 0$ or $C_n = 0$, we can have invariant subspaces and we need to consider the spectral decomposition on such an invariant subspaces, and on its complement if this is also invariant and otherwise on the corresponding quotient space. An example of this will be encountered in Section 3.6.1.

It follows that $\sum_{n=0}^{\infty} p_n(z) y_n$ is a formal eigenfunction of L for the eigenvalue z if p_n satisfies

$$z p_n(z) = C_{n+1} p_{n+1}(z) + B_n p_n(z) + A_{n-1} p_{n-1}(z) \tag{3.31}$$

for $n \in \mathbb{N}$ with the convention $A_{-1} = 0$. In the case $C_n \neq 0$ for $n \geq 1$, we can define $p_0(z) = 1$ and use (3.31) recursively to find $p_n(z)$ as a polynomials of degree n in z. In the case $A_n C_{n+1} > 0$, $B_n \in \mathbb{R}$, for $n \geq 0$, the polynomials p_n are orthogonal with respect to a positive measure on $\mathbb{R}$ by Favard's theorem, see Corollary 3.22. The measure from Favard's theorem then gives information on the action of L on the subspace obtained as the closure of the span of $\{y_n\}_{n=0}^{\infty}$. The special case that the subspace is the whole space on which L acts, is the most interesting case, especially when $\{y_n\}_{n=0}^{\infty}$ constitutes an orthogonal basis. Of particular interest is whether we can match the corresponding Jacobi operator to a well-known class of orthogonal polynomials, e.g., from the $(q\text{-})$Askey scheme.

We illustrate this method by a couple of examples. In the first example in Section 3.6.1, an explicit Schrödinger operator is considered. The Schrödinger operator with the Morse potential is used in modelling potential energy in diatomic molecules, and it is physically relevant since it allows for bound states, which is reflected in the occurrence of an invariant finite-dimensional subspace of the corresponding Hilbert space in Section 3.6.1.

In the second example we use an explicit differential operator for orthogonal polynomials to construct another differential operator suitable for the J-matrix method. We work out the details in a specific case.

In the third example we extend the method to obtain an operator for which we have a 5-term recurrence relation, to which we associate 2×2-matrix valued orthogonal polynomials.

3.6.1 Schrödinger equation with the Morse potential

The Schrödinger equation with the Morse potential is studied by Broad [13] and Diestler [21] in the study of a larger system of coupled equations used in modeling atomic dissocation. The Schrödinger equation with the Morse potential is used to model a two-atom molecule in this larger system. We use the approach as discussed in [48, §3].

The Schrödinger equation with the Morse potential is

$$-\frac{d^2}{dx^2} + q, \qquad q(x) = b^2(e^{-2x} - 2e^{-x}), \tag{3.32}$$

which is an unbounded operator on $L^2(\mathbb{R})$. Here $b > 0$ is a constant. It is a self-adjoint operator with respect to its form domain, see [86, Ch. 5] and $\lim_{x\to\infty} q(x) = 0$, and $\lim_{x\to-\infty} q(x) = +\infty$. Note $\min(q) = -b^2$ so that, by general results in scattering theory, the discrete spectrum is contained in $[-b^2, 0]$ and it consists of isolated points. We show how these isolated occur in this approach.

We look for solutions to $-f''(x) + q(x)f(x) = \gamma^2 f(x)$. Put $z = 2be^{-x}$ so that $x \in \mathbb{R}$ corresponds to $z \in (0,\infty)$, and let $f(x)$ correspond to $\frac{1}{\sqrt{z}}g(z)$. Then

$$g''(z) + \frac{(-\frac{1}{4}z^2 + bz + \gamma^2 + \frac{1}{4})}{z^2}g(z) = 0; \tag{3.33}$$

this is precisely the Whittaker equation with $\kappa = b$, $\mu = \pm i\gamma$. The Whittaker integral transform gives the spectral decomposition for this Schrödinger equation, see e.g., [28, §IV]. In particular, depending on the value of b, the Schrödinger equation has finite discrete spectrum, i.e., bound states – see the Plancherel formula [28, §IV] – and in this case the Whittaker function terminates and can be written as a Laguerre polynomial of type $L_m^{(2b-2m-1)}(x)$, for those $m \in \mathbb{N}$ such that $2b - 2m > 0$. So the spectral decomposition can be done directly using the Whittaker transform.

We now show how the spectral decomposition of three-term recurrence (Jacobi) operators can also be used to find the spectral decomposition of the Schrödinger operator (3.32). The Schrödinger operator is tridiagonal in a basis introduced by Broad [13] and Diestler [21]. Put $N = \#\{n \in \mathbb{N} \mid n < b - \frac{1}{2}\}$, i.e., $N = \lfloor b + \frac{1}{2} \rfloor$, so that $2b - 2N > -1$, and assume for simplicity

$b \notin \frac{1}{2} + \mathbb{N}$. Let $T : L^2(\mathbb{R}) \to L^2((0, \infty); z^{2b-2N} e^{-z} dz)$ be the map $(Tf)(z) = z^{N-b-\frac{1}{2}} e^{\frac{1}{2}z} f(\ln(2b/z))$; then T is unitary, and

$$T\left(-\frac{d^2}{dx^2} + q\right) T^* = L, \qquad L = M_A \frac{d^2}{dz^2} + M_B \frac{d}{dz} + M_C,$$

where M_f denotes the operator of multiplication by f. Here $A(z) = -z^2$, $B(z) = (2N - 2b - 2 + z)z$, $C(z) = -(N - b - \frac{1}{2})^2 + z(1 - N)$. Using the second-order differential equation, see e.g., [47, (4.6.15)], [55, (1.11.5)], [93, (5.1.2)], for the Laguerre polynomials, the three-term recurrence relation for the Laguerre polynomials, see e.g., [47, (4.6.26)], [55, (1.11.3)], [93, (5.1.10)], and the differential-recursion formula

$$x \frac{d}{dx} L_n^{(\alpha)}(x) = n L_n^{(\alpha)}(x) - (n + \alpha) L_{n-1}^{(\alpha)}(x)$$

see [4, Case II], for the Laguerre polynomials we find that this operator is tridiagonalized by the Laguerre polynomials $L_n^{(2b-2N)}$.

Translating this back to the Schrödinger operator we started with, we obtain

$$y_n(x) =$$

$$(2b)^{(b-N+\frac{1}{2})} \sqrt{\frac{n!}{\Gamma(2b - 2N + n + 1)}} e^{-(b-N+\frac{1}{2})x} e^{-be^{-x}} L_n^{(2b-2N)}(2be^{-x})$$

as an orthonormal basis for $L^2(\mathbb{R})$ such that

$$\left(-\frac{d^2}{dx^2} + q\right) y_n =$$
$$- (1 - N + n)\sqrt{(n+1)(2b - 2N + n + 1)}\, y_{n+1}$$
$$+ \left(-(N - b - \frac{1}{2})^2 + (1 - N + n)(2n + 2b - 2N + 1) - n\right) y_n$$
$$- (n - N)\sqrt{n(2b - 2N + n)}\, y_{n-1}. \qquad (3.34)$$

Note that (3.34) is written in a symmetric tridiagonal form.

The space $\mathcal{H}^+$ spanned by $\{y_n\}_{n=N}^\infty$ and the space $\mathcal{H}^-$ spanned by $\{y_n\}_{n=0}^{N-1}$ are invariant with respect to $-\frac{d^2}{dx^2} + q$ which follows from (3.34). Note that $L^2(\mathbb{R}) = \mathcal{H}^+ \oplus \mathcal{H}^-$, $\dim(\mathcal{H}^-) = N$. In particular, there will be discrete eigenvalues, therefore bound states, for the restriction to $\mathcal{H}^-$.

In order to determine the spectral properties of the Schrödinger operator, we first consider its restriction on the finite-dimensional invariant subspace $\mathcal{H}^-$. We look for eigenfunctions $\sum_{n=0}^{N-1} P_n(z) y_n$ and for eigenvalue z, so we need to solve

$$z P_n(z) =$$
$$(N - 1 - n)\sqrt{(n+1)(2b - 2N + n + 1)}\, P_{n+1}(z)$$
$$+ \left(-(N - b - \frac{1}{2})^2 + (1 - N + n)(2n + 2b - 2N + 1) - n\right) P_n(z)$$
$$+ (N - n)\sqrt{n(2b - 2N + n)}\, P_{n-1}(z), \qquad 0 \le n \le N - 1.$$

which corresponds to some orthogonal polynomials on a finite discrete set. These polynomials are expressible in terms of the dual Hahn polynomials, see [47, §6.2], [55, §1.6], and we find that z is of the form $-(b - m - \frac{1}{2})^2$, m a non-negative integer less than $b - \frac{1}{2}$, and

$$P_n(-(b - m - \frac{1}{2})^2) =$$
$$\sqrt{\frac{(2b - 2N + 1)_n}{n!}}\, R_n(\lambda(N - 1 - m); 2b - 2N, 0, N - 1),$$

using the notation of [47, §6.2], [55, §1.6]. Since we have now two expressions for the eigenfunctions of the Schrödinger operator for a specific simple eigenvalue, we obtain, after simplifications,

$$\sum_{n=0}^{N-1} R_n(\lambda(N - 1 - m); 2b - 2N, 0, N - 1) L_n^{(2b-2N)}(z)$$
$$= C z^{N-1-m} L_m^{(2b-2m-1)}(z), \tag{3.35}$$
$$C = (-1)^{N+m+1}\left((N + m - 2b)_{N-1-m} \binom{N-1}{m}\right)^{-1},$$

where the constant C can be determined by e.g., considering leading coefficients on both sides.

We seek formal eigenvectors $\sum_{n=0}^{\infty} P_n(z)\, y_{N+n}(x)$ for the eigenvalue z on the invariant subspace $\mathscr{H}^+$. This leads to the recurrence relation

$$z P_n(z) = -(1 + n)\sqrt{(N + n + 1)(2b - N + n + 1)}\, P_{n+1}(z)$$
$$+ \left(-(N - b - \frac{1}{2})^2 + (1 + n)(2n + 2b + 1) - n - N\right) P_n(z)$$
$$- n\sqrt{(N + n)(2b - N + n)}\, P_{n-1}(z).$$

This corresponds with the three-term recurrence relation for the continuous dual Hahn polynomials, see [55, §1.3], with (a, b, c) replaced by $(b + \frac{1}{2}, N - b + \frac{1}{2}, b - N + \frac{1}{2})$, and note that the coefficients a, b and c are positive. We find, with $z = \gamma^2 \geq 0$

$$P_n(z) = \frac{S_n(\gamma^2; b + \frac{1}{2}, N - b + \frac{1}{2}, b - N + \frac{1}{2})}{n! \sqrt{(N+1)_n (2b - N + 1)_n}}$$

and these polynomials satisfy

$$\int_0^\infty P_n(\gamma^2) P_m(\gamma^2) w(\gamma) \, d\gamma = \delta_{n,m},$$

$$w(\gamma) = \frac{1}{2\pi N! \Gamma(2b - N + 1)}$$

$$\times \left| \frac{\Gamma(b + \frac{1}{2} + i\gamma) \Gamma(N - b + \frac{1}{2} + i\gamma) \Gamma(b - N + \frac{1}{2} + i\gamma)}{\Gamma(2i\gamma)} \right|^2.$$

Note that the series $\sum_{n=0}^\infty P_n(\gamma^2) y_{N+n}$ diverges in $\mathcal{H}^+$ (as a closed subspace of $L^2(\mathbb{R})$). Using the results on spectral decomposition of Jacobi operators as in Section 3.3, we obtain the spectral decomposition of the Schrödinger operator restricted to $\mathcal{H}^+$ as

$$\Upsilon : \mathcal{H}^+ \to L^2((0, \infty); w(\gamma) \, d\gamma), \qquad (\Upsilon y_{N+n})(\gamma) = P_n(\gamma^2),$$

$$\left\langle \left(-\frac{d^2}{dx^2} + q \right) f, g \right\rangle = \int_0^\infty \gamma^2 (\Upsilon f)(\gamma) \overline{(\Upsilon g)(\gamma)} \, w(\gamma) \, d\gamma$$

for $f, g \in \mathcal{H}^+ \subset L^2(\mathbb{R})$ such that f is in the domain of the Schrödinger operator.

In this way we obtain the spectral decomposition of the Schrödinger operator on the invariant subspaces $\mathcal{H}^-$ and $\mathcal{H}^+$, where the space $\mathcal{H}^-$ is spanned by the bound states, i.e., by the eigenfunctions for the negative eigenvalues, and $\mathcal{H}^+$ is the reducing subspace on which the Schrödinger operator has spectrum $[0, \infty)$. The link between the two approaches for the discrete spectrum is given by (3.35). For the continuous spectrum it leads to the fact that the Whittaker integral transform maps Laguerre polynomials to continuous dual Hahn polynomials, and we can interpret (3.35) also in this way. For explicit formulas we refer to [70, (5.14)].

Koornwinder [70] generalizes this to the case of the Jacobi function transform mapping Jacobi polynomials to Wilson polynomials, which in turn has been generalized by Groenevelt [32] to the Wilson function transform, an integral transformation with a $_7F_6$ as kernel, mapping Wilson polynomials to Wilson polynomials, which is at the highest level of the Askey scheme, see Figure 3.1. Note that conversely, we can define a unitary map $U : L^2(\mu) \to L^2(\nu)$ between two weighted L^2-spaces by mapping an orthonormal basis $\{\phi_n\}_{n \in \mathbb{N}}$ of $L^2(\mu)$ to an orthonormal basis $\{\Phi_n\}_{n \in \mathbb{N}}$

of $L^2(v)$. Then we can define formally a map $U_t : L^2(\mu) \to L^2(v)$ by

$$(U_t f)(\lambda) = \int_{\mathbb{R}} f(x) \sum_{k=0}^{\infty} t^k \phi_k(x) \Phi_k(\lambda) \, d\mu(x)$$

and consider convergence as $t \to 1$. Note that the convergence of the (non-symmetric) Poisson kernel $\sum_{k=0}^{\infty} t^k \phi_k(x) \Phi_k(\lambda)$ needs to be studied carefully. In the case of the Hermite functions as eigenfunctions of the Fourier transform, this approach is due to Wiener [97, Ch. 1], in which the Poisson kernel is explicitly known as the Mehler formula. More information on explicit expressions of non-symmetric Poisson kernels for orthogonal polynomials from the q-Askey scheme can be found in [8].

3.6.2 A tridiagonal differential operator

In this section we create tridiagonal operators from explicit well-known operators, and we show in an explicit example how this works. This example is based on [49], and we refer to [48, 50] for more examples and general constructions. Genest et al. [30] have generalised this approach and have obtained the full family of Wilson polynomials in terms of an algebraic interpretation.

Assume now μ and v are orthogonality measures of infinite support for orthogonal polynomials;

$$\int_{\mathbb{R}} P_n(x) P_m(x) \, d\mu(x) = H_n \delta_{n,m}, \qquad \int_{\mathbb{R}} p_n(x) p_m(x) \, dv(x) = h_n \delta_{n,m}.$$

We assume that both μ and v correspond to a determinate moment problem, so that the space $\mathscr{P}$ of polynomials is dense in $L^2(\mu)$ and $L^2(v)$. We also assume that $\int_{\mathbb{R}} f(x) \, d\mu(x) = \int_{\mathbb{R}} f(x) r(x) \, dv(x)$, where r is a polynomial of degree 1, so that the Radon–Nikodym derivative $\frac{dv}{d\mu} = \delta = 1/r$. Then we obtain, using lc (p) for the leading coefficient of a polynomial p,

$$p_n = \frac{\text{lc}(p_n)}{\text{lc}(P_n)} P_n + \text{lc}(r) \frac{h_n}{H_{n-1}} \frac{\text{lc}(P_{n-1})}{\text{lc}(p_n)} P_{n-1} \tag{3.36}$$

by expanding p_n in the basis $\{P_n\}_{n\in\mathbb{N}}$. Indeed, $p_n(x) = \sum_{k=0}^{n} c_k^n P_k(x)$ with

$$c_k^n H_k = \int_{\mathbb{R}} p_n(x) P_k(x) \, d\mu(x) = \int_{\mathbb{R}} p_n(x) P_k(x) r(x) \, dv(x),$$

so that $c_k^n = 0$ for $k < n-1$ by orthogonality of the polynomials $p_n \in L^2(v)$. Then c_n^n follows by comparing leading coefficients, and

$$c_{n-1}^n = \int_{\mathbb{R}} p_n(x) P_k(x) r(x) \, dv(x) = \frac{\text{lc}(P_{n-1} \text{lc}(r)}{\text{lc}(p_n)} h_n.$$

By taking ϕ_n, respectively Φ_n, the corresponding orthonormal polynomials to p_n, respectively P_n, we see that

$$\phi_n = A_n \Phi_n + B_n \Phi_{n-1}, \qquad A_n = \frac{\mathrm{lc}\,(p_n)}{\mathrm{lc}\,(P_n)} \sqrt{\frac{H_n}{h_n}},$$

$$B_n = \mathrm{lc}\,(r) \sqrt{\frac{h_n}{H_{n-1}} \frac{\mathrm{lc}\,(P_{n-1})}{\mathrm{lc}\,(p_n)}}.$$

$$(3.37)$$

We assume the existence of a self-adjoint operator L with domain $\mathscr{D} = \mathscr{P}$ on $L^2(\mu)$ with $LP_n = \Lambda_n P_n$, and so $L\Phi_n = \Lambda_n \Phi_n$, for eigenvalues $\Lambda_n \in \mathbb{R}$. By convention $\Lambda_{-1} = 0$. So this means that we assume that $\{P_n\}_{n \in \mathbb{N}}$ satisfies a bispectrality property, and we can typically take the family $\{P_n\}_n$ from the Askey scheme or its q-analogue, see Figure 3.1 and 3.2.

Lemma 3.67   *The operator $T = r(L + \gamma)$ with domain $\mathscr{D} = \mathscr{P}$ on $L^2(\nu)$ is tridiagonal with respect to the basis $\{\phi_n\}_{n \in \mathbb{N}}$. Here γ is a constant, and r denotes multiplication by the polynomial r of degree 1.*

Proof   Note that $(L + \gamma)\Phi_n = \Lambda_n^\gamma \Phi_n = (\Lambda_n + \gamma)\Phi_n$ and

$$\langle T\phi_n, \phi_m \rangle_{L^2(\nu)}$$
$$= \langle A_n T\Phi_n + B_n T\Phi_{n-1}, A_m \Phi_m + B_m \Phi_{m-1} \rangle_{L^2(\nu)}$$
$$= \langle A_n (L + \gamma)\Phi_n + B_n (L + \gamma)\Phi_{n-1}, A_m \Phi_m + B_m \Phi_{m-1} \rangle_{L^2(\mu)}$$
$$= \Lambda_n^\gamma A_n B_{n+1} \delta_{n+1,m} + (A_n^2 \Lambda_n^\gamma + B_n^2 \Lambda_{n-1}^\gamma)\delta_{n,m} + \Lambda_{n-1}^\gamma A_{n-1} B_n \delta_{n,m+1},$$

so that

$$T\phi_n = a_n \phi_n + b_n \phi_n + a_{n-1}\phi_{n-1},$$

$$a_n = \Lambda_n^\gamma \mathrm{lc}\,(r) \frac{\mathrm{lc}\,(p_n)}{\mathrm{lc}\,(p_{n+1})} \sqrt{\frac{h_{n+1}}{h_n}},$$

$$b_n = \Lambda_n^\gamma \frac{H_n}{h_n} \left(\frac{\mathrm{lc}\,(p_n)}{\mathrm{lc}\,(P_n)} \right)^2 + \Lambda_{n-1}^\gamma \mathrm{lc}\,(r)^2 \frac{h_n}{H_{n-1}} \left(\frac{\mathrm{lc}\,(P_{n-1})}{\mathrm{lc}\,(p_n)} \right)^2. \qquad \square$$

So we need to solve for the orthonormal polynomials $r_n(\lambda)$ satisfying

$$\lambda r_n(\lambda) = a_n r_n(\lambda) + b_n r_n(\lambda) + a_{n-1} r_{n-1}(\lambda),$$

where we assume that we can use the parameter γ in order ensure that $a_n \neq 0$. If $a_n = 0$, then we need to proceed as in Section 3.6.1 and split the space into invariant subspaces.

This is a general set-up to find tridiagonal operators. In general, the three-term recurrence relation of Lemma 3.67 needs not be matched with a known family of orthogonal polynomials, such as e.g., from the Askey

scheme. Let us work out a case where it does, namely for the Jacobi polynomials and the related hypergeometric differential operator. See [49] for
other cases.

For the Jacobi polynomials $P_n^{(\alpha,\beta)}(x)$, we follow the standard notation
[5, 47, 55]. We take the measures μ and ν to be the orthogonality measures for the Jacobi polynomials for parameters $(\alpha+1,\beta)$, and (α,β) respectively. We assume $\alpha,\beta > -1$. So we set $P_n(x) = P_n^{(\alpha+1,\beta)}(x)$, $p_n(x) =$
$P_n^{(\alpha,\beta)}(x)$. This gives

$$h_n = N_n(\alpha) = \frac{2^{\alpha+\beta+1}}{2n+\alpha+\beta+1} \frac{\Gamma(n+\alpha+1)\Gamma(n+\beta+1)}{\Gamma(n+\alpha+\beta+1)n!},$$

$$H_n = N_n(\alpha+1), \; \mathrm{lc}\,(p_n) = l_n(\alpha) = \frac{(n+\alpha+\beta+1)_n}{2^n n!}, \; \mathrm{lc}\,(P_n) = l_n(\alpha+1).$$

Moreover, $r(x) = 1 - x$. Note that we could also have shifted in β, but due
to the symmetry $P_n^{(\alpha,\beta)}(x) = (-1)^n P_n^{(\beta,\alpha)}(-x)$ of the Jacobi polynomials
in α and β it suffices to consider only the shift in α.

The Jacobi polynomials are eigenfunctions of a hypergeometric differential operator,

$$L^{(\alpha,\beta)} f(x) = (1-x^2) f''(x) + \big(\beta - \alpha - (\alpha+\beta+2)x\big) f'(x),$$
$$L^{(\alpha,\beta)} P_n^{(\alpha,\beta)} = -n(n+\alpha+\beta+1) P_n^{(\alpha,\beta)},$$

and we take $L = L^{(\alpha+1,\beta)}$ so that $\Lambda_n = -n(n+\alpha+\beta+2)$. We set $\gamma =$
$-(\alpha+\delta+1)(\beta-\delta+1)$, so that we have the factorisation $\Lambda_n^\gamma = -(n+\alpha+$
$\delta+1)(n+\beta-\delta+1)$. So on $L^2([-1,1],(1-x)^\alpha(1+x)^\beta\,dx)$ we study the
operator $T = (1-x)(L+\gamma)$. Explicitly T is the second-order differential
operator

$$T = (1-x)(1-x^2)\frac{d^2}{dx^2} + (1-x)\big(\beta - \alpha - 1 - (\alpha+\beta+3)x\big)\frac{d}{dx}$$
$$- (1-x)(\alpha+\delta+1)(\beta-\delta+1), \qquad (3.38)$$

which is tridiagonal by construction. Going through the explicit details of
Lemma 3.67 we find the explicit expression for the recursion coefficients

in the three-term realisation of T;

$$a_n = \frac{2(n+\alpha+\delta+1)(n+\beta-\delta+1)}{2n+\alpha+\beta+2}$$

$$\times \sqrt{\frac{(n+1)(n+\alpha+1)(n+\beta+1)(n+\alpha+\beta+1)}{(2n+\alpha+\beta+1)(2n+\alpha+\beta+3)}},$$

$$b_n = -\frac{2(n+\alpha+\delta+1)(n+\beta-\delta+1)(n+\alpha+1)(n+\alpha+\beta+1)}{(2n+\alpha+\beta+1)(2n+\alpha+\beta+2)}$$

$$- \frac{2n(n+\beta)(n+\alpha+\delta+1)(n+\beta-\delta)}{(2n+\alpha+\beta)(2n+\alpha+\beta+1)}.$$

Then the recursion relation from Lemma 3.67 for $\frac{1}{2}T$ is solved by the orthonormal version of the Wilson polynomials [55, §1.1], [54, §9.1],

$$W_n\left(\mu^2; \frac{1}{2}(1+\alpha), \frac{1}{2}(1+\alpha)+\delta, \frac{1}{2}(1-\alpha)+\beta-\delta, \frac{1}{2}(1+\alpha)\right),$$

where the relation between the eigenvalue λ of T and μ^2 is given by $\lambda = -2\left(\frac{\alpha+1}{2}\right)^2 - 2\mu^2$. Using the spectral decomposition of a Jacobi operator as in Section 3.3 proves the following theorem.

Theorem 3.68 *Let $\alpha > -1$, $\beta > -1$, and assume*

$$\gamma = -(\alpha+\delta+1)(\beta-\delta+1) \in \mathbb{R}.$$

The unbounded operator $(T, \mathscr{P})$ defined by (3.38) on

$$L^2([-1,1], (1-x)^\alpha(1+x)^\beta\, dx)$$

with domain the polynomials $\mathscr{P}$ is essentially self-adjoint. The spectrum of the closure $\bar{T}$ is simple and given by

$$\left(-\infty, -\frac{1}{2}(\alpha+1)^2\right)$$

$$\cup \left\{ -\frac{1}{2}(\alpha+1)^2 + 2\left(\frac{1}{2}(1+\alpha)+\delta+k\right)^2 : k \in \mathbb{N}, \frac{1}{2}(1+\alpha)+\delta+k < 0 \right\}$$

$$\cup \left\{ -\frac{1}{2}(\alpha+1)^2 + 2\left(\frac{1}{2}(1-\alpha)+\beta-\delta+l\right)^2 : \right.$$

$$\left. l \in \mathbb{N}, \frac{1}{2}(1-\alpha)+\beta-\delta+l < 0 \right\}$$

where the first set gives the absolutely continuous spectrum and the other sets correspond to the discrete spectrum of the closure of T. The discrete spectrum consists of at most one of these sets, and can be empty.

Note that in Theorem 3.68 we require $\delta \in \mathbb{R}$ or $\Re \delta = \frac{1}{2}(\beta - \alpha)$. In the second case there is no discrete spectrum.

The eigenvalue equation $T f_\lambda = \lambda f_\lambda$ is a second-order differential operator with regular singularities at -1, 1, ∞. In the Riemann–Papperitz notation, see e.g., [94, §5.5], it is

$$\mathscr{P} \left\{ \begin{array}{cccc} -1 & 1 & \infty & \\ 0 & -\frac{1}{2}(1+\alpha) + i\tilde{\lambda} & \alpha + \delta + 1 & x \\ -\beta & -\frac{1}{2}(1+\alpha) + i\tilde{\lambda} & \beta - \delta + 1 & \end{array} \right\}$$

with the reparametrisation $\lambda = -\frac{1}{2}(\alpha+1)^2 - 2\tilde{\lambda}^2$ of the spectral parameter. If $\gamma = 0$, we can exploit this relation and establish a link to the Jacobi function transform mapping (special) Jacobi polynomials to (special) Wilson polynomials, see [70]. We refer to [49] for the details. Going through this procedure and starting with the Laguerre polynomials and taking special values for the additional parameter gives results relating Laguerre polynomials to Meixner polynomials involving confluent hypergeometric functions, i.e., Whittaker functions. This is then related to the results of Section 3.6.1. Genest et al. [30] show how to extend this method in order to find the full 4-parameter family of Wilson polynomials in this way.

3.6.3 J-matrix method with matrix-valued orthogonal polynomials

We generalise the situation of Section 3.3.2 to operators that are 5-diagonal in a suitable basis. By Durán and Van Assche [27], see also e.g., [12, 26], a 5-diagonal recurrence can be written as a three-term recurrence relation for 2×2-matrix-valued orthogonal polynomials. More generally, Durán and Van Assche [27] show that $2N + 1$-diagonal recurrence can be written as a three-term recurrence relation for $N \times N$-matrix-valued orthogonal polynomials, and we leave it to the reader to see how the result of this section can be generalised to $2N + 1$-diagonal operators. The results of this section are based on [38], and we specialise again to the case of the Jacobi polynomials. Another similar example is based on the little q-Jacobi polynomials and other operators which arise as 5-term recurrence operators in a natural way, see [38] for these cases.

In Section 3.6.2 we used known orthogonal polynomials, in particular their orthogonality relations, in order to find spectral information on a differential operator. In this section we generalise the approach of Section 3.6.2 by assuming now that the polynomial r, the inverse of the Radon–Nikodym derivative, is of degree 2. This then leads to a 5-term recurrence relation, see Exercise 3.72. Hence we have an explicit expression for the matrix-valued Jacobi operator. Now we assume that the resulting differential or difference operator leads to an operator of which the spectral decomposition is known. Then we can find from this information the orthogonality measure for the matrix-valued polynomials. This leads to a case of matrix-valued orthogonal polynomials where both the orthogonality measure and the three-term recurrence can be found explicitly.

So let us start with the general set-up. Let T be an operator on a Hilbert space $\mathscr{H}$ of functions, typically a second-order difference or differential operator. We assume that T has the following properties;

(a) T is (a possibly unbounded) self-adjoint operator on $\mathscr{H}$ (with domain D in case T is unbounded);

(b) there exists an orthonormal basis $\{f_n\}_{n=0}^{\infty}$ of $\mathscr{H}$ so that $f_n \in D$ in case T is unbounded and so that there exist sequences $\{a_n\}_{n=0}^{\infty}$, $\{b_n\}_{n=0}^{\infty}$, $\{c_n\}_{n=0}^{\infty}$ of complex numbers with $a_n > 0$, $c_n \in \mathbb{R}$, for all $n \in \mathbb{N}$ so that

$$T f_n = a_n f_{n+2} + b_n f_{n+1} + c_n f_n + \overline{b_{n-1}} f_{n-1} + \overline{a_{n-2}} f_{n-2}. \quad (3.39)$$

Next we assume that we have a suitable spectral decomposition of T. We assume that the spectrum $\sigma(T)$ is simple or at most of multiplicity 2. The double spectrum is contained in $\Omega_2 \subset \sigma(T) \subset \mathbb{R}$, and the simple spectrum is contained in $\Omega_1 = \sigma(T) \setminus \Omega_2 \subset \mathbb{R}$. Consider functions f defined on $\sigma(T) \subset \mathbb{R}$ so that $f|_{\Omega_1} : \Omega_1 \to \mathbb{C}$ and $f|_{\Omega_2} : \Omega_2 \to \mathbb{C}^2$. We let σ be a Borel measure on Ω_1 and $V\rho$ a 2×2-matrix-valued measure on Ω_2 as in [19, §1.2], so $V : \Omega_2 \to M_2(\mathbb{C})$ maps into the positive semi-definite matrices and ρ is a positive Borel measure on Ω_2. We assume V is positive semi-definite ρ-a.e., but not necessarily positive-definite.

Next we consider the weighted Hilbert space $L^2(\mathscr{V})$ of such functions for which

$$\int_{\Omega_1} |f(\lambda)|^2 d\sigma(\lambda) + \int_{\Omega_2} f^*(\lambda) V(\lambda) f(\lambda) d\rho(\lambda) < \infty$$

and we obtain $L^2(\mathscr{V})$ by modding out by the functions of norm zero, see the discussion in Section 3.5.1. The inner product is given by

$$\langle f, g \rangle = \int_{\Omega_1} f(\lambda) \overline{g(\lambda)} d\sigma(\lambda) + \int_{\Omega_2} g^*(\lambda) V(\lambda) f(\lambda) d\rho(\lambda).$$

The final assumption is then

(c) there exists a unitary map $U : \mathscr{H} \to L^2(\mathscr{V})$ so that $UT = MU$, where M is the multiplication operator by λ on $L^2(\mathscr{V})$.

Note that assumption (c) is saying that $L^2(\mathcal{V})$ is the spectral decomposition of T, and since this also gives the spectral decomposition of polynomials in T, we see that all moments exist in $L^2(\mathcal{V})$.

Under the assumptions (a), (b), (c) we link the spectral measure to an orthogonality measure for matrix-valued orthogonal polynomials. Apply U to the 5-term expression (3.39) for T on the basis $\{f_n\}_{n=0}^{\infty}$, so that

$$\lambda(U f_n)(\lambda) = a_n(U f_{n+2})(\lambda) + b_n(U f_{n+1})(\lambda)$$
$$+ c_n(U f_n)(\lambda) + \overline{b_{n-1}}(U f_{n-1})(\lambda) + a_{n-2}(U f_{n-2})(\lambda) \quad (3.40)$$

which is to be interpreted as an identity in $L^2(\mathcal{V})$. When restricted to Ω_1, (3.40) is a scalar identity; and when restricted to Ω_2, the components of $U f(\lambda) = (U_1 f(\lambda), U_2 f(\lambda))^t$ satisfy (3.40).

Working out the details for $N = 2$ of [27], we see that we have to generate the 2×2-matrix-valued polynomials by

$$\lambda P_n(\lambda) = \begin{cases} A_n P_{n+1}(\lambda) + B_n P_n(\lambda) + A_{n-1}^* P_{n-1}(\lambda), & n \geq 1, \\ A_0 P_1(\lambda) + B_0 P_0(\lambda), & n = 0, \end{cases}$$
$$(3.41)$$
$$A_n = \begin{pmatrix} a_{2n} & 0 \\ b_{2n+1} & a_{2n+1} \end{pmatrix}, \qquad B_n = \begin{pmatrix} c_{2n} & b_{2n} \\ \overline{b_{2n}} & c_{2n+1} \end{pmatrix},$$

with initial conditions $P_{-1}(\lambda) = 0$ and $P_0(\lambda)$ is a constant non-singular matrix, which we take to be the identity, so $P_0(\lambda) = I$. Note that A_n is a non-singular matrix and B_n is a Hermitian matrix for all $n \in \mathbb{N}$. Then the $\mathbb{C}^2$-valued functions

$$\mathcal{U}_n(\lambda) = \begin{pmatrix} U f_{2n}(\lambda) \\ U f_{2n+1}(\lambda) \end{pmatrix}, \qquad \mathcal{U}_n^1(\lambda) = \begin{pmatrix} U_1 f_{2n}(\lambda) \\ U_1 f_{2n+1}(\lambda) \end{pmatrix},$$
$$\mathcal{U}_n^2(\lambda) = \begin{pmatrix} U_2 f_{2n}(\lambda) \\ U_2 f_{2n+1}(\lambda) \end{pmatrix},$$

satisfy (3.41) for vectors for $\lambda \in \Omega_1$ in the first case and for $\lambda \in \Omega_2$ in the last cases. Hence,

$$\mathcal{U}_n(\lambda) = P_n(\lambda)\mathcal{U}_0(\lambda), \ \mathcal{U}_n^1(\lambda) = P_n(\lambda)\mathcal{U}_0^1(\lambda), \ \mathcal{U}_n^2(\lambda) = P_n(\lambda)\mathcal{U}_0^2(\lambda),$$
$$(3.42)$$

where the first holds σ-a.e. and the last two hold ρ-a.e. We can now state the orthogonality relations for the matrix-valued orthogonal polynomials.

Theorem 3.69 *With the assumptions* (a), (b), (c) *as given above, the 2×2-matrix-valued polynomials P_n generated by* (3.41) *and $P_0(\lambda) = I$ satisfy*

$$\int_{\Omega_1} P_n(\lambda) W_1(\lambda) P_m(\lambda)^* \, d\sigma(\lambda) + \int_{\Omega_2} P_n(\lambda) W_2(\lambda) P_m(\lambda)^* \, d\rho(\lambda) = \delta_{n,m} I$$

where

$$W_1(\lambda) = \begin{pmatrix} \overline{|U f_0(\lambda)|^2} & U f_0(\lambda)\overline{U f_1(\lambda)} \\ \overline{U f_0(\lambda)}U f_1(\lambda) & |U f_1(\lambda)|^2 \end{pmatrix}, \qquad \sigma\text{-a.e.}$$

$$W_2(\lambda) = \begin{pmatrix} \langle U f_0(\lambda), U f_0(\lambda)\rangle_{V(\lambda)} & \langle U f_0(\lambda), U f_1(\lambda)\rangle_{V(\lambda)} \\ \langle U f_1(\lambda), U f_0(\lambda)\rangle_{V(\lambda)} & \langle U f_1(\lambda), U f_1(\lambda)\rangle_{V(\lambda)} \end{pmatrix}, \qquad \rho\text{-a.e.}$$

and $\langle x, y\rangle_{V(\lambda)} = x^* V(\lambda) y$.

Since we stick to the situation with the assumptions (a), (b), (c), the multiplicity of T cannot be higher than 2. Note that the matrices $W_1(\lambda)$ and $W_2(\lambda)$ are Gram matrices. In particular, $\det(W_1(\lambda)) = 0$ for all λ. So the weight matrix $W_1(\lambda)$ is semi-definite positive with eigenvalues 0 and $\mathrm{tr}(W_1(\lambda)) = |U f_0(\lambda)|^2 + |U f_1(\lambda)|^2 > 0$. Note that

$$\ker(W_1(\lambda)) = \mathbb{C}\begin{pmatrix} \overline{U f_1(\lambda)} \\ -\overline{U f_0(\lambda)} \end{pmatrix} = \begin{pmatrix} U f_0(\lambda) \\ U f_1(\lambda) \end{pmatrix}^{\perp},$$

$$\ker(W_1(\lambda) - \mathrm{tr}(W_1(\lambda))) = \mathbb{C}\begin{pmatrix} U f_0(\lambda) \\ U f_1(\lambda) \end{pmatrix}.$$

Moreover, $\det(W_2(\lambda)) = 0$ if and only if $U f_0(\lambda)$ and $U f_1(\lambda)$ are multiples of each other.

Denoting the integral in Theorem 3.69 as $\langle P_n, P_m\rangle_W$, we see that all the assumptions on the matrix-valued inner product, as in the definition of the Hilbert C^*-module $L^2_C(\mu)$ in Section 3.4.1, are trivially satisfied, except for $\langle Q, Q\rangle_W = 0$ implies $Q = 0$ for a matrix-valued polynomial Q. We can proceed by writing $Q = \sum_{k=1}^n C_k P_k$ for suitable matrices C_k, since the leading coefficient of P_k is non-singular by (3.41). Then by Theorem 3.69 we have $\langle Q, Q\rangle_W = \sum_{k=0}^n C_k C_k^*$ which is a sum of positive-definite elements, which can only give 0 if each of the terms is zero. So $\langle Q, Q\rangle_W = 0$ implies $C_k = 0$ for all k, therefore $Q = 0$.

Proof Start using the unitarity:

$$\delta_{n,m}\begin{pmatrix} 1 & 0 \\ 0 & 1 \end{pmatrix} = \begin{pmatrix} \langle f_{2n}, f_{2m}\rangle_{\mathcal{H}} & \langle f_{2n}, f_{2m+1}\rangle_{\mathcal{H}} \\ \langle f_{2n+1}, f_{2m}\rangle_{\mathcal{H}} & \langle f_{2n+1}, f_{2m+1}\rangle_{\mathcal{H}} \end{pmatrix}$$

$$= \begin{pmatrix} \langle U f_{2n}, U f_{2m}\rangle_{L^2(\gamma)} & \langle U f_{2n}, U f_{2m+1}\rangle_{L^2(\gamma)} \\ \langle U f_{2n+1}, U f_{2m}\rangle_{L^2(\gamma)} & \langle U f_{2n+1}, U f_{2m+1}\rangle_{L^2(\gamma)} \end{pmatrix}.$$

$$(3.43)$$

Split each of the inner products on the right-hand side of (3.43) as a sum over two integrals, one over Ω_1 and the other over Ω_2. First the integral over Ω_1 equals

$$\begin{pmatrix} \int_{\Omega_1} U f_{2n}(\lambda)\overline{U f_{2m}(\lambda)}\,d\sigma(\lambda) & \int_{\Omega_1} U f_{2n}(\lambda)\overline{U f_{2m+1}(\lambda)}\,d\sigma(\lambda) \\ \int_{\Omega_1} U f_{2n+1}(\lambda)\overline{U f_{2m}(\lambda)}\,d\sigma(\lambda) & \int_{\Omega_1} U f_{2n+1}(\lambda)\overline{U f_{2m+1}(\lambda)}\,d\sigma(\lambda) \end{pmatrix}$$

$$= \int_{\Omega_1} \begin{pmatrix} U f_{2n}(\lambda)\overline{U f_{2m}(\lambda)} & U f_{2n}(\lambda)\overline{U f_{2m+1}(\lambda)} \\ U f_{2n+1}(\lambda)\overline{U f_{2m}(\lambda)} & U f_{2n+1}(\lambda)\overline{U f_{2m+1}(\lambda)} \end{pmatrix} d\sigma(\lambda)$$

$$= \int_{\Omega_1} \begin{pmatrix} U f_{2n}(\lambda) \\ U f_{2n+1}(\lambda) \end{pmatrix} \begin{pmatrix} U f_{2m}(\lambda) \\ U f_{2m+1}(\lambda) \end{pmatrix}^* d\sigma(\lambda)$$

$$= \int_{\Omega_1} P_n(\lambda) \begin{pmatrix} U f_0(\lambda) \\ U f_1(\lambda) \end{pmatrix} \begin{pmatrix} U f_0(\lambda) \\ U f_1(\lambda) \end{pmatrix}^* P_m(\lambda)^* d\sigma(\lambda)$$

$$= \int_{\Omega_1} P_n(\lambda) W_1(\lambda) P_m(\lambda)^* d\sigma(\lambda), \tag{3.44}$$

where we have used (3.42). For the integral over Ω_2 we write $U f(\lambda) = (U_1 f(\lambda), U_2 f(\lambda))^t$ and $V(\lambda) = (v_{ij}(\lambda))_{i,j=1}^2$, so that the integral over Ω_2 can be written as (temporarily writing $v_{ij}(\lambda)$ simply as v_{ij})

$$\sum_{i,j=1}^2 \int_{\Omega_2} \begin{pmatrix} U_j f_{2n}(\lambda) v_{ij}\overline{U_i f_{2m}(\lambda)} & U_j f_{2n}(\lambda) v_{ij}\overline{U_i f_{2m+1}(\lambda)} \\ U_j f_{2n+1}(\lambda) v_{ij}\overline{U_i f_{2m}(\lambda)} & U_j f_{2n+1}(\lambda) v_{ij}\overline{U_i f_{2m+1}(\lambda)} \end{pmatrix} d\rho(\lambda)$$

$$= \sum_{i,j=1}^2 \int_{\Omega_2} \begin{pmatrix} U_j f_{2n}(\lambda) \\ U_j f_{2n+1}(\lambda) \end{pmatrix} \begin{pmatrix} U_i f_{2m}(\lambda) \\ U_i f_{2m+1}(\lambda) \end{pmatrix}^* v_{ij}\,d\rho(\lambda)$$

$$= \sum_{i,j=1}^2 \int_{\Omega_2} P_n(\lambda) \begin{pmatrix} U_j f_0(\lambda) \\ U_j f_1(\lambda) \end{pmatrix} \begin{pmatrix} U_i f_0(\lambda) \\ U_i f_1(\lambda) \end{pmatrix}^* P_m(\lambda)^* v_{ij}\,d\rho(\lambda)$$

$$= \int_{\Omega_2} P_n(\lambda) W_2(\lambda) P_m(\lambda)^* d\rho(\lambda), \tag{3.45}$$

where we have used (3.42) again and with

$$\begin{aligned} W_2(\lambda) &= \sum_{i,j=1}^2 \begin{pmatrix} U_j f_0(\lambda) \\ U_j f_1(\lambda) \end{pmatrix} \begin{pmatrix} U_i f_0(\lambda) \\ U_i f_1(\lambda) \end{pmatrix}^* v_{ij}(\lambda) \\ &= \sum_{i,j=1}^2 v_{ij}(\lambda) \begin{pmatrix} U_j f_0(\lambda)\overline{U_i f_0(\lambda)} & U_j f_0(\lambda)\overline{U_i f_1(\lambda)} \\ U_j f_1(\lambda)\overline{U_i f_0(\lambda)} & U_j f_1(\lambda)\overline{U_i f_1(\lambda)} \end{pmatrix} \\ &= \begin{pmatrix} (U f_0(\lambda))^* V(\lambda) U f_0(\lambda) & (U f_1(\lambda))^* V(\lambda) U f_0(\lambda) \\ (U f_0(\lambda))^* V(\lambda) U f_1(\lambda) & (U f_1(\lambda))^* V(\lambda) U f_1(\lambda) \end{pmatrix} \end{aligned} \tag{3.46}$$

and putting (3.44) and (3.45), (3.46) into (3.43) proves the result. $\square$

If we additionally assume T is bounded, so that the measures σ and ρ have compact support, the coefficients in (3.39) and (3.41) are bounded. In this case the corresponding Jacobi operator is bounded and self-adjoint.

Remark 3.70 Assume that $\Omega_1 = \sigma(T)$ or $\Omega_2 = \emptyset$, so that T has simple spectrum. Then

$$\mathcal{L}^2(W_1 d\sigma) = \{f\colon \mathbb{R} \to \mathbb{C}^2 \mid \int_{\mathbb{R}} f(\lambda)^* W_1(\lambda) f(\lambda) \, d\sigma(\lambda) < \infty\} \quad (3.47)$$

has the subspace of null-vectors

$$\mathcal{N} = \{f \in \mathcal{L}^2(W_1 d\sigma) \mid \int_{\mathbb{R}} f(\lambda)^* W_1(\lambda) f(\lambda) \, d\sigma(\lambda) = 0\}$$

$$= \{f \in \mathcal{L}^2(W_1 d\sigma) \mid f(\lambda) = c(\lambda) \begin{pmatrix} U f_1(\lambda) \\ -U f_0(\lambda) \end{pmatrix} \sigma\text{-a.e.}\},$$

where c is a scalar-valued function. In this case $L^2(\mathcal{V}) = \mathcal{L}^2(W_1 d\sigma)/\mathcal{N}$. Note that $\mathcal{U}_n \colon \mathbb{R} \to L^2(W_1 d\sigma)$ is completely determined by $U f_0(\lambda)$, which is a restatement of T having simple spectrum. From Theorem 3.69 we see that, cf. (3.24),

$$\langle P_n(\cdot) v_1, P_m(\cdot) v_2 \rangle_{L^2(W_1 d\sigma)} = \delta_{n,m} \langle v_1, v_2 \rangle$$

so that $\{P_n(\cdot) e_i\}_{i \in \{1,2\}, n \in \mathbb{N}}$ is linearly independent in $L^2(W_1 d\sigma)$ for any basis $\{e_1, e_2\}$ of $\mathbb{C}^2$, cf. (3.24).

We illustrate Theorem 3.69 with an example, and we refer to Groenevelt and the author [38] and [41] for details. We extend the approach of Section 3.6.2 and Lemma 3.67 by now assuming that r is a polynomial of degree 2. Then the relations (3.36) and (3.37) go through, except that it also involves a term P_{n-2}, respectively Φ_{n-2}. Then we find that $r(L+\gamma)$ is a 5-term recurrence operator. Adding a three-term recurrence relation, so $T = r(L+\gamma) + \rho x$, gives a 5-term recurrence operator, see Exercise 3.72. However it is usually hard to establish the assumption that an explicit spectral decomposition of such an operator is available. Moreover, we want to have an example of such an operator where the spectrum of multiplicity 2 is non-trivial.

We do this for the Jacobi polynomials, and we consider $T = T^{(\alpha,\beta;\kappa)}$ defined by

$$T = (1-x^2)^2 \frac{d^2}{dx^2} + (1-x^2)\big(\beta - \alpha - (\alpha + \beta + 4)x\big)\frac{d}{dx}$$

$$+ \frac{1}{4}\big(\kappa^2 - (\alpha + \beta + 3)^2\big)(1 - x^2) \quad (3.48)$$

as an operator in the weighted L^2-space for the Jacobi polynomials; that is, $L^2((-1,1), w^{(\alpha,\beta)})$ with $w^{(\alpha,\beta)}$ the normalised weight function for the

Jacobi polynomials as given below. Here $\alpha, \beta > -1$ and $\kappa \in \mathbb{R}_{\geq 0} \cup i\mathbb{R}_{>0}$. Then we can use (3.38) to obtain

$$T^{(\alpha,\beta;\kappa)} = r\left(L^{(\alpha+1,\beta+1)} + \rho\right), \qquad \rho = \frac{1}{4}\left(\kappa^2 - (\alpha+\beta+3)^2\right),$$

where $r(x) = 1 - x^2$ is, up to a constant, the quotient of the normalised weight functions of the Jacobi polynomial,

$$r(x) = K\frac{w^{(\alpha+1,\beta+1)}(x)}{w^{(\alpha,\beta)}(x)}, \qquad K = \frac{4(\alpha+1)(\beta+1)}{(\alpha+\beta+2)(\alpha+\beta+3)},$$

$$w^{(\alpha,\beta)}(x) = 2^{-\alpha-\beta-1}\frac{\Gamma(\alpha+\beta+2)}{\Gamma(\alpha+1,\beta+1)}(1-x)^\alpha(1+x)^\beta.$$

It is then clear from the analogue of Lemma 3.67 that T is 5-term recurrence relation with respect to Jacobi polynomials

In order to describe the spectral decomposition, we have to introduce some notation. For proofs we refer to Groenevelt and the author [41]. We assume $\beta \geq \alpha$. Let $\Omega_1, \Omega_2 \subset \mathbb{R}$ be given by

$$\Omega_1 = \left(-(\beta+1)^2, -(\alpha+1)^2\right) \quad \text{and} \quad \Omega_2 = \left(-\infty, -(\beta+1)^2\right).$$

We assume $0 \leq \kappa < 1$ or $\kappa \in i\mathbb{R}_{>0}$ for convenience, in order to avoid discrete spectrum of T. When the condition on κ is not valid, we have a discrete spectrum, which occurs with multiplicity one, see [41]. We set

$$\delta_\lambda = i\sqrt{-\lambda - (\alpha+1)^2}, \qquad \lambda \in \Omega_1 \cup \Omega_2,$$

$$\eta_\lambda = i\sqrt{-\lambda - (\beta+1)^2}, \qquad \lambda \in \Omega_2,$$

$$\delta(\lambda) = \sqrt{\lambda + (\alpha+1)^2}, \qquad \lambda \in \mathbb{C} \setminus (\Omega_1 \cup \Omega_2),$$

$$\eta(\lambda) = \sqrt{\lambda + (\beta+1)^2}, \qquad \lambda \in \mathbb{C} \setminus \Omega_2.$$

Here $\sqrt{\cdot}$ denotes the principal branch of the square root. We denote by σ the set $\Omega_2 \cup \Omega_1$. Theorem 3.71 will show that σ is the spectrum of T.

Next we introduce the weight functions that we need to define $L^2(\mathcal{V})$. First we define

$$c(x;y) = \frac{\Gamma(1+y)\Gamma(-x)}{\Gamma(\frac{1}{2}(1+y-x+\kappa))\Gamma(\frac{1}{2}(1+y-x-\kappa))}.$$

With this function we define for $\lambda \in \Omega_1$

$$v(\lambda) = \frac{1}{c(\delta_\lambda;\eta(\lambda))c(-\delta_\lambda;\eta(\lambda))}.$$

For $\lambda \in \Omega_2$ we define the matrix-valued weight function $V(\lambda)$ by

$$V(\lambda) = \begin{pmatrix} 1 & v_{12}(\lambda) \\ v_{21}(\lambda) & 1 \end{pmatrix},$$

with

$$\begin{aligned}
v_{21}(\lambda) &= \frac{c(\eta_\lambda; \delta_\lambda)}{c(-\eta_\lambda; \delta_\lambda)} \\
&= \frac{\Gamma(-\eta_\lambda)\Gamma(\frac{1}{2}(1+\delta_\lambda+\eta_\lambda+\kappa))\Gamma(\frac{1}{2}(1+\delta_\lambda+\eta_\lambda-\kappa))}{\Gamma(\eta_\lambda)\Gamma(\frac{1}{2}(1+\delta_\lambda-\eta_\lambda+\kappa))\Gamma(\frac{1}{2}(1+\delta_\lambda-\eta_\lambda-\kappa))},
\end{aligned}$$

and $v_{12}(\lambda) = \overline{v_{21}(\lambda)}$.

Now we are ready to define the Hilbert space $L^2(\mathcal{V})$. It consists of functions that are $\mathbb{C}^2$-valued on Ω_2 and $\mathbb{C}$-valued on Ω_1. The inner product on $L^2(\mathcal{V})$ is given by

$$\langle f, g \rangle_{\mathcal{V}} = \frac{1}{2\pi D} \int_{\Omega_2} g(\lambda)^* V(\lambda) f(\lambda) \frac{d\lambda}{-i\eta_\lambda} + \frac{1}{2\pi D} \int_{\Omega_1} f(\lambda)\overline{g(\lambda)}v(\lambda)\frac{d\lambda}{-i\delta_\lambda},$$

where $D = \frac{4\Gamma(\alpha+\beta+2)}{\Gamma(\alpha+1,\beta+1)}$.

Next we introduce the integral transform $\mathscr{F}$. For $\lambda \in \Omega_1$ and $x \in (-1,1)$ we define

$$\begin{aligned}
\varphi_\lambda(x) = &\left(\frac{1-x}{2}\right)^{-\frac{1}{2}(\alpha-\delta_\lambda+1)}\left(\frac{1+x}{2}\right)^{-\frac{1}{2}(\beta-\eta(\lambda)+1)} \\
&\times {}_2F_1\left(\begin{matrix} \frac{1}{2}(1+\delta_\lambda+\eta(\lambda)-\kappa), \frac{1}{2}(1+\delta_\lambda+\eta(\lambda)+\kappa) \\ 1+\eta(\lambda) \end{matrix}; \frac{1+x}{2}\right).
\end{aligned}$$

By Euler's transformation, see e.g., [5, (2.2.7)], we have the symmetry $\delta_\lambda \leftrightarrow -\delta_\lambda$. Furthermore, we define for $\lambda \in \Omega_2$ and $x \in (-1,1)$,

$$\begin{aligned}
\varphi_\lambda^{\pm}(x) = &\left(\frac{1-x}{2}\right)^{-\frac{1}{2}(\alpha-\delta_\lambda+1)}\left(\frac{1+x}{2}\right)^{-\frac{1}{2}(\beta\mp\eta_\lambda+1)} \\
&\times {}_2F_1\left(\begin{matrix} \frac{1}{2}(1+\delta_\lambda\pm\eta_\lambda-\kappa), \frac{1}{2}(1+\delta_\lambda\pm\eta_\lambda+\kappa) \\ 1\pm\eta_\lambda \end{matrix}; \frac{1+x}{2}\right).
\end{aligned}$$

Observe that $\overline{\varphi_\lambda^+(x)} = \varphi_\lambda^-(x)$, again by Euler's transformation. Now, let $\mathscr{F}$ be the integral transform defined by

$$(\mathscr{F}f)(\lambda) = \begin{cases} \int_{-1}^{1} f(x)\begin{pmatrix} \varphi_\lambda^+(x) \\ \varphi_\lambda^-(x) \end{pmatrix} w^{(\alpha,\beta)}(x)\,dx, & \lambda \in \Omega_2, \\ \int_{-1}^{1} f(x)\varphi_\lambda(x)w^{(\alpha,\beta)}(x)\,dx, & \lambda \in \Omega_1, \end{cases}$$

for all $f \in \mathcal{H}$ such that the integrals converge. The following result says that $\mathcal{F}$ is the required unitary operator U intertwining T with multiplication.

Theorem 3.71 *The transform $\mathcal{F}$ extends uniquely to a unitary operator $\mathcal{F} \colon \mathcal{H} \to L^2(\mathcal{V})$ such that $\mathcal{F}T = M\mathcal{F}$, where $M \colon L^2(\mathcal{V}) \to L^2(\mathcal{V})$ is the unbounded multiplication operator given by $(Mg)(\lambda) = \lambda g(\lambda)$ for almost all $\lambda \in \sigma$.*

The proof of Theorem 3.71 is based on the fact that the eigenvalue equation $Tf_\lambda = \lambda f_\lambda$ can be solved in terms of hypergeometric functions since it is a second-order differential equation with regular singularities at three points. Having sufficiently many solutions available gives the opportunity to find the Green kernel, and therefore the resolvent operator, from which one derives the spectral decomposition; see [41] for details.

Now we want to apply Theorem 3.69 for the polynomials generated by (3.41). For this it suffices to write down explicitly the coefficients a_n, b_n and c_n in the 5-term recurrence realisation of the operator T, cf. Exercise 3.72, and to calculate the matrix entries in the weight matrices of Theorem 3.69.

The coefficients a_n, b_n and c_n follow by keeping track of the method of Exercise 3.72, and this worked out in Exercise 3.74. This then makes the matrix entries in the three-term recurrence relation (3.41) completely explicit.

It remains to calculate the matrix entries of the weight functions in Theorem 3.69. In [41] these functions are calculated in terms of $_3F_2$-functions.

3.6.4 Exercises

Exercise 3.72   Generalise the situation of Section 3.6.2 to the case where the polynomial r is of degree 2. Show that in this case the analogue of (3.36) and (3.37) involve three terms on the right-hand side. Show now that the operator $T = r(L + \gamma) + \tau x$ is a 5-term operator in the bases $\{\phi_n\}_{n \in \mathbb{N}}$ of $L^2(v)$. Here r, respectively x, denotes multiplication by r, respectively x, and γ, τ are constants.

Exercise 3.73   Show that (3.41) and (3.42) hold starting from (3.40).

Exercise 3.74   (a) Show that

$$\phi_n = \alpha_n \Phi_n + \beta_n \Phi_{n-1} + \gamma_n \Phi_{n-2},$$

where ϕ_n, respectively Φ_n, are the orthonormalised Jacobi polynomials

$P_n^{(\alpha,\beta)}$, respectively $P_n^{(\alpha+1,\beta+1)}$, and where

$$\alpha_n = \frac{2}{\sqrt{K}} \frac{1}{2n+\alpha+\beta+2}$$

$$\times \sqrt{\frac{(\alpha+n+1)(\beta+n+1)(n+\alpha+\beta+1)(n+\alpha+\beta+2)}{(\alpha+\beta+2n+1)(\alpha+\beta+2n+3)}},$$

$$\beta_n = (-1)^n \frac{2}{\sqrt{K}} \frac{(\beta-\alpha)\sqrt{n(n+\alpha+\beta+1)}}{(\alpha+\beta+2n)(\alpha+\beta+2n+2)},$$

$$\gamma_n = -\frac{2}{\sqrt{K}} \frac{1}{2n+\alpha+\beta} \sqrt{\frac{n(n-1)(\alpha+n)(\beta+n)}{(\alpha+\beta+2n-1)(\alpha+\beta+2n+1)}}.$$

Here K as in the definition of $r(x)$.

(b) Show that

$$a_n = K\alpha_n\gamma_{n+2}(\Lambda_n+\rho),$$

$$b_n = K\alpha_n\beta_{n+1}(\Lambda_n+\rho) + K\beta_n\gamma_{n+1}(\Lambda_{n+1}+\rho),$$

$$c_n = K\alpha_n^2(\Lambda_n+\rho) + K\beta_n^2(\Lambda_{n-1}+\rho) + K\gamma_n^2(\Lambda_{n-2}+\rho),$$

where $\Lambda_n = -n(n+\alpha+\beta+3)$, ρ as in the definition of $T = T^{(\alpha,\beta;\kappa)}$ and $\alpha_n, \beta_n, \gamma_n$ as in (a).

3.7 Appendix: The spectral theorem

In this appendix we recall some facts from functional analysis with emphasis on the spectral theorem. There are many sources for this appendix, or parts of it, see e.g., [23, 77, 82, 85, 87, 92, 96], but many other sources are available.

3.7.1 Hilbert spaces and operators

A vector space $\mathscr{H}$ over $\mathbb{C}$ is an inner product space if there exists a mapping $\langle \cdot, \cdot \rangle \colon \mathscr{H} \times \mathscr{H} \to \mathbb{C}$ such that for all $u, v, w \in \mathscr{H}$ and for all $a, b \in \mathbb{C}$ we have (i) $\langle av+bw, u \rangle = a\langle v, u \rangle + b\langle w, u \rangle$, (ii) $\langle u, v \rangle = \overline{\langle v, u \rangle}$, and (iii) $\langle v, v \rangle \geq 0$ and $\langle v, v \rangle = 0$ if and only if $v = 0$. With the inner product we associate the norm $\|v\| = \|v\|_{\mathscr{H}} = \sqrt{\langle v, v \rangle}$, and the topology from the corresponding metric $d(u, v) = \|u - v\|$. The standard inequality is the Cauchy–Schwarz inequality; $|\langle u, v \rangle| \leq \|u\|\|v\|$. A Hilbert space $\mathscr{H}$ is a complete inner product space, i.e., for any Cauchy sequence $\{x_n\}_n$ in $\mathscr{H}$, i.e., for all $\varepsilon > 0$ there exists $N \in \mathbb{N}$ such that for all $n, m \geq N$ $\|x_n - x_m\| < \varepsilon$, there exists an element $x \in \mathscr{H}$ such that x_n converges to x. In these notes all Hilbert spaces are separable, i.e., there exists a denumerable set of basis vectors. The Cauchy–Schwarz inequality can be extended to the Bessel inequality; for an orthonormal sequence $\{f_i\}_{i \in I}$ in $\mathscr{H}$, i.e., $\langle f_i, f_j \rangle = \delta_{i,j}$,

$$\sum_{i \in I} |\langle x, f_i \rangle|^2 \leq \|x\|^2.$$

Example 3.75 (i) The finite-dimensional inner product space $\mathbb{C}^N$ with its standard inner product is a Hilbert space.

(ii) $\ell^2(\mathbb{Z})$, the space of square summable sequences $\{a_k\}_{k \in \mathbb{Z}}$, and $\ell^2(\mathbb{N})$, the space of square summable sequences $\{a_k\}_{k \in \mathbb{N}}$, are Hilbert spaces. The inner product is given by $\langle \{a_k\}, \{b_k\} \rangle = \sum_{k \in \mathbb{N}} a_k \overline{b_k}$. An orthonormal basis is given by the sequences $\{e_k\}$ defined by $(e_k)_l = \delta_{k,l}$, so we identify $\{a_k\}$ with $\sum_{k \in \mathbb{N}} a_k e_k$.

(iii) We consider a positive Borel measure μ on the real line $\mathbb{R}$ such that all moments exist, i.e., $\int_{\mathbb{R}} |x|^m \, d\mu(x) < \infty$ for all $m \in \mathbb{N}$. Without loss of generality we assume that μ is a probability measure, $\int_{\mathbb{R}} d\mu(x) = 1$. By $L^2(\mu)$ we denote the space of square integrable functions on $\mathbb{R}$, i.e., $\int_{\mathbb{R}} |f(x)|^2 \, d\mu(x) < \infty$. Then $L^2(\mu)$ is a Hilbert space (after identifying two functions f and g for which $\int_{\mathbb{R}} |f(x) - g(x)|^2 \, d\mu(x) = 0$) with respect to the inner product $\langle f, g \rangle = \int_{\mathbb{R}} f(x) \overline{g(x)} \, d\mu(x)$. In the case that μ is a finite sum of discrete Dirac measures, we find that $L^2(\mu)$ is finite dimensional.

(iv) For two Hilbert spaces $\mathscr{H}_1$ and $\mathscr{H}_2$ we can take its algebraic tensor product $\mathscr{H}_1 \otimes \mathscr{H}_2$ and equip it with an inner product defined on simple tensors by

$$\langle v_1 \otimes v_2, w_1 \otimes w_2 \rangle = \langle v_1, w_1 \rangle_{\mathscr{H}_1} \langle v_2, w_2 \rangle_{\mathscr{H}_2}.$$

Taking its completion gives the Hilbert space $\mathscr{H}_1 \hat{\otimes} \mathscr{H}_2$.

An operator T from a Hilbert space $\mathscr{H}$ into another Hilbert space $\mathscr{K}$ is linear if for all $u, v \in \mathscr{H}$ and for all $a, b \in \mathbb{C}$ we have $T(au + bv) = aT(u) + bT(v)$. An operator T is bounded if there exists a constant M such that $\|Tu\|_{\mathscr{K}} \leq M\|u\|_{\mathscr{H}}$ for all $u \in \mathscr{H}$. The smallest M for which this holds is the norm, denoted by $\|T\|$, of T. A bounded linear operator is continuous. The adjoint of a bounded linear operator $T \colon \mathscr{H} \to \mathscr{K}$ is a map $T^* \colon \mathscr{K} \to \mathscr{H}$ with $\langle Tu, v \rangle_{\mathscr{K}} = \langle u, T^*v \rangle_{\mathscr{H}}$. We call $T \colon \mathscr{H} \to \mathscr{H}$ self-adjoint if $T^* = T$. $T^* \colon \mathscr{K} \to \mathscr{H}$ is unitary if $T^*T = 1_{\mathscr{H}}$ and $TT^* = 1_{\mathscr{K}}$. A projection $P \colon \mathscr{H} \to \mathscr{H}$ is a self-adjoint bounded operator such that $P^2 = P$.

An operator $T \colon \mathscr{H} \to \mathscr{K}$ is compact if the closure of the image of the unit ball $B_1 = \{v \in \mathscr{H} \mid \|v\| \leq 1\}$ under T is compact in $\mathscr{K}$. In the case $\mathscr{K}$ is finite-dimensional any bounded operator $T \colon \mathscr{H} \to \mathscr{K}$ is compact, and slightly more general, any operator which has finite rank, i.e., its range is finite dimensional, is compact. Moreover, any compact operator can be approximated in the operator norm by finite-rank operators.

3.7.2 Hilbert C*-modules

For more information on Hilbert C*-modules, see e.g., Lance [75]. The space $B(\mathcal{H})$ of bounded linear operators $T\colon \mathcal{H} \to \mathcal{H}$ is a $*$-algebra, where the $*$-operation is given by the adjoint, satisfying $\|TS\| \leq \|T\|\|S\|$ and $\|T^*T\| = \|T\|^2$. With the operator-norm, $B(\mathcal{H})$ is a metric space. Moreover, $B(\mathcal{H})$ is a $*$-algebra, and the norm and the algebra structure are compatible. A C*-algebra is a $*$-invariant subalgebra of $B(\mathcal{H})$, which is closed in the metric topology. Examples of a C*-algebra are $B(\mathcal{H})$ and the space of all compact operators $T\colon \mathcal{H} \to \mathcal{H}$. We only need $M_N(\mathbb{C}) = B(\mathbb{C}^N)$, the space of all linear maps from $\mathbb{C}^N$ to itself, as an example of a C*-algebra. An element $a \in A$ in a C*-algebra A is positive if $a = b^*b$ for some element $b \in A$, and we use the notation $a \geq 0$. This notation is extended to $a \geq b$ meaning $(a - b) \geq 0$. In the case of $A = M_N(\mathbb{C})$, $T \geq 0$ means that T corresponds to a positive semi-definite matrix, i.e., $\langle Tx, x\rangle \geq 0$ for all $x \in \mathbb{C}^N$. The positive-definite matrices form the cone $P_N(\mathbb{C})$ in $M_N(\mathbb{C})$. We say T is a positive matrix or a positive-definite matrix if $\langle Tx, x\rangle > 0$ for all $x \in \mathbb{C}^N \setminus \{0\}$. Note that terminology concerning positivity in C*-algebras and matrix algebras does not coincide, and we follow the latter, see [46].

A Hilbert C*-module E over the (unital) C*-algebra A is a left A-module E equipped with an A-valued inner product $\langle \cdot, \cdot \rangle\colon E \times E \to A$ so that for all $v, w, u \in E$ and all $a, b \in A$

$$\langle av + bw, u\rangle = a\langle v, u\rangle + b\langle w, u\rangle, \qquad \langle v, w\rangle = \langle w, v\rangle^*,$$
$$\langle v, v\rangle \geq 0 \quad \text{and} \quad \langle v, v\rangle = 0 \Leftrightarrow v = 0,$$

and E is complete with respect to the norm $\|v\| = \sqrt{\|\langle v, v\rangle\|}$. The analogue of the Cauchy–Schwarz inequality then reads

$$\langle v, w\rangle\langle w, v\rangle \leq \|\langle w, w\rangle\| \langle v, v\rangle, \qquad v, w \in E,$$

and the analogue of the Bessel inequality

$$\sum_{i \in I} \langle v, f_i\rangle\langle f_i, v\rangle \leq \langle v, v\rangle, \qquad v \in E,$$

for $(f_i)_{i \in I}$ an orthonormal set in E, i.e., $\langle f_i, f_j\rangle = \delta_{i,j} \in A$. (Here we use that A is unital.)

3.7.3 Unbounded operators

We are also interested in unbounded linear operators. In that case we denote $(T, \mathscr{D}(T))$, where $\mathscr{D}(T)$, the domain of T, is a linear subspace of $\mathscr{H}$ and $T: \mathscr{D}(T) \to \mathscr{H}$. Then T is densely defined if the closure of $\mathscr{D}(T)$ equals $\mathscr{H}$. All unbounded operators that we consider in these notes are densely defined. If the operator $(T - z)$, $z \in \mathbb{C}$, has an inverse $R(z) = (T - z)^{-1}$ which is densely defined and is bounded, so that $R(z)$, the resolvent operator, extends to a bounded linear operator on $\mathscr{H}$, then we call z a regular value. The set of all regular values is the resolvent set $\rho(T)$. The complement of the resolvent set $\rho(T)$ in $\mathbb{C}$ is the spectrum $\sigma(T)$ of T. The point spectrum is the subset of the spectrum for which $T - z$ is not one-to-one. In this case there exists a vector $v \in \mathscr{H}$ such that $(T - z)v = 0$, and z is an eigenvalue. The continuous spectrum consists of the points $z \in \sigma(T)$ for which $T - z$ is one-to-one, but for which $(T - z)\mathscr{H}$ is dense in $\mathscr{H}$, but not equal to $\mathscr{H}$. The remaining part of the spectrum is the residual spectrum. For self-adjoint operators, both bounded and unbounded, the spectrum only consists of the discrete and continuous spectrum.

The resolvent operator is defined in the same way for a bounded operator. For a bounded operator T the spectrum $\sigma(T)$ is a compact subset of the disk of radius $\|T\|$. Moreover, if T is self-adjoint, then $\sigma(T) \subset \mathbb{R}$, so that $\sigma(T) \subset [-\|T\|, \|T\|]$ and the spectrum consists of the point spectrum and the continuous spectrum.

3.7.4 The spectral theorem for bounded self-adjoint operators

A resolution of the identity, say E, of a Hilbert space $\mathscr{H}$ is a projection valued Borel measure on $\mathbb{R}$ such that for all Borel sets $A, B \subseteq \mathbb{R}$ we have (i) $E(A)$ is a self-adjoint projection, (ii) $E(A \cap B) = E(A)E(B)$, (iii) $E(\emptyset) = 0$, $E(\mathbb{R}) = \mathbf{1}_{\mathscr{H}}$, (iv) $A \cap B = \emptyset$ implies $E(A \cup B) = E(A) + E(B)$, and (v) for all $u, v \in \mathscr{H}$ the map $A \mapsto E_{u,v}(A) = \langle E(A)u, v \rangle$ is a complex Borel measure.

A generalisation of the spectral theorem for matrices is the following theorem for compact self-adjoint operators, see, e.g, [96, VI.3].

Theorem 3.76 (Spectral theorem for compact operators) *Let $T: \mathscr{H} \to \mathscr{H}$ be a compact self-adjoint linear map, then there exists a sequence of orthonormal vectors $(f_i)_{i \in I}$ such that $\mathscr{H}$ is the orthogonal direct sum of $\mathrm{Ker}(T)$ and the subspace spanned by $(f_i)_{i \in I}$ and there exists a sequence $(\lambda_i)_{i \in I}$ of non-zero real numbers converging to 0 so that*

$$Tv = \sum_{i \in I} \lambda_i \langle v, f_i \rangle f_i.$$

Here I is at most countable, since we assume $\mathscr{H}$ to be separable. In the case I is finite, the fact that the sequence $(\lambda_i)_{i \in I}$ is a null-sequence is automatic.

The following theorem is the corresponding statement for bounded self-adjoint operators, see [23, §X.2], [85, §12.22].

Theorem 3.77 (Spectral theorem) *Let $T : \mathscr{H} \to \mathscr{H}$ be a bounded self-adjoint linear map, then there exists a unique resolution of the identity such that $T = \int_{\mathbb{R}} t \, dE(t)$, i.e., $\langle Tu, v \rangle = \int_{\mathbb{R}} t \, dE_{u,v}(t)$. Moreover, E is supported on the spectrum $\sigma(T)$, which is contained in the interval $[-\|T\|, \|T\|]$. Moreover, any of the spectral projections $E(A)$, $A \subset \mathbb{R}$ a Borel set, commutes with T.*

A more general theorem of this kind holds for normal operators, i.e., for those operators satisfying $T^*T = TT^*$.

For the case of a compact operator, we have in the notation of Theorem 3.76 that for λ_i the spectral measure evaluated at $\{\lambda_k\}$ is the orthogonal projection on the corresponding eigenspace;

$$E(\{\lambda_k\})v = \sum_{i \in I; \lambda_i = \lambda_k} \langle v, f_i \rangle f_i.$$

Using the spectral theorem, we define for any continuous function f on the spectrum $\sigma(T)$ the operator $f(T)$ by $f(T) = \int_{\mathbb{R}} f(t) \, dE(t)$, i.e., $\langle f(T)u, v \rangle = \int_{\mathbb{R}} f(t) \, dE_{u,v}(t)$. Then $f(T)$ is bounded operator with norm equal to the supremum norm of f on the spectrum of T, i.e., $\|f(T)\| = \sup_{x \in \sigma(T)} |f(x)|$. This is known as the functional calculus for self-adjoint operators. In particular, for $z \in \rho(T)$ we see that $f \colon x \mapsto (x - z)^{-1}$ is continuous on the spectrum, and the corresponding operator is just the resolvent operator $R(z)$. The functional calculus can be extended to measurable functions, but then $\|f(T)\| \leq \sup_{x \in \sigma(T)} |f(x)|$.

The spectral measure can be obtained from the resolvent operators by the Stieltjes–Perron inversion formula, see [23, Thm. X.6.1].

Theorem 3.78 *The spectral measure of the open interval $(a,b) \subset \mathbb{R}$ is given by*

$$E_{u,v}\big((a,b)\big) = \lim_{\delta \downarrow 0} \lim_{\varepsilon \downarrow 0} \frac{1}{2\pi i} \int_{a+\delta}^{b-\delta} \langle R(x+i\varepsilon)u, v \rangle - \langle R(x-i\varepsilon)u, v \rangle \, dx.$$

The limit holds in the strong operator topology, i.e., $T_n x \to Tx$ for all $x \in \mathscr{H}$.

Note that the right-hand side of Theorem 3.78 is like the Cauchy integral formula, where we integrate over a rectangular contour.

3.7.5 Unbounded self-adjoint operators

Let $(T, \mathscr{D}(T))$, with $\mathscr{D}(T)$ the domain of T, be a densely-defined unbounded operator on $\mathscr{H}$. We can now define the adjoint operator $(T^*, \mathscr{D}(T^*))$ as follows. First write

$$\mathscr{D}(T^*) = \{v \in \mathscr{H} \mid u \mapsto \langle Tu, v \rangle \text{ is continuous on } \mathscr{D}(T)\}.$$

By the density of $\mathscr{D}(T)$ the map $u \mapsto \langle Tu, v \rangle$ for $v \in \mathscr{D}(T^*)$ extends to a continuous linear functional $\omega \colon \mathscr{H} \to \mathbb{C}$, and by the Riesz representation theorem there exists a unique $w \in \mathscr{H}$ such that $\omega(u) = \langle u, w \rangle$ for all $u \in \mathscr{H}$. Now the adjoint T^* is defined by $T^*v = w$, so that

$$\langle Tu, v \rangle = \langle u, T^*v \rangle \quad \text{for all } u \in \mathscr{D}(T), \text{ for all } v \in \mathscr{D}(T^*).$$

If T and S are unbounded operators on $\mathscr{H}$, then T extends S, notation $S \subset T$, if $\mathscr{D}(S) \subset \mathscr{D}(T)$ and $Sv = Tv$ for all $v \in \mathscr{D}(S)$. Two unbounded operators S and T are equal, $S = T$, if $S \subset T$ and $T \subset S$, or S and T have the same domain and act in the same way. In terms of the graph

$$\mathscr{G}(T) = \{(u, Tu) \mid u \in \mathscr{D}(T)\} \subset \mathscr{H} \times \mathscr{H}$$

we see that $S \subset T$ if and only if $\mathscr{G}(S) \subset \mathscr{G}(T)$. An operator T is closed if its graph is closed in the product topology of $\mathscr{H} \times \mathscr{H}$. The adjoint of a densely defined operator is a closed operator, since the graph of the adjoint is given as

$$\mathscr{G}(T^*) = \{(-Tu, u) \mid u \in \mathscr{D}(T)\}^\perp,$$

for the inner product $\langle (u, v), (x, y) \rangle = \langle u, x \rangle + \langle v, y \rangle$ on $\mathscr{H} \times \mathscr{H}$, see [85, §13.8].

A densely defined operator is symmetric if $T \subset T^*$, or,

$$\langle Tu, v \rangle = \langle u, Tv \rangle, \quad \text{for all } u, v \in \mathscr{D}(T).$$

A densely defined operator is self-adjoint if $T = T^*$, so that a self-adjoint operator is closed. The spectrum of an unbounded self-adjoint operator is contained in $\mathbb{R}$. Note that $\mathscr{D}(T) \subset \mathscr{D}(T^*)$, so that $\mathscr{D}(T^*)$ is a dense subspace and taking the adjoint once more gives $(T^{**}, \mathscr{D}(T^{**}))$ as the minimal closed extension of $(T, \mathscr{D}(T))$, i.e., any densely defined symmetric operator has a closed extension. We have $T \subset T^{**} \subset T^*$. We say that the densely defined symmetric operator is essentially self-adjoint if its closure is self-adjoint, i.e., if $T \subset T^{**} = T^*$.

In general, a densely defined symmetric operator T might not have self-adjoint extensions. This can be measured by the deficiency indices. Define, for $z \in \mathbb{C} \backslash \mathbb{R}$, the eigenspace

$$N_z = \{v \in \mathscr{D}(T^*) \mid T^*v = zv\}.$$

Then $\dim N_z$ is constant for $\Im z > 0$ and for $\Im z < 0$, [23, Thm. XII.4.19], and we put $n_+ = \dim N_i$ and $n_- = \dim N_{-i}$. The pair (n_+, n_-) are the deficiency indices for the densely defined symmetric operator T. Note that if T^* commutes with complex conjugation, then we automatically have $n_+ = n_-$. Here complex conjugation is an antilinear mapping $f = \sum_n f_n e_n$ to $\sum_n \overline{f_n} e_n$, where $\{e_n\}_n$ is an orthonormal basis of the separable Hilbert space $\mathscr{H}$. Note furthermore that if T is self-adjoint then $n_+ = n_- = 0$, since a self-adjoint operator cannot have non-real eigenvalues. Now the following holds, see [23, §XII.4].

Proposition 3.79 *Let* $(T, \mathscr{D}(T))$ *be a densely defined symmetric operator.*

(i) $\mathscr{D}(T^*) = \mathscr{D}(T^{**}) \oplus N_i \oplus N_{-i}$, *as an orthogonal direct sum with respect to the graph norm for* T^* *from* $\langle u, v \rangle_{T^*} = \langle u, v \rangle + \langle T^*u, T^*v \rangle$. *As a direct sum,* $\mathscr{D}(T^*) = \mathscr{D}(T^{**}) + N_z + N_{\bar{z}}$ *for general* $z \in \mathbb{C} \backslash \mathbb{R}$.

(ii) *Let* U *be an isometric bijection* $U \colon N_i \to N_{-i}$ *and define* $(S, \mathscr{D}(S))$ *by*

$$\mathscr{D}(S) = \{u + v + Uv \mid u \in \mathscr{D}(T^{**}), \ v \in N_i\}, \quad Sw = T^*w,$$

then $(S, \mathscr{D}(S))$ *is a self-adjoint extension of* $(T, \mathscr{D}(T))$, *and every self-adjoint extension of* T *arises in this way.*

In particular, T has self-adjoint extensions if and only if the deficiency indices are equal; $n_+ = n_-$. However, T has a self-adjoint extension to a bigger Hilbert space in case the deficiency indices are unequal, see e.g., [87, Prop. 3.17, Cor. 13.4], but we will not take this into account. T^{**} is a closed symmetric extension of T. We can also characterise the domains of the self-adjoint extensions of T using the sesquilinear form

$$B(u, v) = \langle T^*u, v \rangle - \langle u, T^*v \rangle, \qquad u, v \in \mathscr{D}(T^*),$$

then $\mathscr{D}(S) = \{u \in \mathscr{D}(T^*) \mid B(u, v) = 0, \text{ for all } v \in \mathscr{D}(S)\}$.

3.7.6 The spectral theorem for unbounded self-adjoint operators

With all the preparations of the previous subsection the Spectral Theorem 3.77 goes through in the unbounded setting, see [23, §XII.4], [85, Ch. 13].

Theorem 3.80 (Spectral theorem) *Let* $T \colon \mathscr{D}(T) \to \mathscr{H}$ *be an unbounded self-adjoint linear map, then there exists a unique resolution of the identity* *such that* $T = \int_{\mathbb{R}} t \, dE(t)$, *i.e.,* $\langle Tu, v \rangle = \int_{\mathbb{R}} t \, dE_{u,v}(t)$ *for* $u \in \mathscr{D}(T)$, $v \in \mathscr{H}$. *Moreover, E is supported on the spectrum* $\sigma(T)$, *which is contained in* $\mathbb{R}$. *For any bounded operator S that satisfies* $ST \subset TS$ *we have* $E(A)S = SE(A)$, $A \subset \mathbb{R}$ *a Borel set. Moreover, the Stieltjes–Perron inversion formula of Theorem 3.78 remains valid;*

$$E_{u,v}\big((a,b)\big) = \lim_{\delta \downarrow 0} \lim_{\varepsilon \downarrow 0} \frac{1}{2\pi i} \int_{a+\delta}^{b-\delta} \langle R(x+i\varepsilon)u, v \rangle - \langle R(x-i\varepsilon)u, v \rangle \, dx.$$

As in the case of bounded self-adjoint operators we can now define $f(T)$ for any measurable function f by

$$\langle f(T)u, v \rangle = \int_{\mathbb{R}} f(t) \, dE_{u,v}(t), \qquad u \in \mathscr{D}(f(T)), \ v \in \mathscr{H},$$

where $\mathscr{D}(f(T)) = \{u \in \mathscr{H} \mid \int_{\mathbb{R}} |f(t)|^2 \, dE_{u,u}(t) < \infty\}$ is the domain of $f(T)$. This makes $f(T)$ into a densely defined closed operator. In particular, if $f \in L^\infty(\mathbb{R})$, then $f(T)$ is a continuous operator, by the closed graph theorem. This in particular applies to $f(x) = (x-z)^{-1}$, $z \in \rho(T)$, which gives the resolvent operator.

3.8 Hints and answers for selected exercises

Exercise 3.10. See, e.g., [57, Lemma (3.3.3)].

Exercise 3.11. See, e.g., the proof of Proposition 3.39 or [57, Proposition (3.4.2)].

Exercise 3.14. See, e.g., [56].

Exercise 3.15. See [10, p. 583].

Exercise 3.24. See [57].

Exercise 3.47. See, e.g., [19, 45] or Section 3.5.1.

Exercise 3.50. See [12].

Exercise 3.52. See, e.g., [57, §3.1], Van Assche [47, §22.1] for comparable calculations.

Exercise 3.54. See, e.g., [6, §1].

Exercise 3.55. Mimick the proof of the Carleman condition for the scalar case, see [2, Ch. 1].

Exercise 3.66. See Tirao and Zurrián [95].

References

[1] V. Adamyan, Y. Berezansky, I. Gohberg, M. Gorbachuk, V. Gorbachuk, A. Kochubei, H. Langer, G. Popov, *Modern Analysis and Applications. The Mark Krein Centenary Conference*, Vol. 1: Operator theory and related topics, Vol. 2: Differential operators and mechanics. Operator Theory: Advances and Appl. 190, 191, Birkhäuser, Basel, 2009.

[2] N.I. Akhiezer, *The Classical Moment Problem and Some Related Questions in Analysis*, Hafner Publishing Co. New York, 1965.

[3] N. Aldenhoven, E. Koelink, P. Román, Matrix-valued orthogonal polynomials related to the quantum analogue of $(SU(2) \times SU(2), diag)$, *Ramanujan J.* (2017) 43, 243–311.

[4] W.A. Al-Salam, T.S. Chihara, Another characterization of the classical orthogonal polynomials, *SIAM J. Math. Anal.* 3 (1972) 65–70.

[5] G.E. Andrews, R.A. Askey, R. Roy, *Special Functions*, Cambridge University Press, Cambridge, 1999.

[6] A.I. Aptekarev, E.M. Nikishin, The scattering problem for a discrete Sturm–Liouville operator. *Mat. USSR Sbornik* 49 (1984), 325–355.

[7] R. Askey, Continuous q-Hermite polynomials when $q > 1$. In *q-Series and Partitions*, D. Stanton (ed), IMA Vol. Math. Appl. 18, Springer-Verlag, New York, 1989, pp. 151–158.

[8] R.A. Askey, M. Rahman, S.K. Suslov, On a general q-Fourier transformation with nonsymmetric kernels *J. Comput. Appl. Math.* 68 (1996), 25–55.

[9] R. Askey, J. Wilson, Some basic hypergeometric orthogonal polynomials that generalize Jacobi polynomials, *Mem. Amer. Math. Soc.* 54 (1985), no. 319.

[10] J.M. Berezanskiĭ, *Expansions in Eigenfunctions of Selfadjoint Operators*, Transl. Math. Monographs 17, Amer. Math. Soc., Providence RI, 1968.

[11] C. Berg, Markov's theorem revisited, *J. Approx. Theory* 78 (1994), 260–275.

[12] C. Berg, The matrix moment problem. In *Coimbra Lecture Notes on Orthogonal Polynomials*, A.J.P.L. Branquinho, A.P. Foulquié Moreno (eds), Nova Science Publishers Inc., New York, 2008, pp. 1–57

[13] J.T. Broad, Extraction of continuum properties from L^2 basis set matrix representations of the Schrödinger equation: the Sturm sequence polynomials and Gauss quadrature. In *Numerical Integration of Differential Equations and Large Linear Systems*, J. Hinze (ed), LNM 968, Springer-Verlag, Berlin–New York, 1982, pp. 53–70

[14] H. Buchwalter, G. Cassier, La paramétrisation de Nevanlinna dans le problème des moments de Hamburger, *Exposition. Math.* 2 (1984), 155–178.

[15] J.S. Christiansen, *Indeterminate Moment Problems within the Askey scheme*, PhD, University of Copenhagen, 2004.

[16] J.S. Christiansen, E. Koelink, Self-adjoint difference operators and classical solutions to the Stieltjes–Wigert moment problem, *J. Approx. Theory* 140 (2006), 1–26.

[17] J.S. Christiansen, E. Koelink, Self-adjoint difference operators and symmetric Al-Salam–Chihara polynomials, *Constr. Approx.* 28 (2008), 199–218.

[18] N. Ciccoli, E. Koelink, T.H. Koornwinder, q-Laguerre polynomials and big q-Bessel functions and their orthogonality relations, *Meth. Appl. Anal.* **6** (1999), 109–127.

[19] D. Damanik, A. Pushnitski, B. Simon, The analytic theory of matrix orthogonal polynomials, *Surveys in Approx. Th.* **4** (2008), 1–85.

[20] P. Deift, *Orthogonal Polynomials and Random Matrices: a Riemann–Hilbert Approach*, Courant Lecture Notes in Mathetics 3. New York University Courant Institute of Mathematical Sciences, American Mathematical Society, Providence RI, 1999.

[21] D.J. Diestler, The discretization of continuous infinite sets of coupled ordinary linear differential equations: application to the collision-induced dissociation of a diatomic molecule by an atom. In *Numerical Integration of Differential Equations and Large Linear Systems*, J. Hinze (ed), LNM 968, Springer-Verlag, Berlin–New York, 1982, pp. 40–52.

[22] J. Dieudonné, *History of Functional Analysis*, North-Holland Math. Stud. **49**, North-Holland, Amsterdam–New York, 1981.

[23] N. Dunford, J.T. Schwartz, *Linear Operators II: Spectral Theory*, Interscience Publisers, John Wiley & Sons, New York–London, 1963.

[24] A.J. Durán, *Exceptional orthogonal polynomials via Krall discrete polynomials*, this volume.

[25] A.J. Durán, Ratio asymptotics for orthogonal matrix polynomials, *J. Approx. Theory* **100** (1999), 304–344.

[26] A.J. Durán, P. López-Rodrıguez, Orthogonal matrix polynomials. In *Laredo Lectures on Orthogonal Polynomials and Special Functions*, R. Álvarez-Nodarse, F. Marcellán, W. Van Assche (eds), Nova Science Publishers Inc., New York, 2004, pp. 13-44

[27] A.J. Durán, W. Van Assche, Orthogonal matrix polynomials and higher-order recurrence relations, *Linear Algebra Appl.* **219** (1995), 261–280.

[28] J. Faraut, Un théorème de Paley–Wiener pour la transformation de Fourier sur un espace riemannien symétrique de rang un, *J. Funct. Anal.* **49** (1982), 230–268.

[29] G. Gasper, M. Rahman, *Basic Hypergeometric Series*, 2nd ed., Cambridge University Press, Cambridge, 2004.

[30] V.X. Genest, M.E.H. Ismail, L. Vinet, A. Zhedanov, Tridiagonalization of the hypergeometric operator and the Racah-Wilson algebra, *Proc. Amer. Math. Soc.* **144** (2016), 4441–4454.

[31] J.S. Geronimo, Scattering theory and matrix orthogonal polynomials on the real line, *Circuits Systems Signal Process.* **1** (1982), 471–495.

[32] W. Groenevelt, The Wilson function transform *Int. Math. Res. Not.* **2003** (2003), 2779–2817.

[33] W. Groenevelt, Laguerre functions and representations of $\mathfrak{su}(1,1)$, *Indagationes Math. N.S.* **14** (2003), 329–352.

[34] W. Groenevelt, *Tensor Product Representations and Special Functions*, PhD thesis, Delft University of Technology, 2004.

[35] W. Groenevelt, Bilinear summation formulas from quantum algebra representations, *Ramanujan J.* **8** (2004), 383–416.

[36] W. Groenevelt, The vector-valued big q-Jacobi transform, *Constr. Approx.* **29** (2009), 85–127.

[37] W. Groenevelt, Coupling coefficients for tensor product representations of quantum $SU(2)$, *J. Math. Phys.* **55** (2014), 101702, 35 pp.

[38] W. Groenevelt, M.E.H. Ismail, E. Koelink, Spectral decomposition and matrix-valued orthogonal polynomials, *Adv. Math.* **244** (2013), 91–105.

[39] W. Groenevelt, E. Koelink, Meixner functions and polynomials related to Lie algebra representations, *J. Phys. A: Math. Gen.* **35** (2002), 65–85.

[40] W. Groenevelt, E. Koelink, The indeterminate moment problem for the q-Meixner polynomials, *J. Approx. Theory* **163** (2011), 838–863.

[41] W. Groenevelt, E. Koelink, A hypergeometric function transform and matrix-valued orthogonal polynomials, *Constr. Approx.* **38** (2013), 277–309.

[42] W. Groenevelt, E. Koelink, J. Kustermans, The dual quantum group for the quantum group analog of the normalizer of $SU(1,1)$ in $SL(2,\mathbb{C})$, *Int. Math. Res. Not.* **2010** (2010), no. 7, 1167–1314.

[43] W. Groenevelt, E. Koelink, H. Rosengren, *Continuous Hahn functions as Clebsch–Gordan coefficients*. In *Theory and Applications of Special Functions. A Volume Dedicated to Mizan Rahman* M. E. H. Ismail, E. Koelink (eds), Developments in Mathematics, Vol. 13, Springer-Verlag, New York, 2005, pp. 221-284

[44] F.A. Grünbaum, I. Pacharoni, J. Tirao, Matrix valued spherical functions associated to the complex projective plane, *J. Funct. Anal.* **188** (2002), 350–441.

[45] F.A. Grünbaum, J. Tirao, The algebra of differential operators associated to a weight matrix, *Integral Eq. Operator Theory* **58** (2007), 449–475.

[46] R.A. Horn, C.R. Johnson, *Matrix Analysis*, Cambridge University Press, Cambridge, 1985.

[47] M.E.H. Ismail, *Classical and Quantum Orthogonal Polynomials in One Variable*, Cambridge University Press, Cambridge, 2009.

[48] M.E.H. Ismail, E. Koelink, The J-matrix method, *Adv. in Appl. Math.* **46** (2011), 379–395.

[49] M.E.H. Ismail, E. Koelink, Spectral properties of operators using tridiagonalization, *Anal. Appl. (Singap.)* **10** (2012), 327–343.

[50] M.E.H. Ismail, E. Koelink, Spectral analysis of certain Schrödinger operators, *SIGMA Symmetry Integrability Geom. Methods Appl.* **8** (2012), Paper 061, 19 pp.

[51] M.E.H. Ismail, D.R. Masson, q-Hermite polynomials, biorthogonal rational functions, and q-beta integrals, *Trans. Amer. Math. Soc.* **346** (1994), 63–116.

[52] T. Kakehi, Eigenfunction expansion associated with the Casimir operator on the quantum group $SU_q(1,1)$, *Duke Math. J.* **80** (1995), 535–573.

[53] T.H. Kjeldsen, The early history of the moment problem, *Historia Math.* **20** (1993), 19–44.

[54] R. Koekoek, P.A. Lesky, R.F. Swarttouw, *Hypergeometric Orthogonal Polynomials and their q-Analogues*, Springer-Verlag, Berlin–New York, 2010.

[55] R. Koekoek, R.F. Swarttouw, The Askey scheme of hypergeometric orthogonal polynomials and its q-analogue. Available online at http://aw. twi.tudelft.nl/~koekoek/askey.html, Report 98–17, Technical University Delft, 1998.

[56] E. Koelink, One-parameter orthogonality relations for basic hypergeometric series, *Indag. Math. (N.S.)* **14** (2003), 423–443.

[57] E. Koelink, Spectral theory and special functions. In *Laredo Lectures on Orthogonal Polynomials and Special Functions*, R. Álvarez-Nodarse, F. Marcellán, W. Van Assche (eds), Nova Science Publishers Inc., New York, 2004, pp. 45–84

[58] E. Koelink, J. Kustermans, A locally compact quantum group analogue of the normalizer of $SU(1,1)$ in $SL(2,\mathbb{C})$, *Comm. Math. Phys.* **233** (2003), 231–296.

[59] E. Koelink, M. van Pruijssen, P. Román, Matrix-valued orthogonal polynomials related to $(\mathrm{SU}(2) \times \mathrm{SU}(2), \mathrm{diag})$, *Int. Math. Res. Not.* **2012** (2012), 5673–5730.

[60] E. Koelink, M. van Pruijssen, P. Román, Matrix valued orthogonal polynomials related to $(\mathrm{SU}(2) \times \mathrm{SU}(2), \mathrm{diag})$, II, *Publ. RIMS Kyoto* **49** (2013), 271–312.

[61] E. Koelink, A.M. de los Ríos, P. Román, Matrix-valued Gegenbauer-type polynomials, *Constr. Approx.* **46** (2017), 459–487.

[62] E. Koelink, P. Román, Orthogonal vs. non-orthogonal reducibility of matrix-valued measures, *SIGMA Symmetry Integrability Geom. Methods Appl.* **12** (2016), Paper 008, 9 pp.

[63] E. Koelink, J.V. Stokman, The big q-Jacobi function transform, *Constr. Approx.* **19** (2003), 191–235.

[64] E. Koelink, J.V. Stokman, with an appendix by M. Rahman, Fourier transforms on the quantum $SU(1,1)$ group, *Publ. RIMS Kyoto* **37** (2001), 621–715.

[65] E. Koelink, J.V. Stokman, The Askey–Wilson function transform scheme. In *Special Functions 2000: Current Perspective and Future Directions*, J. Bustoz, M.E.H. Ismail, S.K. Suslov (eds), NATO Science Series II, Vol. 30, Kluwer Academic Publishers, Dordrecht, 2001, pp. 221–241

[66] E. Koelink, J.V. Stokman, The Askey–Wilson function transform, *Intern. Math. Res. Notices* **2001**, 22, 1203-1227.

[67] H.T. Koelink, J. Van Der Jeugt, Convolutions for orthogonal polynomials from Lie and quantum algebra representations, *SIAM J. Math. Anal.* **29** (1998), 794–822.

[68] H.T. Koelink, J. Van Der Jeugt, Bilinear generating functions for orthogonal polynomials, *Constr. Approx.* **15** (1999), 481–497.

[69] T.H. Koornwinder, *Jacobi functions and analysis on noncompact semisimple Lie groups*. In *Special Functions: Group-Theoretical Aspects and Applications*, R.A. Askey, T.H. Koornwinder, W. Schempp (eds), Math. Appl., Reidel Publishing Co., Dordrecht, 1984, pp. 1–85

[70] T.H. Koornwinder, Special orthogonal polynomial systems mapped onto each other by the Fourier–Jacobi transform. In *Orthogonal Polynomials and*

Applications, C. Brezinski, A. Draux, A.P. Magnus, P. Maroni, A. Ronveaux (eds), LNM 1171, Springer-Verlag, Berlin–New York, 1985 pp. 174–183.

[71] T.H. Koornwinder, Matrix elements of irreducible representations of $SU(2) \times SU(2)$ and vector-valued orthogonal polynomials, *SIAM J. Math. Anal.* **16** (1985), 602–613.

[72] M.G. Kreĭn, The fundamental propositions of the theory of representations of Hermitian operators with deficiency index (m,m), *Ukrain. Mat. Žurnal* **1** (1949), 3–66. English translation in *AMS Translations ser. 2* **97** (1970), 75–143.

[73] M. Kreĭn, Infinite J-matrices and a matrix-moment problem, *Doklady Akad. Nauk SSSR (N.S.)* **69** (1949), 125–128. (English translation by W. Van Assche at arXiv:1606.07754 [math.CA].)

[74] J. Labelle, *Tableau d'Askey*. In *Orthogonal Polynomials and Applications (Bar-le-Duc, 1984)*, C. Brezinski, A. Draux, A.P. Magnus, P. Maroni, A. Ronveaux (eds), LNM 1171, Springer-Verlag, Berlin–New York, 1985, pp. xxxvi–xxxvii.

[75] E.C. Lance, *Hilbert C^*-modules. A Toolkit for Operator Algebraists*, London Math. Soc. Lecture Note Series 210, Cambridge University Press, Cambridge, 1995.

[76] H.J. Landau, The classical moment problem: Hilbertian proofs, *J. Funct. Anal.* **38** (1980), 255–272.

[77] P.D. Lax, *Functional Analysis*, Wiley-Interscience, New York 2002.

[78] D.R. Masson, J. Repka, Spectral theory of Jacobi matrices in $\ell^2(\mathbb{Z})$ and the $\mathfrak{su}(1,1)$ Lie algebra, *SIAM J. Math. Anal.* **22** (1991), 1131–1146.

[79] A.F. Monna, *Functional Analysis in Historical Perspective*, John Wily & Sons, New York, 1973.

[80] Yu.A. Neretin, Some continuous analogues of the expansion in Jacobi polynomials, and vector-valued orthogonal bases, *Funct. Anal. Appl.* **39** (2005), 106–119.

[81] I. Pacharoni, I. Zurrián, Matrix Gegenbauer polynomials: the 2×2 fundamental cases, *Constr. Approx.* **43** (2016), 253–271.

[82] M. Reed, B. Simon, *Methods of Modern Mathematical Physics. I. Functional Analysis*, Academic Press, New York–London, 1972.

[83] M. Rosenberg, The square-integrability of matrix-valued functions with respect to a non-negative Hermitian measure, *Duke Math. J.* **31** (1964), 291–298.

[84] W. Rudin, *Real and Complex Analysis*, McGraw-Hill, New York–Toronto, 1966.

[85] W. Rudin, *Functional Analysis*, McGraw-Hill, New York–Toronto, 1973.

[86] M. Schechter, *Operator Methods in Quantum Mechanics*, North-Holland Publishing Co., New York–Amsterdam, 1981.

[87] K. Schmüdgen, *Unbounded Self-adjoint Operators on Hilbert Space*, GTM 265, Springer-Verlag, New York, 2012.

[88] J.A. Shohat, J.D. Tamarkin, *The Problem of Moments*, Math. Surveys 2, American Mathematical Society, Providence RI, 1943.

[89] B. Simon, The classical moment problem as a self-adjoint finite difference operator, *Adv. Math.* **137** (1998), 82–203.

[90] B. Simon, *Szegő's Theorem and its Descendants. Spectral Theory for L² perturbations of orthogonal polynomials*, Princeton University Press, Princeton NJ, 2011.

[91] T.J. Stieltjes, Recherches sur les fractions continues, *Annales de la Faculté des Sciences de Toulouse* **8** (1894), J.1–122, **9** (1895), A.1–47. Reprinted in *Œuvres Complètes – Collected Papers, vol. II*, G. van Dijk (ed), Springer-Verlag, Berlin–New York, 1993, pp. 406–570.

[92] M.H. Stone, *Linear Transformations in Hilbert Space*, AMS Colloq. Publ. **15**, American Mathematical Society, Providence RI, 1932.

[93] G. Szegő, *Orthogonal Polynomials*, 4th ed., AMS Colloquium Publ. **23**, American Mathematical Society, Providence RI, 1975.

[94] N.M. Temme, *Special Functions*, Wiley–Interscience, New York, 1996.

[95] J. Tirao, I. Zurrián. Reducibility of matrix weights, *Ramanujan J.* **45**, 349–374, (2018).

[96] D. Werner, *Funktionalanalyis*, 4th ed., Springer-Verlag, Berlin, 2002.

[97] N. Wiener, *The Fourier Integral and Certain of its Applications*, Cambridge University Press, Cambridge, 1933.

4

Elliptic Hypergeometric Functions

Hjalmar Rosengren

Abstract: The theory of hypergeometric functions is of central importance to special functions and orthogonal polynomials. It gives a unified framework for most special functions of classical mathematical physics (e.g., Bessel functions, Hermite polynomials and spherical harmonics). Connections between hypergeometric functions and mathematical fields such as representation theory and combinatorics continues to be an active area of research.

A (classical) hypergeometric series can be defined as a finite or infinite series $\sum_n a_n$ such that $a_{n+1}/a_n = f(n)$ for some rational function f. This classical theory is in fact the bottom level of a three-step hierarchy. When f is a trigonometric function, the series is called basic hypergeometric and when f is an elliptic function it is called elliptic hypergeometric. At all three levels, there are also hypergeometric functions defined by integrals rather than series.

It is interesting to contrast the historical development of basic and elliptic hypergeometric series. Simple examples of basic hypergeometric series appear in the work of Euler, and the theory was gradually developed during the 19th and 20th Century. By contrast, there is no hint about elliptic hypergeometric series in the literature until 1988. Then, Date et al. found examples of elliptic hypergeometric series in the context of higher-spin versions of Baxter's eight-vertex-solid-on-solid model. After that, very little happened until 1997, when Frenkel and Turaev gave the first explicit definition of elliptic hypergeometric series. Since the year 2000, the field has expanded rapidly, with more than 150 scientific papers devoted to mathematical and physical aspects of elliptic hypergeometric functions. One particularly intriguing development was the discovery by Dolan and Osborn in 2009 that certain indices of four-dimensional quantum field theories can be expressed as elliptic hypergeometric integrals. This has already led to a lot of work, including a large number of new conjectured integral evaluations.

The purpose of our lectures is to give an elementary introduction to the most fundamental aspects of elliptic hypergeometric functions. We will also give some indication of their historical origin in statistical mechanics. We will explain how elliptic hypergeometric functions can be used to construct biorthogonal rational functions, generalizing the famous Askey scheme of orthogonal polynomials. Apart from some general mathematical maturity, the only prerequisites will be elementary linear algebra and complex function theory.

Introduction

Various physical models and mathematical objects come in three levels: rational, trigonometric and elliptic. A classical example is Weierstrass's theorem, which states that a meromorphic one-variable function f satisfying an algebraic addition theorem; that is,

$$P\big(f(w), f(z), f(w+z)\big) \equiv 0, \qquad (4.1)$$

for some polynomial P, is either rational, trigonometric or elliptic. Here, trigonometric means that $f(z) = g(q^z)$ for some rational function g, where q is a fixed number. Writing $q = e^{2i\pi\eta}$, one may express f in terms of trigonometric functions. Elliptic means that f has two independent periods. If we let one of these periods tend to infinity, elliptic solutions of (4.1) degenerate to trigonometric ones. Letting the remaining period tend to infinity, we recover rational solutions.

Further examples of the hierarchy rational – trigonometric – elliptic are abundant in the context of classical and quantum integrable systems. Integrability is closely related to exact solvability, which means that some physically interesting quantities can be computed exactly. The meaning of the word "exactly" is loose but the answer may, for instance, involve hypergeometric functions. From this perspective, it is not surprising that there is a hierarchy of rational, trigonometric and elliptic hypergeometric functions. What is perhaps more surprising is that only the first two levels were known classically, with fundamental contributions by mathematicians such as Euler, Gauss, Cauchy and Heine. Elliptic hypergeometric functions appeared much later, first in the work of Date et al. [8] from 1988 and more explicitly in the 1997 paper [12] by Frenkel and Turaev. These authors only consider elliptic hypergeometric functions defined by finite sums. An important step forward was taken by Spiridonov [35, 37] who introduced elliptic hypergeometric integrals. Since the turn of the millenium, development has been rapid; in June 2017, the on-line bibliography [30] contained 186 entries. A new wave of interest from physicists was initiated by Dolan and Osborn [10], who found that elliptic hypergeometric integrals appear in the context of four-dimensional quantum field theories.

The purpose of the present notes is to give an elementary introduction to elliptic hypergeometric functions. They were written for the summer school OPSF-S6 on orthogonal polynomials and special functions, but I hope that they can be useful also in other contexts. I focus on motivating and exemplifying the main ideas, rather than giving a comprehensive survey of relevant results. The required background knowledge is modest and should be covered by a first course in complex analysis and some basic notions from linear and abstract algebra. Previous acquaintance with special functions will make the material easier to digest, but is not required.

Section 4.1 provides a brief introduction to elliptic functions. The presentation may seem idiosyncratic to some readers, but I believe it is close to the thinking of many contemporary researchers on elliptic integrable systems. Most textbooks follow a combination of Jacobi's and Weierstrass's approaches, which hides the elegance of the theory by cumbersome and (at least for our purposes) useless notation. My philosophy has been to completely avoid notation for specific elliptic functions. Another point is to consistently work with expressions for theta functions and elliptic functions as infinite products rather than series. In my opinion, this is more natural and often simplifies the theory.

The main part of the text is Section 4.2, where I give an introduction to elliptic hypergeometric sums and integrals. Although a large part of the literature deals with multivariable functions, I have decided to restrict to the one-variable theory. The main results are the Frenkel–Turaev summation and Spiridonov's elliptic beta integral evaluation. To give an indication of further results, I also present a quadratic summation and a Karlsson–Minton-type summation. Finally, in Section 4.3 I briefly explain the historical origin of elliptic hypergeometric functions in the context of solvable lattice models. In particular, I give a new proof of the fact that fused Boltzmann weights for Baxter's elliptic solid-on-solid model can be expressed as elliptic hypergeometric sums.

As they are based on a one-week course, the present lecture notes are very limited in scope. Let me provide some suggestions for further reading. More extensive introductions to elliptic hypergeometric functions are given in Chapter 11 of the textbook [13] and in the survey [39]. For multivariable elliptic hypergeometric sums, a natural starting point would be

[27], where some of the more accessible results are derived in an elementary manner much in the spirit of the present notes. Rains [25, 26] goes much further, introducing elliptic extensions of Okounkov's interpolation polynomials and Koornwinder–Macdonald polynomials. A succinct but rather comprehensive overview of multivariable elliptic hypergeometric functions is given in [32]. The reader interested in relations to the Sklyanin algebra and other elliptic quantum groups could start with [9, 19, 28]. One emerging research area is "elliptic combinatorics", where combinatorial objects are dressed with elliptic weight functions, see e.g., [7, 34]. In mathematical physics, there is much on-going activity on relations to four-dimensional supersymmetric quantum field theories. For someone with my own mathematical background, the literature is hard to get into, but I recommend the reader to have a look at [42, 43], where many (> 100) intriguing new integral identities are conjectured. A related topic is connections between elliptic hypergeometric integrals and two-dimensional lattice models with continuous spin, see [6, 40] for a start. Some of these recent applications in physics are briefly surveyed in [41]. Naturally, the above selection is biased by my own taste and interests. A more complete list of references can be found in [30].

Acknowledgements I would like to thank the organizers of OPSF-S6 for inviting me. I am grateful Gaurav Bhatnagar and Linnea Hietala, as well as the anonymous referees, for many useful comments on the manuscript. Finally, I thank all the students who followed the lectures and contributed to the course.

4.1 Elliptic functions

4.1.1 Definitions

The classical definition of an *elliptic function* is a meromorphic function f on $\mathbb{C}$ with two periods η and τ; that is,

$$f(z+\eta) = f(z+\tau) = f(z), \qquad z \in \mathbb{C}. \tag{4.2}$$

To avoid trivialities, one assumes that η and τ are non-zero and $\tau/\eta \notin \mathbb{R}$. Possibly interchanging η and τ, we may assume that $\mathrm{Im}(\tau/\eta) > 0$. Finally, after the change of variables $z \mapsto \eta z$, we may take $\eta = 1$. Thus, it is enough to consider meromorphic functions satisfying

$$f(z+1) = f(z+\tau) = f(z), \qquad z \in \mathbb{C}, \tag{4.3}$$

where $\mathrm{Im}(\tau) > 0$.

This "additive" definition goes back to Abel's memoir from 1827 and previous unpublished work of Gauss. We will mostly work with an equivalent, "multiplicative", definition. Note first that, if f is a meromorphic function satisfying $f(z+1) = f(z)$, then we can introduce a new function g by $f(z) = g(e^{2i\pi z})$. Then, g is meromorphic on the punctured plane $\mathbb{C}^* = \mathbb{C} \setminus \{0\}$. The periodicity $f(z + \tau) = f(z)$ is equivalent to $g(px) = g(x)$, where $p = e^{2i\pi\tau}$. Thus, we can alternatively define an *elliptic function* as a meromorphic function g on $\mathbb{C}^*$ such that $g(px) = g(x)$ for all x, where the *period* p satisfies $0 < |p| < 1$. We will distinguish the two definitions by using the terms *additively elliptic* and *multiplicatively elliptic*, respectively.

One can also give a coordinate-free definition of an elliptic function as an analytic function from a complex torus (compact Riemann surface of genus one) to a complex sphere (compact Riemann surface of genus zero). Our two definitions then correspond to two distinct choices of a complex coordinate on the torus. For the additive definition, we realize the torus as a parallelogram with opposite edges identified, for the multiplicative definition as an annulus with the inner and outer boundary circles identified.

Exercise 4.1 What can you say about functions satisfying (4.2) when $\tau/\eta \in \mathbb{R}$?

Exercise 4.2 What can you say about meromorphic functions with *three* additive periods?

4.1.2 Theta functions

We want to think of elliptic functions as analogues of rational functions, which can be factored as

$$f(z) = C\frac{(z-a_1)\cdots(z-a_m)}{(z-b_1)\cdots(z-b_n)}. \tag{4.4}$$

We will see that there is an analogous result for elliptic functions, where the building blocks (analogues of first degree polynomials) are known as *theta functions*.

How can we find an elliptic analogue of (4.4)? We expect that the individual factors on the right should correspond to zeroes and poles of f. In particular, the analogue of the building block $z - 0$ should vanish at $z = 0$. If we want to construct a solution to (4.3), it is natural to assume that it vanishes at the whole lattice $\mathbb{Z} + \tau\mathbb{Z}$. In multiplicative language (that is, writing $x = e^{2i\pi z}$, $p = e^{2i\pi\tau}$), we are looking for a function vanishing for $x \in p^{\mathbb{Z}}$. A naive way to construct such a function would be as an infinite product

$$\cdots(x-p^{-2})(x-p^{-1})(x-1)(x-p)(x-p^2)\cdots = \prod_{k=-\infty}^{\infty}(x-p^k).$$

However, this product diverges. For convergence, the factors should tend to 1 as $k \to \pm\infty$, but in fact they tend to x as $k \to \infty$ and behave as $-p^k$ when $k \to -\infty$. It is therefore natural to normalize the product by dividing factors with large k by x and factors with large negative k by $-p^k$. The details of how this is done are not important; we will make these substitutions for $k > 0$ and $k \le 0$, respectively, denoting the resulting function $\theta(x;p)$. That is,

$$\theta(x;p) = \prod_{k=-\infty}^{0}\left(1-\frac{x}{p^k}\right)\prod_{k=1}^{\infty}\left(1-\frac{p^k}{x}\right) = \prod_{k=0}^{\infty}(1-xp^k)\left(1-\frac{p^{k+1}}{x}\right).$$

Equivalently, in the standard notation

$$(a;p)_\infty = \prod_{k=0}^{\infty}(1-ap^k), \qquad (a_1,\ldots,a_m;p)_\infty = (a_1;p)_\infty\cdots(a_m;p)_\infty,$$

we have[1]

$$\theta(x;p) = (x,p/x;p)_\infty.$$

It will be convenient to use the shorthand notation

$$\theta(a_1,\ldots,a_m;p) = \theta(a_1;p)\cdots\theta(a_m;p)$$

as well as

$$\theta(ax^{\pm};p) = \theta(ax;p)\theta(a/x;p). \tag{4.5}$$

Note that the *trigonometric limit* $\tau \to i\infty$ corresponds to $p \to 0$. Then, our theta function reduces to the first degree polynomial $\theta(x;0) = 1-x$.

As all readers may not be so comfortable with infinite products, we give a direct proof of the following fact.

Lemma 4.3   *For $|p| < 1$, $(x;p)_\infty$ is an entire function of x with zeroes precisely at $x \in p^{\mathbb{Z}_{\le 0}}$.*

Proof   We start from the Taylor expansion

$$\log\frac{1}{1-x} = \sum_{n=1}^{\infty}\frac{x^n}{n}, \qquad |x| < 1,$$

---

[1] The reader who is more familiar with the four classical Jacobi theta functions should have a look at Exercise 4.44.

which gives

$$1 - x = \exp\left(-\sum_{n=1}^{\infty} \frac{x^n}{n}\right), \qquad |x| < 1.$$

Fixing x, pick N so that $|xp^{N+1}| < 1$. We can then write

$$(x;p)_{\infty} = \prod_{j=0}^{N}(1 - xp^j)\exp\left(-\sum_{j=N+1}^{\infty}\sum_{n=1}^{\infty}\frac{(xp^j)^n}{n}\right).$$

As the double series converges absolutely we may change the order of summation and obtain

$$(x;p)_{\infty} = \prod_{j=0}^{N}(1 - xp^j)\exp\left(-\sum_{n=1}^{\infty}\frac{x^n p^{(N+1)n}}{n(1-p^n)}\right).$$

The stated properties are then obvious. $\square$

We have the following immediate consequence.

Corollary 4.4 *The theta function $\theta(x;p)$ is analytic for $x \neq 0$ and has zeroes precisely at $x \in p^{\mathbb{Z}}$.*

Note that the theta function is *not* elliptic. In fact,

$$\frac{\theta(px;p)}{\theta(x;p)} = \frac{(px,1/x;p)_{\infty}}{(x,p/x;p)_{\infty}} = \frac{1 - 1/x}{1 - x} = -\frac{1}{x}.$$

This relation,

$$\theta(px;p) = -x^{-1}\theta(x;p), \tag{4.6}$$

is called *quasi-periodicity* of the theta function. More generally,

$$\theta(p^k x;p) = (-1)^k p^{-\binom{k}{2}}x^{-k}\theta(x;p), \qquad k \in \mathbb{Z}. \tag{4.7}$$

Another useful identity is

$$\theta(1/x;p) = -x^{-1}\theta(x;p).$$

Exercise 4.5 Prove that $\prod_{n=2}^{N}(1 - 1/n) \to 0$ as $N \to \infty$. (This shows that one has to be a little bit careful when showing that $\theta(x;p) \neq 0$ for $x \notin p^{\mathbb{Z}}$; it does not just follow from the fact that the factors are non-zero and tend to 1.)

Exercise 4.6 Prove (4.7).

Exercise 4.7 Show that $\theta(x^2;p^2) = \theta(x,-x;p)$ and $\theta(x;p) = \theta(x,px;p^2)$. Deduce the duplication formula

$$\theta(x^2;p) = \theta(x,-x,\sqrt{p}x,-\sqrt{p}x;p). \tag{4.8}$$

Exercise 4.8 Show that

$$\theta(-1, \sqrt{p}, -\sqrt{p}; p) = 2, \tag{4.9}$$

first using (4.8) and then by direct manipulation of infinite products.

Exercise 4.9 Show that $\theta(x; p)$ has the Laurent expansion[2]

$$\theta(x; p) = \frac{1}{(p; p)_\infty} \sum_{n=-\infty}^{\infty} (-1)^n p^{\binom{n}{2}} x^n.$$

4.1.3 Factorization of elliptic functions

We will now show that elliptic functions can be factored in terms of theta functions. We first recall the following elementary fact, which is easily proved by expanding f as a Laurent series.

Lemma 4.10 *If f is analytic on $\mathbb{C}^*$ and $f(px) = Cf(x)$ for some $C \in \mathbb{C}$ and $|p| < 1$, then $f(x) = Dx^N$ for some $D \in \mathbb{C}$ and $N \in \mathbb{Z}$.*

The following result will be useful.

Lemma 4.11 *Let f be multiplicatively elliptic with period p. Then, f has as many poles as zeroes, counted with multiplicity, in each period annulus $A = \{x; pr \le |x| < r\}$.*

Proof We can assume that there are no zeroes or poles at ∂A; otherwise we just vary r slightly. By the argument principle, if N is the number of zeroes and P the number of poles inside A, then

$$N - P = \int_{\partial A} \frac{f'(x)}{f(x)} \frac{dx}{2\pi i}.$$

Here, the inner boundary circle should be oriented clockwise and the outer circle counter-clockwise. To compare the two components we change x to px, as well as the orientation, at the inner boundary. Since $f(px) = f(x)$ gives $pf'(px) = f'(x)$, it follows that

$$N - P = \int_{|x|=r} \left(\frac{f'(x)}{f(x)} - \frac{pf'(px)}{f(px)} \right) \frac{dx}{2\pi i} = 0.$$

$\square$

We can now obtain the following fundamental result.

---

[2] This is known as Jacobi's triple product identity. A neat way to compute the prefactor is to compare the cases $x = i\sqrt{p}$ and $x = \sqrt{p}$, see [1, §10.4].

Theorem 4.12 *Any multiplicatively elliptic function f with period p can be factored as*

$$f(x) = C\frac{\theta(x/a_1,\ldots,x/a_n;p)}{\theta(x/b_1,\ldots,x/b_n;p)},\qquad(4.10a)$$

where $C \in \mathbb{C}$ and $a_j, b_j \in \mathbb{C}^*$ are subject to the condition

$$a_1 \cdots a_n = b_1 \cdots b_n.\qquad(4.10b)$$

Proof Pick a period annulus A such that f has no zeroes or poles at the boundary. By Lemma 4.11, there are as many zeroes as poles inside A (counted with multiplicity); denote them $a_1,\ldots,a_n$ and $b_1,\ldots,b_n$, respectively. All zeroes and poles are then of the form $p^{\mathbb{Z}}a_j$ and $p^{\mathbb{Z}}b_j$. Thus, by Corollary 4.4,

$$g(x) = f(x)\frac{\theta(x/b_1,\ldots,x/b_n;p)}{\theta(x/a_1,\ldots,x/a_n;p)}$$

is analytic for $x \neq 0$. Using (4.6) it follows that $g(px) = Dg(x)$, where $D = b_1 \cdots b_n/a_1 \cdots a_n$. By Lemma 4.10, $g(x) = Cx^N$ and $D = p^N$ for some $C \in \mathbb{C}$ and $N \in \mathbb{Z}$. We have now proved that

$$f(x) = Cx^N\frac{\theta(x/a_1,\ldots,x/a_n;p)}{\theta(x/b_1,\ldots,x/b_n;p)},\qquad p^N a_1 \cdots a_n = b_1 \cdots b_n.$$

Replacing a_1 by $p^{-N}a_1$ and using (4.7) we arrive at (4.10). $\square$

Note that the limit $p \to 0$ of (4.10) does not give all rational functions but only those of the form

$$f(x) = C\frac{(x-a_1)\cdots(x-a_n)}{(x-b_1)\cdots(x-b_n)},\qquad a_1 \cdots a_n = b_1 \cdots b_n;$$

that is, all rational functions such that $f(0) = f(\infty)$. From this perspective, it is natural to consider any function of the form

$$C\frac{\theta(x/a_1,\ldots,x/a_m;p)}{\theta(x/b_1,\ldots,x/b_n;p)}\qquad(4.11)$$

as "kind of elliptic". Indeed, such functions are sometimes called *elliptic functions of the third kind*. The special case $m = n$ is then called *elliptic functions of the second kind* and the true elliptic functions, satisfying in addition $a_1 \cdots a_n = b_1 \cdots b_n$, are *elliptic functions of the first kind*. Moreover, the special case $n = 0$ (corresponding to polynomials in the trigonometric limit) is referred to as *higher order theta functions*, or simply *theta functions*.

We state an extension of Theorem 4.12 to elliptic functions of the third kind, but leave the proof to the reader.

Theorem 4.13 *Let f be a meromorphic function on $\mathbb{C}^*$ satisfying $f(px) = tx^{-k}f(x)$, where $k \in \mathbb{Z}$, $t \in \mathbb{C}^*$ and $0 < |p| < 1$. Then, $f(x)$ can be factored as in (4.11), where $m = n + k$ and $(-1)^k a_1 \cdots a_m / b_1 \cdots b_n = t$.*

The special case of higher order theta functions is as follows.

Corollary 4.14 *Let f be an analytic function on $\mathbb{C}^*$ satisfying the equation $f(px) = tx^{-k}f(x)$, where $k \in \mathbb{Z}$, $t \in \mathbb{C}^*$ and $0 < |p| < 1$. Then, $k \geq 0$ and*

$$f(x) = C\theta(x/a_1, \ldots, x/a_k; p), \qquad (-1)^k a_1 \cdots a_k = t.$$

Exercise 4.15 Show that an elliptic function assumes each value (including ∞) an equal number of times in each period annulus $|pr| \leq |z| < r$.

Exercise 4.16 Prove Theorem 4.13.

4.1.4 The three-term identity

From the viewpoint of elliptic hypergeometric series, the most fundamental result on elliptic functions is a certain three-term relation for theta functions due to Weierstrass.[3] To motivate this relation, consider the space V of analytic functions on $\mathbb{C}^*$ satisfying $f(px) = f(x)/px^2$. By Corollary 4.14, it consists of functions of the form $C\theta(xa, px/a; p) = C\theta(ax^{\pm}; p)$ (recall the notation (4.5)). As it is described by two parameters, we expect that $\dim V = 2$, so V should have a basis of the form $\theta(bx^{\pm}; p)$ and $\theta(cx^{\pm}; p)$. Thus, we should be able to write

$$\theta(ax^{\pm}; p) = B\theta(bx^{\pm}; p) + C\theta(cx^{\pm}; p). \tag{4.12}$$

If we put $x = c$ we get $B = \theta(ac^{\pm}; p)/\theta(bc^{\pm}; p)$, provided that the denominator is non-zero. Similarly, $C = \theta(ab^{\pm}; p)/\theta(cb^{\pm}; p)$. Clearing the denominator, we are led to Weierstrass's identity

$$\theta(ax^{\pm}, bc^{\pm}; p) = \theta(bx^{\pm}, ac^{\pm}; p) + \frac{a}{c}\theta(cx^{\pm}, ba^{\pm}; p). \tag{4.13}$$

Although it is not hard to make the above argument rigorous, let us give an independent proof of (4.13) from scratch. Let $f(x)$ denote the difference of the left-hand and right-hand sides in (4.13). We may assume that all parameters are generic. It is clear that $f(c) = f(c^{-1}) = 0$. Since $f(px) = f(x)/px^2$, f vanishes at $c^{\pm}p^{\mathbb{Z}}$. It follows that $g(x) = f(x)/\theta(cx, c/x; p)$

---

[3] It has also been attributed to Riemann, but that seems incorrect [20].

is analytic for $x \neq 0$. Moreover, $g(px) = g(x)$. By Liouville's theorem[4], a non-constant elliptic function must have poles, so g is a constant. But since we also have $f(b) = 0$, that constant must be zero. Hence, f is identically zero.

Exercise 4.17 Deduce from (4.13) that $\theta(bx^{\pm}; p)$ and $\theta(cx^{\pm}; p)$ form a basis for the space V if and only if bc, $b/c \notin p^{\mathbb{Z}}$; in particular, $\dim V = 2$.

Exercise 4.18 Prove that the "elliptic number" $[z] = e^{-i\pi z}\theta(e^{2\pi i z}; e^{2\pi i \tau})$ satisfies

$$[z+a][z-a][b+c][b-c]$$
$$= [z+b][z-b][a+c][a-c] + [z+c][z-c][b+a][b-a]. \quad (4.14)$$

Deduce as limit cases that the same identity holds for the "trigonometric number" $[z] = \sin(z)$ and the "rational number" $[z] = z$.[5]

Exercise 4.19 Show that the trigonometric and rational numbers in the previous exercise satisfy

$$[b+c][b-c] = [a+c][a-c] + [b+a][b-a], \quad (4.15)$$

but that this is *not* true for the elliptic numbers (one way to see this is to shift one of the variables by τ).

4.1.5 Even elliptic functions

It will be useful to understand the structure of even additively elliptic functions. Note that if $f(z) = g(e^{2i\pi z})$ is such a function, then the corresponding multiplicatively elliptic function g satisfies $g(1/x) = g(x)$. By slight abuse of terminology, we will use the word *even* also for the latter type of symmetry.

Lemma 4.20 *Let g be an even multiplicatively elliptic function; that is, g is meromorphic on $\mathbb{C}^*$ and satisfies*

$$g(px) = g(1/x) = g(x). \quad (4.16)$$

Then, if $a^2 \in p^{\mathbb{Z}}$, the multiplicity of a as a zero or pole of g is even.

---

[4] In complex analysis you have probably learned that Liouville's theorem says that entire bounded functions are constant. What Liouville in fact proved was the weaker statement that entire elliptic functions are constant (this follows from Theorem 4.12). The generalization to bounded functions is due to Cauchy.

[5] Any entire function satisfying (4.14) is of one of these three forms, up to the transformations $[z] \mapsto ae^{bz^2}[cz]$, see [47, Ex. 20.38].

Proof It follows from (4.16) that $g(ax) = g(1/ax) = g(a/x)$. Suppose the Laurent expansion of g near a starts as $C(x-a)^j$. Then $(ax-a)^j \sim (a/x-a)^j$ as $x \to 1$, which is only possible for j even.   □

We can now give a counterpart of Theorem 4.12 for even elliptic functions.

Proposition 4.21 *Any even multiplicatively elliptic function g can be factored as*

$$g(x) = C\frac{\theta(c_1 x^{\pm}, \ldots, c_m x^{\pm}; p)}{\theta(d_1 x^{\pm}, \ldots, d_m x^{\pm}; p)}, \tag{4.17}$$

where $C \in \mathbb{C}$ and $c_j, d_j \in \mathbb{C}^*$.

Proof We first factor g as in (4.10). Since $g(1/a_1) = g(a_1) = 0$, we must have $a_1 a_j \in p^{\mathbb{Z}}$ for some j. Consider first the case $j \neq 1$. It then follows that $\theta(x/a_1, x/a_j; p) \sim \theta(a_1 x^{\pm}; p)$, where $\sim$ means equality up to a factor of the form Cx^k. If $j = 1$, we have from Lemma 4.20 that the multiplicity of a_1 as a zero of g is even. If the multiplicity is $2l$, this leads to a factor $\sim \theta(a_1 x^{\pm}; p)^l$. The same argument applies to the poles of g, so in conclusion we find that

$$g(x) = Cx^k \frac{\theta(c_1 x^{\pm}, \ldots, c_m x^{\pm}; p)}{\theta(d_1 x^{\pm}, \ldots, d_m x^{\pm}; p)}$$

for some integer k. It is easy to check that (4.16) holds if and only if $k = 0$.   □

This factorization has the following important consequence.

Proposition 4.22 *Let*

$$X(x) = \frac{\theta(ax^{\pm}; p)}{\theta(bx^{\pm}; p)}, \tag{4.18}$$

where $ab, a/b \notin p^{\mathbb{Z}}$. Then X generates the field of even multiplicatively elliptic functions; that is, any such function g is of the form $g(x) = p(X(x))$, with p a rational function.

Proof Starting from (4.17), we order the parameters so that $c_j \in bp^{\mathbb{Z}}$ if and only if $k+1 \leq j \leq m$ and $d_j \in bp^{\mathbb{Z}}$ if and only if $l+1 \leq j \leq m$. Using (4.7), we can assume that the remaining parameters c_j and d_j equal b, so that

$$g(x) = C\theta(bx^{\pm}; p)^{l-k} \frac{\theta(c_1 x^{\pm}, \ldots, c_k x^{\pm}; p)}{\theta(d_1 x^{\pm}, \ldots, d_l x^{\pm}; p)}.$$

On the other hand, (4.13) gives

$$X(x) - X(c) = \frac{a\theta(ba^{\pm}, cx^{\pm}; p)}{c\theta(bc^{\pm}, bx^{\pm}; p)}. \tag{4.19}$$

It follows that

$$g(x) = D\frac{(X(x) - X(c_1)) \cdots (X(x) - X(c_k))}{(X(x) - X(d_1)) \cdots (X(x) - X(d_l))},$$

where D is a non-zero constant. $\square$

That the field of even elliptic functions is generated by a single element can also be understood geometrically. Such functions live on the quotient S of the torus by the relation $x = x^{-1}$ (or, additively, $z = -z$). It can be shown that S is a sphere, so there must exist a holomorphic bijection (known as a uniformizing map) X from S to the Riemann sphere $\mathbb{C} \cup \{\infty\}$. The meromorphic functions on the Riemann sphere are simply the rational functions. Consequently, the meromorphic functions on S are precisely the rational functions in $X(x)$.

Exercise 4.23 By drawing pictures, convince yourself that the quotient S discussed in the text is a topological sphere.

Exercise 4.24 Generalize Proposition 4.21 to elliptic functions of the third kind.

Exercise 4.25 Deduce from (4.19) that

$$X'(x) = \frac{a(p; p)^2_{\infty}\theta(ba^{\pm}, x^2; p)}{x^2\theta(bx^{\pm}; p)^2}.$$

Exercise 4.26 Suppose that f is analytic on $\mathbb{C}^*$ and satisfies $f(px) = f(x)/x^n p^{2n}$ and $f(1/x) = f(x)$, where $n \in \mathbb{Z}_{\geq 0}$. Show that

$$f(x) = \theta(bx^{\pm}; p)^n p(X(x)),$$

where p is a unique polynomial of degree at most n and X is as in (4.18). In particular, the space of such functions has dimension $n + 1$.

4.1.6 Interpolation and partial fractions

Lagrange interpolation expresses a polynomial of degree $n - 1$ in terms of its values at n distinct points. If the points are $(y_j)_{j=1}^n$, one introduces the polynomials $(p_j)_{j=1}^n$ by

$$p_j(x) = \prod_{k=1, k\neq j}^n (x - y_k).$$

Note that $p_j(y_k) \neq 0$ if and only if $k = j$. Thus, if

$$p(x) = \sum_{j=1}^{n} c_j p_j(x), \qquad (4.20)$$

then $c_k = p(y_k)/p_k(y_k)$. In particular, choosing p as the zero polynomial it follows that $(p_j)_{j=1}^{n}$ are linearly independent. Counting dimensions, they form a basis for the polynomials of degree at most $n - 1$, so any such polynomial p can be expanded as in (4.20). This yields the interpolation formula

$$p(x) = \sum_{j=1}^{n} p(y_j) \prod_{k=1, k \neq j}^{n} \frac{x - y_k}{y_j - y_k}.$$

If we let $p(x) = \prod_{k=1}^{n-1}(x - z_k)$ and divide by $\prod_{k=1}^{n}(x - y_k)$ we get the partial fraction expansion

$$\frac{\prod_{k=1}^{n-1}(x - z_k)}{\prod_{k=1}^{n}(x - y_k)} = \sum_{j=1}^{n} \frac{\prod_{k=1}^{n-1}(y_j - z_k)}{\prod_{k=1, k \neq j}^{n}(y_j - y_k)} \cdot \frac{1}{x - y_j}, \qquad (4.21)$$

which is useful for integrating rational functions.

Lagrange interpolation also works for theta functions; in fact, we have already seen an example in §4.1.4. It may seem natural to replace the polynomials p_j with the theta functions $f_j(x) = \prod_{k \neq j} \theta(x/y_k; p)$. However, these functions satisfy different quasi-periodicity relations and thus don't span a very natural space. Instead, we take

$$f_j(x) = \theta(tx/y_j; p) \prod_{k \neq j} \theta(x/y_k; p).$$

If we first let $p = 0$ and then $t = 0$ we recover the polynomials p_j. We then have the following fact.

Proposition 4.27 *Let $t, y_1, \ldots, y_n \in \mathbb{C}^*$ be such that neither t nor y_j/y_k for $j \neq k$ is in $p^{\mathbb{Z}}$. Let V be the space of functions that are analytic for $x \neq 0$ and satisfy $f(px) = (-1)^n y_1 \cdots y_n t^{-1} x^{-n} f(x)$. Then, any $f \in V$ is uniquely determined by the values $f(y_1), \ldots, f(y_n)$ and given by*

$$f(x) = \sum_{j=1}^{n} f(y_j) \frac{\theta(tx/y_j; p)}{\theta(t; p)} \prod_{k=1, k \neq j}^{n} \frac{\theta(x/y_k; p)}{\theta(y_j/y_k; p)}. \qquad (4.22)$$

Proof With f_j as defined above, it is easy to see that $f_j \in V$. By the conditions on the parameters, $f_j(y_k) \neq 0$ if and only if $k = j$. Take now $f \in V$ and consider

$$g(x) = f(x) - \sum_{j=1}^{n} \frac{f(y_j)}{f_j(y_j)} f_j(x).$$

Then, $g \in V$ and g vanishes at $x = y_1, \ldots, y_n$. By quasi-periodicity, it vanishes at $p^{\mathbb{Z}} y_j$, so

$$h(x) = \frac{g(x)}{\theta(x/y_1, \ldots, x/y_n; p)}$$

is analytic on $\mathbb{C}^*$. Moreover, $h(px) = h(x)/t$. By Lemma 4.10, we can write $h(x) = Cx^N$ with $N \in \mathbb{Z}$. Since $t \notin p^{\mathbb{Z}}$, we must have $C = 0$ and consequently

$$f(x) = \sum_{j=1}^{n} \frac{f(y_j)}{f_j(y_j)} f_j(x).$$

Writing this out explicitly gives (4.22). □

Note that it follows that $(f_j)_{j=1}^n$ form a basis for V and, in particular, that $\dim V = n$. This is expected since, by Corollary 4.14, any $f \in V$ is of the form $f(x) = C \prod_{k=1}^n \theta(x/z_k)$, where $tz_1 \cdots z_n = y_1 \cdots y_n$, and is thus described by n free parameters. Inserting this factorization in Proposition 4.27 and dividing by $\prod_{k=1}^n \theta(x/y_k)$, we obtain the elliptic partial fraction expansion

$$\prod_{k=1}^{n} \frac{\theta(x/z_k; p)}{\theta(x/y_k; p)} = \sum_{j=1}^{n} \frac{\prod_{k=1}^{n} \theta(y_j/z_k; p)}{\prod_{k=1, k \neq j}^{n} \theta(y_j/y_k; p)} \cdot \frac{\theta(xy_1 \cdots y_n/y_j z_1 \cdots z_n; p)}{\theta(y_1 \cdots y_n/z_1 \cdots z_n, x/y_j; p)}.$$
(4.23)

If we let $x = z_n$, cancel all factors involving z_n and then introduce a new variable $z_n = y_1 \cdots y_n/z_1 \cdots z_{n-1}$, we obtain the elegant identity[6]

$$\sum_{j=1}^{n} \frac{\prod_{k=1}^{n} \theta(y_j/z_k; p)}{\prod_{k=1, k \neq j}^{n} \theta(y_j/y_k; p)} = 0, \qquad y_1 \cdots y_n = z_1 \cdots z_n, \qquad (4.24)$$

which is in fact equivalent to (4.23).

We will also need another elliptic partial fraction expansion, connected with even elliptic functions. Namely,

$$\frac{\prod_{k=1}^{n-1} \theta(xz_k^{\pm}; p)}{\prod_{k=1}^{n} \theta(xy_k^{\pm}; p)} = \sum_{j=1}^{n} \frac{\prod_{k=1}^{n-1} \theta(y_j z_k^{\pm}; p)}{\theta(xy_j^{\pm}; p) \prod_{k=1, k \neq j}^{n} \theta(y_j y_k^{\pm}; p)}.$$
(4.25)

The proof is left to the reader as Exercise 4.30. The special case $x = z_1$ is

$$\sum_{j=1}^{n} \frac{y_j \prod_{k=2}^{n-1} \theta(y_j z_k^{\pm}; p)}{\prod_{k=1, k \neq j}^{n} \theta(y_j y_k^{\pm}; p)} = 0, \qquad n \geq 2. \qquad (4.26)$$

Again, this is equivalent to the general case.

---

[6] The earliest reference I have found is [45, p. 46].

Exercise 4.28   Show that (4.23) is equivalent to (4.24) (with n replaced by $n+1$).

Exercise 4.29   Show that the case $n = 2$ of (4.23) and (4.25) are both equivalent to Weierstrass's identity (4.13).

Exercise 4.30   Give two proofs of (4.25). First, imitate the proof of (4.23), using the basis $f_j(x) = \prod_{k\neq j} \theta(y_k x^\pm; p)$ for an appropriate space of theta functions. Second, substitute $x = X(x)$, $y_k = X(y_k)$, $z_k = X(z_k)$ in (4.21), where X is as in (4.18).

Exercise 4.31   Show in two ways that, for $a_1 \cdots a_n b_1 \cdots b_{n+2} = 1$,

$$
\begin{aligned}
x^{-n-1} &\theta(a_1 x, \ldots, a_n x, b_1 x, \ldots, b_{n+2} x; p) \\
&- x^{n+1} \theta(a_1 x^{-1}, \ldots, a_n x^{-1}, b_1 x^{-1}, \ldots, b_{n+2} x^{-1}; p) \\
&= \frac{(-1)^n x \theta(x^{-2}; p)}{a_1 \cdots a_n} \sum_{k=1}^{n} \prod_{j=1}^{n+2} \theta(a_k b_j; p) \prod_{j=1, j\neq k}^{n} \frac{\theta(a_j x^\pm; p)}{\theta(a_k/a_j; p)}. \quad (4.27)
\end{aligned}
$$

First, prove that (4.27) is equivalent to (4.23), then prove it directly by viewing it as an interpolation formula for functions in x.

Exercise 4.32   Let V be the vector space of functions satisfying the conditions of Corollary 4.14. By expanding the elements of V as Laurent series and using Exercise 4.9, show that the functions $f_j(x) = x^j \theta(-p^j x^n/t; p^n)$, $j = 1, \ldots, n$, form a basis for V. (This gives an independent proof that $\dim V = n$.)

Exercise 4.33   Use (4.23) to prove Frobenius's determinant evaluation[7]

$$
\begin{aligned}
\det_{1\leq i,j\leq n} &\left(\frac{\theta(t x_i y_j; p)}{\theta(x_i y_j; p)} \right) \\
&= \frac{\theta(t; p)^{n-1} \theta(t x_1 \cdots x_n y_1 \cdots y_n; p) \prod_{1\leq i<j\leq n} x_j y_j \theta(x_i/x_j, y_i/y_j; p)}{\prod_{i,j=1}^{n} \theta(x_i y_j; p)}.
\end{aligned}
$$

4.1.7 Modularity and elliptic curves

We have now presented a minimum of material on elliptic functions needed for the remainder of these notes. We proceed to discuss some topics that are more peripheral to our main purpose, but so central in other contexts that we cannot ignore them completely. We start with the following important fact.

---

[7] See [17] for applications to multivariable elliptic hypergeometric series.

Theorem 4.34 *If τ and τ' are in the upper half-plane, the corresponding complex tori $E_\tau = \mathbb{C}/(\mathbb{Z} + \tau\mathbb{Z})$ and $E_{\tau'}$ are equivalent as Riemann surfaces if and only if $\tau' = (a\tau + b)/(c\tau + d)$ for some integers a, b, c, d with $ad - bc = 1$. If that is the case, then $\phi(z) = z/(c\tau + d)$ gives an equivalence $E_\tau \to E_{\tau'}$.*

In this context, τ and τ' are called *moduli*[8] and the map $\tau \mapsto \tau'$ a *modular transformation*. The rule for composing such maps is the same as for multiplying the matrices $\left(\begin{smallmatrix} a & b \\ c & d \end{smallmatrix}\right)$, which form the *modular group* $\mathrm{SL}(2, \mathbb{Z})$. (More precisely, the group of modular transformations is isomorphic to $\mathrm{SL}(2, \mathbb{Z})/\{\pm 1\}$.) The lemma states that we can parametrize the space of complex tori by identifying any two moduli related by a modular transformation; the corresponding quotient of the upper half-plane is called the moduli space. The term moduli space is nowadays used more generally for any space parametrizing geometric objects, but the origin of the term comes from the special case considered here.

Proof of Theorem 4.34   Let ϕ be an invertible analytic map from E_τ to $E_{\tau'}$. Equivalently, ϕ is an invertible entire function such that

$$\phi(z + \mathbb{Z} + \tau\mathbb{Z}) = \phi(z) + \mathbb{Z} + \tau'\mathbb{Z}.$$

Since any invertible entire function has the form $\phi(z) = Cz + D, C \neq 0$, we get

$$C(\mathbb{Z} + \tau\mathbb{Z}) = \mathbb{Z} + \tau'\mathbb{Z}. \tag{4.28}$$

Choosing 1 in the right-hand side gives $C = 1/(c\tau + d)$ for some integers c, d. If we then choose τ' in the right-hand side we find that $\tau' = (a\tau + b)/(c\tau + d)$ for some integers a, b. Interchanging the roles of τ and τ' gives $\tau = (A\tau + B)/(C\tau + D)$ for integers A, B, C, D, where necessarily

$$\begin{pmatrix} a & b \\ c & d \end{pmatrix} \begin{pmatrix} A & B \\ C & D \end{pmatrix} = \begin{pmatrix} 1 & 0 \\ 0 & 1 \end{pmatrix}.$$

Taking determinants gives $(ad - bc)(AD - BC) = 1$ and therefore $ad - bc = \pm 1$. We may compute

$$\mathrm{Im}(\tau') = \frac{ad - bc}{|c\tau + d|^2} \mathrm{Im}(\tau)$$

and conclude that $ad - bc = 1$. The final statement is easy to check. $\square$

Since meromorphic functions on E_τ can be factored in terms of theta

---

[8] The word modulus is also used for other quantities parametrized by τ, cf. Exercise 4.45.

functions, one would expect that $\theta(e^{2i\pi z}; e^{2i\pi\tau})$ transforms nicely under the modular group. Indeed, if $\left(\begin{smallmatrix} a & b \\ c & d \end{smallmatrix}\right) \in \mathrm{SL}(2,\mathbb{Z})$, then

$$
e^{-i\pi z/(c\tau+d)}\,\theta\!\left(e^{2i\pi z/(c\tau+d)}; e^{2i\pi(a\tau+b)/(c\tau+d)}\right)
$$
$$
= C e^{i\pi c z^2/(c\tau+d)-i\pi z}\,\theta\!\left(e^{2i\pi z}; e^{2i\pi\tau}\right), \tag{4.29}
$$

where $C = C(a,b,c,d;\tau)$ is independent of z. To see this, simply observe that if $f(z)$ is the quotient of the left-hand and right-hand sides, then f is an entire function, as both sides vanish precisely at $\mathbb{Z} + \tau\mathbb{Z}$. Moreover, f is periodic with periods 1 and τ so, by Liouville's theorem, f is constant.

Explicit expressions for the constant in (4.29) exist but are somewhat complicated [23, §80]. Usually, one is content with giving it for the transformations $\tau \mapsto \tau + 1$ and $\tau \mapsto -1/\tau$, which are known to generate the modular group. In the first case, $C = 1$ trivially; the second case is treated in Exercise 4.36.

In Section 4.1.5 we showed that the field of even elliptic functions is generated by a single element. A slight extension of the argument gives an analogous statement for general elliptic functions.

Theorem 4.35 *The field of all elliptic functions, with periods 1 and τ, is isomorphic to the quotient field $\mathbb{C}(X,Y)/(Y^2 - X(X-1)(X-\lambda))$, where $\lambda = \lambda(\tau)$ is a certain function (the lambda invariant).*[9]

Proof We will only sketch the proof. Consider first an odd multiplicatively elliptic function g; that is, $g(px) = -g(1/x) = g(x)$. Let $h(x) = xg(x)/\theta(x^2; p)$. Then, $h(1/x) = h(x)$ and $h(px) = p^2 x^4 h(x)$. As in the proof of Proposition 4.21 (cf. Exercise 4.24), it follows that

$$
g(x) = C \frac{\theta(x^2, c_1 x^{\pm}, \ldots, c_{m-2} x^{\pm}; p)}{x\,\theta(d_1 x^{\pm}, \ldots, d_m x^{\pm}; p)}.
$$

If we take $X(x)$ as in Proposition 4.22 and $Y(x) = \theta(x^2; p)/x\theta(bx^{\pm}; p)^2$, it follows that $g(x) = Y(x)p(X(x))$, where p is a rational function. As any function is the sum of an even and an odd function, and the even elliptic functions are rational in X, it follows that the field of all elliptic functions is generated by X and Y.[10]

Let us now consider the possible relations between X and Y. Since Y^2 is even, it is a rational function of X. If we choose a and b in the definition

---

[9] There are six possible choices for λ. The standard one is derived in Exercise 4.39:

$$
\lambda(\tau) = 16\sqrt{p}\,\frac{(-p;p)_\infty^8}{(-\sqrt{p};p)_\infty^8} = 16q - 128q^2 + 704q^3 - \cdots, \qquad q = \sqrt{p} = e^{i\pi\tau}.
$$

[10] By Exercise 4.25, we can always choose $Y(x) = xX'(x)$.

of X and Y as two of the numbers $\{1,-1,\sqrt{p},-\sqrt{p}\}$, then it follows from the proof of Proposition 4.22 that

$$Y^2 = CX(X-A)(X-B),$$

where A and B are the values of X at the remaining two numbers and $C \neq 0$. By rescaling X and Y, this can be reduced to $Y^2 = X(X-1)(X-\lambda)$. It remains to prove that there are no further algebraic relations between X and Y. To this end, suppose that $P(X(x),Y(x)) \equiv 0$ for some rational P. We can write

$$P(X,Y) = Q(X,Y^2) + YR(X,Y^2)$$

with Q and R rational. Let $E(X) = X(X-1)(X-\lambda)$. Taking the even and odd part of $P(X(x),Y(x))$, it follows that

$$Q(X(x),E(X(x))) = R(X(x),E(X(x))) = 0.$$

But, by the proof of Proposition 4.22, the zero elliptic function can only be obtained from the zero rational function, so $Q(X,E(X)) = R(X,E(X)) = 0$. By the factor theorem, $Q(X,Y)$ and $R(X,Y)$ are divisible by $Y - E(X)$, therefore $P(X,Y)$ is divisible by $Y^2 - E(X)$. $\qquad\square$

Theorem 4.35 shows that a complex torus is an algebraic variety, known as an *elliptic curve*.

Exercise 4.36 Compute the constant $C(\tau)$ in (4.29) for the transformation $\tau \mapsto -1/\tau$. Indeed, show that

$$e^{-i\pi z/\tau}\theta(e^{2i\pi z/\tau};e^{-2i\pi/\tau}) = -ie^{i\pi(\tau+\tau^{-1})/6+i\pi z^2/\tau-i\pi z}\theta(e^{2i\pi z};e^{2i\pi\tau}).$$

One way is to first specialize z to the three values $1/2$, $\tau/2$ and $(\tau+1)/2$. Then apply (4.9) to get an expression for $C(\tau)^3$ and finally let $\tau = i$ to find the correct branch of the cubic root.[11]

Exercise 4.37 By letting $z \to 0$ in Exercise 4.36, prove that *Dedekind's eta function* $\eta(\tau) = p^{1/24}(p;p)_\infty$, $p = e^{2i\pi\tau}$, satisfies $\eta(\frac{-1}{\tau}) = \sqrt{-i\tau}\eta(\tau)$. Using the fact that the modular group is generated by $\tau \mapsto \tau+1$ and $\tau \mapsto$

---

[11] This is the famous imaginary transformation of Jacobi. It is often proved in more complicated ways, by authors insisting on using the series representation for θ given in Exercise 4.9. The elementary proof sketched here is taken from the classic textbook [47].

$\frac{-1}{\tau}$, conclude that the *modular discriminant* $\Delta(\tau) = p(p;p)_\infty^{24}$ satisfies[12]

$$\Delta\left(\frac{a\tau+b}{c\tau+d}\right) = (c\tau+d)^{12}\Delta(\tau), \qquad \begin{pmatrix} a & b \\ c & d \end{pmatrix} \in \mathrm{SL}(2,\mathbb{Z}). \qquad (4.30)$$

Exercise 4.38 Let $\psi(x) = \sum_{n=1}^{\infty} e^{-n^2\pi x}$. Combining the previous two problems with Exercise 4.9, show that $\psi(1/x) = \sqrt{x}\psi(x) + (1-\sqrt{x})/2$. Next, show that

$$\pi^{-s/2}\Gamma(s/2)\zeta(s) = \int_0^\infty \psi(x)x^{s/2-1}\,dx, \qquad \mathrm{Re}(s) > 1,$$

where

$$\Gamma(s) = \int_0^\infty e^{-x}x^{s-1}\,dx, \qquad \zeta(s) = \sum_{n=1}^\infty \frac{1}{n^s}.$$

Deduce that

$$\pi^{-s/2}\Gamma(s/2)\zeta(s) = -\frac{1}{s} - \frac{1}{1-s} + \int_1^\infty \psi(x)\left(x^{s/2} + x^{(1-s)/2}\right)\frac{dx}{x}.$$

It follows that the analytic continuation of the left-hand side is invariant under $s \mapsto 1-s$.[13]

Exercise 4.39 Let

$$X(x) = \frac{4\sqrt{p}(-p;p)_\infty^4}{(-\sqrt{p};p)_\infty^4}\,\frac{\theta(-x^\pm;p)}{\theta(-\sqrt{p}x^\pm;p)}.$$

Using Exercise 4.25, show that

$$(xX'(x))^2 = -CX(X-1)(X-\lambda),$$

where

$$C = (p;p)_\infty^4(-\sqrt{p};p)_\infty^8, \qquad \lambda = 16\sqrt{p}\frac{(-p;p)_\infty^8}{(-\sqrt{p};p)_\infty^8}.$$

This gives an explicit expression for the lambda invariant in Theorem 4.35.

Exercise 4.40 Consider a pendulum released from an angle ϕ_0. Its subsequent movement is described by the initial value problem

$$\phi'' + \frac{g}{l}\sin\phi = 0, \qquad \phi(0) = \phi_0, \qquad \phi'(0) = 0.$$

---

[12] More generally, a *cusp form* is an analytic function on the upper half-plane that vanishes at $p = 0$ ($\tau = i\infty$) and satisfies (4.30) with 12 replaced by an arbitrary positive integer (the weight). The modular discriminant is the simplest cusp form in the sense that any cusp form of weight 12 is proportional to Δ and any cusp form of smaller weight vanishes identically [2].

[13] This is Riemann's original proof of the functional equation for Riemann's zeta function.

where ϕ is the displacement angle, g the gravitational acceleration and l the length of the pendulum. Let $Y = (1 - \cos\phi)/2 = \sin^2(\phi/2)$. Show that

$$(Y')^2 = \frac{4g}{l} Y(Y-1)(Y-Y_0), \qquad Y(0) = Y_0 = \sin^2(\phi_0/2).$$

Deduce that $Y(t) = X(e^{i\mu t})$, where X is as in Exercise 4.39 with $\mu = \sqrt{4g/lC}$ and p defined implicitly by $\lambda = Y_0$. In particular, deduce that the period of the pendulum is $2\pi\sqrt{lC/g}$.[14]

Exercise 4.41 Show that if $\lambda = \lambda(\tau)$ is as in Theorem 4.35 and

$$\tilde{\lambda}(\tau) = \lambda\left(\frac{a\tau + b}{c\tau + d}\right), \qquad \begin{pmatrix} a & b \\ c & d \end{pmatrix} \in \mathrm{SL}(2, \mathbb{Z}),$$

then

$$\tilde{\lambda} \in \left\{ \lambda, \ \frac{1}{1-\lambda}, \ \frac{\lambda-1}{\lambda}, \ \frac{1}{\lambda}, \ \frac{\lambda}{\lambda-1}, \ 1-\lambda \right\}.$$

Exercise 4.42 Using the previous exercise, show that $j(\tau) = 256(1 - \mu)^3/\mu^2$, where $\mu = \lambda(1-\lambda)$, is invariant under $\mathrm{SL}(2, \mathbb{Z})$.[15]

4.1.8 Comparison with classical notation

For the benefit of the reader who wants to compare our presentation with the classical approaches of Weierstrass or Jacobi (see, e.g., [47]) we provide a few exercises as starting points.

Exercise 4.43 In Weierstrass's theory of elliptic functions, the fundamental building block is the $\wp$-function

$$\wp(z; \tau) = \frac{1}{z^2} + \sum_{m,n \in \mathbb{Z}, \, (m,n) \neq (0,0)} \left(\frac{1}{(z+m+n\tau)^2} - \frac{1}{(m+n\tau)^2} \right).$$

Show that $\wp$ is an even additively elliptic function with periods 1 and τ and poles precisely at $\mathbb{Z} + \tau\mathbb{Z}$. Using Proposition 4.21, deduce that

$$\wp(z; \tau) = C\frac{\theta(ax^\pm; p)}{\theta(x^\pm; p)}, \qquad x = e^{2i\pi z}, \qquad p = e^{2i\pi\tau},$$

---

[14] One can give an elegant explanation of the fact that the pendulum is described by elliptic functions as follows. Assume that ϕ is analytic in t and consider $\psi(t) = \pi - \phi(it)$. Then, ψ satisfies the same initial value problem as ϕ but with ϕ_0 replaced by $\pi - \phi_0$. Since we expect that ϕ has a real period, so should ψ, which leads to an imaginary period for ϕ.

[15] This is Klein's famous invariant

$$j(\tau) = p^{-1} + 744 + 196884p + \cdots, \qquad p = e^{2i\pi\tau}.$$

One can show that it generates the field of modular functions; that is, the analytic functions on the upper half-plane that are invariant under $\mathrm{SL}(2, \mathbb{Z})$ [2].

for some constants C and a (depending on τ). Deduce from Theorem 4.35 that any additively elliptic function with periods 1 and τ is a rational function of $\wp(z;\tau)$ and its z-derivative $\wp'(z;\tau)$.

Exercise 4.44   Jacobi's theta functions are defined by the Fourier series[16]

$$\theta_1(z|\tau) = 2 \sum_{n=0}^{\infty} (-1)^n q^{(n+1/2)^2} \sin((2n+1)z),$$

$$\theta_2(z|\tau) = 2 \sum_{n=0}^{\infty} q^{(n+1/2)^2} \cos((2n+1)z),$$

$$\theta_3(z|\tau) = 1 + 2 \sum_{n=1}^{\infty} q^{n^2} \cos(2nz),$$

$$\theta_4(z|\tau) = 1 + 2 \sum_{n=1}^{\infty} (-1)^n q^{n^2} \cos(2nz),$$

where $q = e^{i\pi\tau}$. Using Exercise 4.9, show that these functions are related to $\theta(x;p)$ by

$$\theta_1(z|\tau) = iq^{1/4} e^{-iz} (q^2;q^2)_\infty \theta(e^{2iz};q^2),$$
$$\theta_2(z|\tau) = q^{1/4} e^{-iz} (q^2;q^2)_\infty \theta(-e^{2iz};q^2),$$
$$\theta_3(z|\tau) = (q^2;q^2)_\infty \theta(-qe^{2iz};q^2)_\infty,$$
$$\theta_4(z|\tau) = (q^2;q^2)_\infty \theta(qe^{2iz};q^2)_\infty.$$

Exercise 4.45   Fixing τ in the upper half-plane and $q = e^{i\pi\tau}$, let

$$k = \frac{\theta_2(0|\tau)^2}{\theta_3(0|\tau)^2} = 4q^{1/2} \frac{(-q^2;q^2)_\infty^4}{(-q;q^2)_\infty^4}.$$

The parameter k is called the *modulus* in Jacobi's theory of elliptic functions.[17] This theory is based on the functions

$$\mathrm{sn}(u,k) = \frac{\theta_3(0|\tau)\theta_1(z|\tau)}{\theta_2(0|\tau)\theta_4(z|\tau)},$$

$$\mathrm{cn}(u,k) = \frac{\theta_4(0|\tau)\theta_2(z|\tau)}{\theta_2(0|\tau)\theta_4(z|\tau)},$$

$$\mathrm{dn}(u,k) = \frac{\theta_4(0|\tau)\theta_3(z|\tau)}{\theta_3(0|\tau)\theta_4(z|\tau)},$$

where k is as above and $u = \theta_3(0|\tau)^2 z$. Prove that the trigonometric limits

---

[16] The reader is warned that there are several slightly different versions of these definitions; we follow the conventions of [47].

[17] Note that k is related to the lambda invariant of Exercise 4.39 by $\lambda = k^2$ (where $p = q^2$).

of these functions are

$$\operatorname{sn}(u,0) = \sin(u), \qquad \operatorname{cn}(u,0) = \cos(u), \qquad \operatorname{dn}(u,0) = 1.$$

Using, for instance, Weierstrass's identity (4.13), show that

$$\operatorname{sn}(u,k)' + \operatorname{cn}(u,k)^2 = k^2 \operatorname{sn}(u,k)^2 + \operatorname{dn}(u,k)'^2 = 1.$$

Note that

$$\operatorname{sn}(u,k)^2 = C \frac{\theta(e^{\pm 2iz};q^2)}{\theta(qe^{\pm 2iz};q^2)}$$

is of the form (4.18) up to a change of variables. As in Exercise 4.39, show that $y(u) = \operatorname{sn}(u,k)^2$ satisfies the differential equation

$$(y')^2 = 4y(1-y)(1-k^2 y).$$

Deduce that[18]

$$\frac{d}{du}\operatorname{sn}(u,k) = \sqrt{(1 - \operatorname{sn}(u,k)^2)(1 - k^2 \operatorname{sn}(u,k)^2)} = \operatorname{cn}(u,k)\operatorname{dn}(u,k).$$

4.2 Elliptic hypergeometric functions

4.2.1 Three levels of hypergeometry

A *classical hypergeometric series* is a series $\sum_k c_k$ such that c_{k+1}/c_k is a rational function of k. Examples include all the standard Taylor series encountered in calculus. For instance,

$$e^z = \sum_{k=0}^{\infty} \frac{z^k}{k!}, \qquad \sin z = \sum_{k=0}^{\infty} \frac{(-1)^k z^{2k+1}}{(2k+1)!}, \qquad \arctan z = \sum_{k=0}^{\infty} \frac{(-1)^k z^{2k+1}}{2k+1}, \tag{4.31}$$

have termwise ratio

$$\frac{z}{k+1}, \qquad -\frac{z^2}{(2k+2)(2k+3)}, \qquad -\frac{(2k+1)z^2}{2k+3},$$

---

[18] At least formally, it follows that the inverse of $u \mapsto \operatorname{sn}(u,k)$ is given by

$$\operatorname{sn}^{-1}(x) = \int_0^x \frac{dt}{\sqrt{(1-t^2)(1-k^2 t^2)}}.$$

This is known as the incomplete elliptic integral of the first kind. The related incomplete elliptic integral of the second kind,

$$\int_0^x \sqrt{\frac{1-k^2 t^2}{1-t^2}}\, dt,$$

appears when one tries to compute the arc-length of an ellipse. This is the somewhat far-fetched historical reason for the terminology "elliptic function."

respectively. Hypergeometric series can also be finite; an example is the binomial sum

$$\sum_{k=0}^{n} \binom{n}{k} z^k,$$

with termwise ratio

$$\frac{(n-k)z}{k+1}. \tag{4.32}$$

This also holds at the boundary of the summation range in the sense that, if we define $c_{n+1} = c_{-1} = 0$, then the ratio c_{k+1}/c_k vanishes for $k = n$ and is infinite for $k = -1$, in agreement with (4.32).

If $\sum_k c_k$ is classical hypergeometric, we can factor the termwise quotient as

$$f(k) = \frac{c_{k+1}}{c_k} = z\frac{(a_1+k)\cdots(a_r+k)}{(b_1+k)\cdots(b_{s+1}+k)}. \tag{4.33}$$

As was just explained, if we want to consider sums supported on $k \geq 0$, it is natural to assume that f has a pole at -1. Thus, we take $b_{s+1} = 1$. This is no restriction, as we recover the general case if in addition $a_r = 1$. Iterating (4.33) then gives

$$c_k = c_0\frac{(a_1)_k\cdots(a_r)_k}{k!(b_1)_k\cdots(b_s)_k}z^k,$$

where

$$(a)_k = a(a+1)\cdots(a+k-1).$$

Thus, any classical hypergeometric series, supported on $k \geq 0$, is a constant multiple of

$$_rF_s\left(\begin{matrix} a_1,\ldots,a_r \\ b_1,\ldots,b_s \end{matrix}; z\right) = \sum_{k=0}^{\infty} \frac{(a_1)_k\cdots(a_r)_k}{k!(b_1)_k\cdots(b_s)_k}z^k.$$

If $a_r = -n$ is a non-negative integer, this reduces to

$$_rF_s\left(\begin{matrix} a_1,\ldots,a_{r-1},-n \\ b_1,\ldots,b_s \end{matrix}; z\right) = \sum_{k=0}^{n} \frac{(a_1)_k\cdots(a_{r-1})_k(-n)_k}{k!(b_1)_k\cdots(b_s)_k}z^k,$$

which is the general form of a finite hypergeometric sum.

In the 19th century, it became apparent that there is a natural generalization of classical hypergeometric series called *basic hypergeometric series*. For these, the ratio c_{k+1}/c_k is a rational function of q^k for some fixed q (known as the base). If we write $q = e^{2i\pi\eta}$, the quotient can be expressed as a trigonometric function of η. Thus, another possible name would be "trigonometric hypergeometric series".

In the late 1980s, mathematical physicists discovered the even more general *elliptic hypergeometric series* while studying Baxter's elliptic solid-on-solid model [8]. For these, the termwise ratio c_{k+1}/c_k is an additively elliptic function of k. Partly because of convergence issues, the theory of infinite elliptic hypergeometric series is not well developed, so we will focus on finite sums. If one wants to consider elliptic extensions of infinite series, then it is usually better to define them by integrals, see §4.2.10.

The restriction to finite sums causes a slight problem of terminology. Namely, given any finite sum $\sum_{k=0}^{n} c_k$, one can obviously find a rational function f assuming the finitely many values $f(k) = c_{k+1}/c_k$. The same is true for trigonometric and elliptic functions. So it would seem that *all* finite sums are both hypergeometric, basic hypergeometric and elliptic hypergeometric. Although this is correct in principle, it is never a problem in practice. The sums we will consider depend on parameters, including n, in a way which will make the elliptic nature of the termwise ratio obvious. In particular, the degree of the relevant elliptic function (number of zeroes and poles up to periodicity) is independent of n.

Exercise 4.46 Express the Taylor series (4.31) in hypergeometric notation.

4.2.2 Elliptic hypergeometric sums

In this section, we discuss elliptic hypergeometric sums in general. As we explain in §4.2.4, this is *not* a natural class of functions. Sums appearing in practice, because of relations to physical models and/or interesting mathematical properties, are of more restricted types, such as the very well-poised sums defined in §4.2.4.

We want to consider sums $\sum_{k=0}^{n} c_k$ such that $c_{k+1}/c_k = f(k)$ for some additively elliptic function f. We denote the periods $1/\eta$ and τ/η and introduce the parameters $p = e^{2i\pi\tau}$, $q = e^{2i\pi\eta}$. We may then write $f(k) = g(q^k)$, where $g(px) = g(x)$. As usual, we assume that $0 < |p| < 1$. To avoid some (potentially interesting) degenerate situations, we will in addition assume that 1, η and τ are linearly independent over $\mathbb{Q}$ or, equivalently,

$$q^k \neq p^l, \qquad k, l \in \mathbb{Z}, \quad (k,l) \neq (0,0). \tag{4.34}$$

By Theorem 4.12, we can write

$$f(k) = \frac{c_{k+1}}{c_k} = z \frac{\theta(a_1 q^k, \ldots, a_{m+1} q^k; p)}{\theta(b_1 q^k, \ldots, b_{m+1} q^k; p)}, \qquad a_1 \cdots a_{m+1} = b_1 \cdots b_{m+1}.$$

$$\tag{4.35}$$

Let us introduce the *elliptic shifted factorial*

$$(a;q,p)_k = \prod_{j=0}^{k-1} \theta(aq^j;p),$$

for which we employ the condensed notation

$$(a_1,\ldots,a_m;q,p)_k = (a_1;q,p)_k \cdots (a_m;q,p)_k,$$

$$(ax^{\pm};q,p)_k = (ax;q,p)_k(a/x;q,p)_k. \tag{4.36}$$

Then, iterating (4.35) gives

$$c_k = c_0 \frac{(a_1,\ldots,a_{m+1};q,p)_k}{(b_1,\ldots,b_{m+1};q,p)_k} z^k.$$

Similarly as for classical hypergeometric sums, it is natural to require that $f(n) = 0$ and $f(-1) = \infty$. For this reason, we take $a_{m+1} = q^{-n}$ and $b_{m+1} = q$. This is no restriction, as we can recover the general case by choosing in addition $a_m = q$ and $b_m = q^{-n}$. We conclude that, up to an inessential prefactor, the general form of a finite elliptic hypergeometric sum is

$$_{m+1}E_m \left(\begin{matrix} q^{-n}, a_1,\ldots,a_m \\ b_1,\ldots,b_m \end{matrix} ; q, p; z \right) = \sum_{k=0}^n \frac{(q^{-n},a_1,\ldots,a_m;q,p)_k}{(q,b_1,\ldots,b_m;q,p)_k} z^k, \tag{4.37}$$

where the parameters should satisfy the *balancing condition*[19]

$$q^{-n} a_1 \cdots a_m = qb_1 \cdots b_m. \tag{4.38}$$

Note that (4.34) is equivalent to requiring that $(q;q,p)_k \neq 0$ for $k \geq 0$, so that the sum is well-defined for generic b_j.

Exercise 4.47   Show that

$$(a;q,p)_{n+k} = (a;q,p)_n(aq^n;q,p)_k,$$

$$(a;q,p)_{n-k} = \frac{(-1)^k q^{\binom{k}{2}}(a;q,p)_n}{(aq^{n-1})^k(q^{1-n}/a;q,p)_k},$$

$$(a;q,p)_k = (-1)^k a^k q^{\binom{k}{2}}(q^{1-k}/a;q,p)_k,$$

$$(p^m a;q,p)_n = \frac{(-1)^{mn}}{a^{mn} p^{n\binom{m}{2}} q^{m\binom{n}{2}}} (a;q,p)_n.$$

These identities are used routinely when manipulating elliptic hypergeometric series.

---

[19] The reader familiar with basic hypergeometric series may be uncomfortable with the position of q in this identity. We will return to this point at the end of §4.2.3.

Exercise 4.48 Show that

$$
{}_{m+1}E_m \left(\begin{matrix} q^{-n}, a_1, \ldots, a_m \\ b_1, \ldots, b_m \end{matrix} ; q, p; z \right)
$$

$$
= \frac{(-z)^n}{q^{\binom{n+1}{2}}} \frac{(a_1, \ldots, a_m; q)_n}{(b_1, \ldots, b_m; q)_n} {}_{m+1}E_m \left(\begin{matrix} q^{-n}, q^{1-n}/b_1, \ldots, q^{1-n}/b_m \\ q^{1-n}/a_1, \ldots, q^{1-n}/a_m \end{matrix} ; q, p; \frac{1}{z} \right).
$$

Exercise 4.49 Let $\left(\begin{smallmatrix} a & b \\ c & d \end{smallmatrix} \right) \in \mathrm{SL}(2, \mathbb{Z})$. Using (4.29), prove that

$$
{}_{m+1}E_m \left(\begin{matrix} q^{-n}, a_1, \ldots, a_m \\ b_1, \ldots, b_m \end{matrix} ; q, p; z \right) = {}_{m+1}E_m \left(\begin{matrix} \tilde{q}^{-n}, \tilde{a}_1, \ldots, \tilde{a}_m \\ \tilde{b}_1, \ldots, \tilde{b}_m \end{matrix} ; \tilde{q}, \tilde{p}; \tilde{z} \right),
$$

where the parameters are related by

$$
a_j = e^{2i\pi\eta\alpha_j}, \qquad b_j = e^{2i\pi\eta\beta_j}, \qquad p = e^{2i\pi\tau}, \qquad q = e^{2i\pi\eta},
$$

$$
\tilde{a}_j = e^{\frac{2i\pi\eta\alpha_j}{(c\tau+d)}}, \qquad \tilde{b}_j = e^{\frac{2i\pi\eta\beta_j}{(c\tau+d)}}, \qquad \tilde{p} = e^{\frac{2i\pi(a\tau+b)}{(c\tau+d)}}, \qquad \tilde{q} = e^{\frac{2i\pi\eta}{(c\tau+d)}},
$$

$$
\tilde{z} = \exp \left(\frac{i\pi c\eta^2}{c\tau+d} \left(1 - n^2 + \sum_{j=1}^m (\beta_j^2 - \alpha_j^2) \right) \right) z,
$$

and we assume[20] (without loss of generality in view of (4.38))

$$
-n + \sum_{j=1}^m \alpha_j = 1 + \sum_{j=1}^m \beta_j.
$$

4.2.3 The Frenkel–Turaev sum

Disregarding for a moment the condition (4.38), the case $p = 0$ of (4.37) is the basic hypergeometric sum

$$
{}_{m+1}\phi_m \left(\begin{matrix} q^{-n}, a_1, \ldots, a_m \\ b_1, \ldots, b_m \end{matrix} ; q; z \right) = \sum_{k=0}^n \frac{(q^{-n}, a_1, \ldots, a_m; q)_k}{(q, b_1, \ldots, b_m; q)_k} z^k, \qquad (4.39)
$$

where

$$
(a; q)_k = (a; q, 0)_k = (1 - a)(1 - aq) \cdots (1 - aq^{k-1}).
$$

Browsing the standard textbook [13], one will find a large (perhaps overwhelming) number of identities for such sums, such as the q-Saalschütz summation

$$
{}_3\phi_2 \left(\begin{matrix} q^{-n}, a, b \\ c, abc^{-1}q^{1-n} \end{matrix} ; q; q \right) = \frac{(c/a, c/b; q)_n}{(c, c/ab; q)_n} \qquad (4.40)
$$

[20] If in addition $n^2 + \sum_j \alpha_j^2 = 1 + \sum_j \beta_j^2$, then $\tilde{z} = z$. In this situation, ${}_{m+1}E_m$ is known as a *modular hypergeometric sum*.

and the Jackson summation

$$_8\phi_7\left(\begin{array}{c} a,q\sqrt{a},-q\sqrt{a},q^{-n},b,c,d,e \\ \sqrt{a},-\sqrt{a},aq^{n+1},aq/b,aq/c,aq/d,aq/e \end{array};q;q\right)$$
$$= \frac{(aq,aq/bc,aq/bd,aq/cd;q)_n}{(aq/b,aq/c,aq/d,aq/bcd;q)_n}, \qquad a^2q^{n+1}=bcde. \quad (4.41)$$

If we let a, d and e tend to zero in (4.41), in such a way that a/d and a/e is fixed, we recover (4.40). In fact, (4.41) is a top result in a hierarchy of summations, containing (4.40) and many other results as limit cases. In [13], the theory of basic hypergeometric series is built from bottom up, starting with the simplest results like the q-binomial theorem, using these to prove intermediate results like (4.40) and eventually proceeding to (4.41) and beyond. For elliptic hypergeometric series, such an approach is impossible since, as a rule of thumb, only "top level results" extend to the elliptic level. This may be the reason why it took so long before elliptic hypergeometric functions were discovered.

Let us return to (4.40) and explain why it does not exist at the elliptic level. As is emphasized in Ismail's lectures at the present summer school [15], it is natural to view (4.40) as a connection formula for the "Askey–Wilson monomials"

$$\phi_n(y;a) = (ax,a/x;q)_n, \qquad y = \frac{x+x^{-1}}{2}$$

(equivalently, $x = e^{i\theta}$ and $y = \cos\theta$). Indeed, if we replace a by ax and b by a/x, (4.40) takes the form

$$\phi_n(y;c/a) = \sum_{k=0}^n C_k\,\phi_k(y;a).$$

One might try to obtain an elliptic extension by considering

$$(cx^{\pm}/a;q,p)_n = \sum_{k=0}^n C_k(ax^{\pm};q,p)_k.$$

However, such an expansion cannot exist, as the terms satisfy different quasi-periodicity conditions. For instance, the case $n = 1$ can be written

$$C_0 = \theta(cx^{\pm}/a;p) - C_1\theta(ax^{\pm};p).$$

Here, both terms on the right satisfy $f(px) = f(x)/px^2$, wheras the left-hand side is independent of x. This is absurd (except in the trivial cases $c \in p^{\mathbb{Z}}$ and $c \in a^2p^{\mathbb{Z}}$, when we may take $C_0 = 0$).

Let us now consider (4.41) as a connection formula. Replacing b by bx and c by b/x, it can be written as

$$(aqx^{\pm}/bd;q)_n = \sum_{k=0}^{n} C_k(bx^{\pm};q)_k(q^{-n}bx^{\pm}/a;q)_{n-k}.$$

This seems more promising to extend to the elliptic level. Changing parameters, we will consider the expansion problem

$$h_n(x;a) = \sum_{k=0}^{n} C_k^n h_k(x;b)h_{n-k}(x;c), \qquad (4.42)$$

where $h_n(x;a) = (ax^{\pm};q,p)_n$. Here, each term satisfies the same quasi-periodicity relation. Following [29], we will compute the coefficients in (4.42) by induction on n, in analogy with a standard proof of the binomial theorem.

When $n = 0$, (4.42) holds trivially with $C_0^0 = 1$. Assuming that (4.42) holds for fixed n, we write

$$h_{n+1}(x;a) = \theta(aq^n x^{\pm};p) \sum_{k=0}^{n} C_k^n h_k(x;b)h_{n-k}(x;c).$$

To proceed, we must split

$$\theta(aq^n x^{\pm};p) = B\theta(bq^k x^{\pm};p) + C\theta(cq^{n-k} x^{\pm};p).$$

By (4.13), such a splitting exists for generic parameters and is given by

$$B = \frac{\theta(acq^{2n-k}, aq^k/c;p)}{\theta(bcq^n, bq^{2k-n}/c;p)}, \qquad C = -\frac{bq^{2k-n}\theta(abq^{n+k}, aq^{n-k}/b;p)}{c\,\theta(bcq^n, bq^{2k-n}/c;p)}$$

(note that we do not need to remember the exact form of (4.13); we can simply compute B and C by substituting $x = cq^{n-k}$ and $x = bq^k$). Then, (4.42) holds with n replaced by $n+1$ and

$$C_k^{n+1} = \frac{\theta(acq^{2n-k+1}, aq^{k-1}/c;p)}{\theta(bcq^n, bq^{2k-2-n}/c;p)} C_{k-1}^n - \frac{bq^{2k-n}\theta(abq^{n+k}, aq^{n-k}/b;p)}{c\,\theta(bcq^n, bq^{2k-n}/c;p)} C_k^n. \qquad (4.43)$$

This is an elliptic extension of Pascal's triangle. We claim that the solution is the "elliptic binomial coefficient"

$$C_k^n = q^{n-k}\frac{\theta(q^{2k-n}b/c;p)(q, ab, ac;q,p)_n\,(a/c;q,p)_k(a/b;q,p)_{n-k}}{\theta(b/c;p)(bc;q,p)_n(q, ab, qb/c;q,p)_k(q, ac, qc/b;q,p)_{n-k}}. \qquad (4.44)$$

Indeed, plugging (4.44) into (4.43) we obtain after cancellation

$$\theta(q^{n+1}, abq^n, acq^n, q^{2k-n-1}b/c; p)$$
$$= \theta(q^k, abq^{k-1}, acq^{2n-k+1}, q^k b/c; p)$$
$$+ q^k \theta(q^{n+1-k}, abq^{n+k}, acq^{n-k}, q^{k-n-1}b/c; p).$$

The reader should check that this is again an instance of (4.13).

Plugging (4.44) into (4.42), simplifying and changing parameters yields the following result, which is the most fundamental and important fact about elliptic hypergeometric sums. It was first obtained by Frenkel and Turaev [12], but with additional restrictions on the parameters it occurs somewhat implicitly in [8].

Theorem 4.50   *When $a^2 q^{n+1} = bcde$, we have the summation formula*

$$\sum_{k=0}^{n} \frac{\theta(aq^{2k}; p)}{\theta(a; p)} \frac{(a, q^{-n}, b, c, d, e; q, p)_k}{(q, aq^{n+1}, aq/b, aq/c, aq/d, aq/e; q, p)_k} q^k$$

$$= {}_{10}E_9\left(\begin{matrix} a, q\sqrt{a}, -q\sqrt{a}, q\sqrt{pa}, -q\sqrt{pa}, q^{-n}, b, c, d, e \\ \sqrt{a}, -\sqrt{a}, \sqrt{pa}, -\sqrt{pa}, aq^{n+1}, aq/b, aq/c, aq/d, aq/e \end{matrix}; q, p; q\right)$$

$$= \frac{(aq, aq/bc, aq/bd, aq/cd; q, p)_n}{(aq/b, aq/c, aq/d, aq/bcd; q, p)_n}. \tag{4.45}$$

Here, we used Exercise 4.7 to write

$$\frac{\theta(aq^{2k}; p)}{\theta(a; p)} = \frac{\theta(\sqrt{a}q^k, -\sqrt{a}q^k, \sqrt{pa}q^k, -\sqrt{pa}q^k; p)}{\theta(\sqrt{a}, -\sqrt{a}, \sqrt{pa}, -\sqrt{pa}; p)}$$

$$= \frac{(q\sqrt{a}, -q\sqrt{a}, q\sqrt{pa}, -q\sqrt{pa}; p, q)_k}{(\sqrt{a}, -\sqrt{a}, \sqrt{pa}, -\sqrt{pa}; p, q)_k}. \tag{4.46}$$

Note that the case $p = 0$,

$$\frac{1 - aq^{2k}}{1 - a} = \frac{(q\sqrt{a}, -q\sqrt{a}; q)_k}{(\sqrt{a}, -\sqrt{a}; q)_k}$$

contains two factors instead of four, so the ${}_{10}E_9$ reduces to the ${}_8\phi_7$ sum in (4.41). This explains an apparent discrepancy between the balancing conditions for basic and elliptic hypergeometric series. Namely, the balancing condition for (4.37) is $q^{-n}a_1 \cdots a_m = qb_1 \cdots b_m$, whereas (4.39) is called balanced if $q^{1-n}a_1 \cdots a_m = b_1 \cdots b_m$. With these definitions, the left-hand sides of (4.41) and (4.45) are both balanced. The shift by q^2 comes from the additional numerator parameters $\pm q\sqrt{pa}$ and denominator parameters $\pm\sqrt{pa}$, which become invisible in the trigonometric limit.

Exercise 4.51 Show that the case $n = 1$ of (4.40) is equivalent to the trigonometric case of (4.15) and the case $n = 1$ of (4.41) to the trigonometric case of (4.14). (This gives another explanation for why the Saalschütz summation does not exist at the elliptic level.)

Exercise 4.52 In (4.45), multiply a, d and e by $\sqrt{p}$ and then let $p \to 0$. Show that you obtain (4.40) in the limit. (In this sense, the Saalschütz summation *does* exist at the elliptic level.)

Exercise 4.53 As we have remarked, (4.41) contains (4.40) as a limit case. Why can't we take the corresponding limit of (4.45) when $p \neq 0$?

Exercise 4.54 Assuming (4.34), show that the $(h_k(x;a)h_{n-k}(x;b))_{k=0}^n$ are linearly independent functions for generic a and b. Combining this with the dimension count in Exercise 4.26, deduce *a priori* that the expansion (4.42) exists generically.

Exercise 4.55 Show that

$$\lim_{q \to 1} \frac{(q^{-n};q,p)_k}{(q;q,p)_k} = (-1)^k \binom{n}{k}$$

and verify that the limit $q \to 1$ of (4.45), all other parameters being fixed, is equivalent to the classical binomial theorem.

4.2.4 Well-poised and very well-poised sums

In concrete examples, the interesting variable in (4.37) is not z; indeed, it is usually fixed to $z = q$. It is more fruitful to consider (4.37) as a function of the parameters a_j and b_j. When viewed in this way, (4.37) is not very natural, as the terms have different quasi-periodicity. For instance, consider the behaviour under the simultaneous shift $a_1 \mapsto pa_1$, $b_1 \mapsto pb_1$ (which preserves (4.38)). By Exercise 4.47, the kth term in (4.37) is then multiplied by $(b_1/a_1)^k$, which in general depends on k. By contrast, it is easy to see that each term in the sum

$$\sum_{k=0}^n \frac{(x_0x_1, -x_0/x_1; q, p)_k}{(-x_0x_1, x_0/x_1; q, p)_k} z^k$$

is invariant under either of the transformations $x_0 \mapsto px_0$ and $x_1 \mapsto px_1$. This is an example of what Spiridonov [36] calls a *totally elliptic* hypergeometric sum. To describe all such sums explicitly seems to be an open problem. We will be content with giving an interesting special case.

Proposition 4.56   *The sum*

$$\sum_k \frac{(x_0x_1,\ldots,x_0x_{m+1};q,p)_k}{(x_0/x_1,\ldots,x_0/x_{m+1};q,p)_k} z^k, \tag{4.47}$$

subject to the balancing condition $x_1^2 \cdots x_{m+1}^2 = 1$, is totally elliptic in the sense that each term is invariant under the simultaneous shifts $x_j \mapsto p^{\alpha_j} x_j$, where $\alpha_0, \ldots, \alpha_{m+1}$ are integers subject to $\alpha_1 + \cdots + \alpha_{m+1} = 0$.

Proof   By Exercise 4.47, under such a shift the kth term is multiplied by

$$\prod_{j=1}^{m+1} \frac{(-1)^{\alpha_0-\alpha_j}(x_0/x_j)^{\alpha_0-\alpha_j} p^{k\binom{\alpha_0-\alpha_j}{2}} q^{(\alpha_0-\alpha_j)\binom{k}{2}}}{(-1)^{\alpha_0+\alpha_j}(x_0x_j)^{\alpha_0+\alpha_j} p^{k\binom{\alpha_0+\alpha_j}{2}} q^{(\alpha_0+\alpha_j)\binom{k}{2}}}$$

$$= \prod_{j=1}^{m+1} \frac{1}{x_0^{2\alpha_j} x_j^{2\alpha_0} p^{k(2\alpha_0-1)\alpha_j} q^{2\alpha_j\binom{k}{2}}} = 1.$$

$\square$

Following [36], we call (4.47) a *well-poised* elliptic hypergeometric sum. Note that convergence is not an issue, as Proposition 4.56 is just a statement about the individual terms. As a first step towards considering finite sums, suppose that the summation is over non-negative integers and that $x_0/x_{m+1} = q$. Writing $a = x_0^2/q$ and $b_j = x_j x_0$, with $j = 1, \ldots, m$, we arrive at the series

$$\sum_{k=0}^{\infty} \frac{(a, b_1, \ldots, b_m; q, p)_k}{(q, aq/b_1, \ldots, aq/b_m; q, p)_k} z^k, \tag{4.48}$$

where the balancing condition is $b_1^2 \cdots b_m^2 = q^{m+1} a^{m-1}$.

The Frenkel–Turaev sum (4.45) is of the form (4.48). Indeed, the factor (4.46) can alternatively be written

$$\frac{\theta(aq^{2k}; p)}{\theta(a; p)} = \frac{(-1)^k}{q^k} \frac{(q\sqrt{a}, -q\sqrt{a}, q\sqrt{a/p}, -q\sqrt{ap}; p, q)_k}{(\sqrt{a}, -\sqrt{a}, \sqrt{ap}, -\sqrt{a/p}; p, q)_k}, \tag{4.49}$$

which looks less symmetric but fits the well-poised pattern. Well-poised series containing the factor (4.49) are called *very well-poised*. Clearly, any well-poised series can be considered as very well-poised, since one can artificially introduce this factor and then remove it again by choosing $(b_1, \ldots, b_4) = (\sqrt{a}, -\sqrt{a}, \sqrt{ap}, -\sqrt{a/p})$. Typically, the z-variable in a very well-poised series is fixed to q. We will therefore use the notation

$$_{m+1}V_m(a;b_1,\ldots,b_{m-4};q,p)$$
$$= \sum_{k=0}^{\infty} \frac{\theta(aq^{2k};p)}{\theta(a;p)} \frac{(a,b_1,\ldots,b_{m-4};q,p)_k}{(q,aq/b_1,\ldots,aq/b_{m-4};q)_k} q^k.$$

The balancing condition for this series is

$$b_1^2 \cdots b_{m-4}^2 = q^{m-7} a^{m-5}.$$

As before, because of convergence problems one usually restricts to the case when $b_{m-4} = q^{-n}$, when n is a non-negative integer.

Exercise 4.57   Show that the sum

$$\sum_k \frac{(x_1^2,\ldots,x_m^2;q,p)_k}{(-x_1^2,\ldots,-x_m^2;q,p)_k} z^k$$

is totally elliptic in the sense of being invariant under any shift $x_j \mapsto px_j$. (This shows that not all totally elliptic series are well-poised.)

Exercise 4.58   Show that the well-poised series (4.47) is modular in the sense explained in the footnote to Exercise 4.49.

Exercise 4.59   By reversing the order of summation, show that

$$_{m+1}V_m(q^{-n};b_1,\ldots,b_{m-4};q,p) = 0$$

if n is even and the balancing condition holds.

4.2.5 The sum $_{12}V_{11}$

Historically, the first examples of elliptic hypergeometric functions were sums of the form $_{12}V_{11}$ (see [8] and §4.3.5). Again following [29], we will see how such sums appear from a natural generalization of the expansion (4.42).

Let us write $C_k^n(a,b,c)$ for the coefficients (depending also on p and q) given in (4.44) and appearing in the expansion

$$h_n(x;a) = \sum_{k=0}^{n} C_k^n(a,b,c)h_k(x;b)h_{n-k}(x;c). \qquad (4.50)$$

We will study more general coefficients $R_k^l(a,b,c,d;n)$ appearing in the expansion

$$h_k(x;a)h_{n-k}(x;b) = \sum_{l=0}^{n} R_k^l(a,b,c,d;n)h_l(x;c)h_{n-l}(x;d). \qquad (4.51)$$

By Exercise 4.54, these coefficients exist uniquely for generic parameter values.

One can obtain explicit expressions for R_k^l by iterating (4.50). For instance, writing

$$h_k(x;a)h_{n-k}(x;b) = \sum_{j=0}^{k} C_j^k(a,c,bq^{n-k})\, h_j(x;c)h_{n-j}(x;b)$$

$$= \sum_{j=0}^{k}\sum_{m=0}^{n-j} C_j^k(a,c,bq^{n-k})C_m^{n-j}(b,cq^j,d)\, h_{j+m}(x;c)h_{n-j-m}(x;d)$$

gives

$$R_k^l(a,b,c,d;n) = \sum_{j=0}^{\min(k,l)} C_j^k(a,c,bq^{n-k})C_{l-j}^{n-j}(b,cq^j,d).$$

Plugging in the expression (4.44) and simplifying one finds that

$$R_k^l(a,b,c,d;n) = q^{l(l-n)} \times$$

$$\frac{(q;q,p)_n(ac,a/c;q,p)_k(q^{n-l}bd,b/d;q,p)_l(bc,b/c;q,p)_{n-k}(b/c;q,p)_{n-l}}{(cd,b/c;q,p)_n(q,bc,q^{l-n}c/d;q,p)_l(q,q^{-l}d/c;q,p)_{n-l}}$$

$$\times\, {}_{12}V_{11}(q^{-n}c/b;q^{-k},q^{-l},q^{k-n}a/b,q^{l-n}c/d,cd,q^{1-n}/ab,qc/b;q,p).$$

$$(4.52)$$

The uniqueness of the expansion (4.51) implies many non-trivial properties of ${}_{12}V_{11}$-sums. For instance, it follows that

$$R_k^l(a,b,c,d;n) = R_{n-k}^l(b,a,c,d;n).$$

Writing this out explicitly, we get the identity

$${}_{12}V_{11}(q^{-n}c/b;q^{-k},q^{-l},q^{k-n}a/b,q^{l-n}c/d,cd,q^{1-n}/ab,qc/b;q,p)$$

$$= \frac{(a/d,q^{n-l}ad,bc,q^{n-l}b/c;q,p)_l}{(b/d,q^{n-l}bd,ac,q^{n-l}a/c;q,p)_l}$$

$$\times\, {}_{12}V_{11}(q^{-n}c/a;q^{k-n},q^{-l},q^{-k}b/a,q^{l-n}c/d,cd,q^{1-n}/ab,qc/a;q,p).$$

$$(4.53)$$

Note that the left-hand side is a sum supported on $[0,\min(k,l)]$ whereas the right-hand side is supported on $[0,\min(n-k,l)]$. Thus, both sides still terminate if k and n are replaced by continuous parameters, as long as l is a non-negative integer. It is natural to ask whether (4.53) still holds in that case. Indeed, it does, which is the content of the following result due to Frenkel and Turaev [12] in general and to Date et al. [8] under some additional restrictions on the parameters. As the case $p = 0$ is due to Bailey, it is called the elliptic Bailey transformation.

Theorem 4.60 *Let n be a non-negative integer, $bcdefg = q^{n+2}a^3$ and $\lambda = a^2 q/bcd$. Then,*

$$_{12}V_{11}(a;b,c,d,e,f,g,q^{-n};q,p) = \frac{(aq,aq/ef,\lambda q/e,\lambda q/f;q,p)_n}{(aq/e,aq/f,\lambda q,\lambda q/ef;q,p_n}$$
$$\times \, _{12}V_{11}(\lambda;\lambda b/a,\lambda c/a,\lambda d/a,e,f,g,q^{-n};q,p). \quad (4.54)$$

In the following two exercises we give two proofs of Theorem 4.60, one that uses (4.53) and one that doesn't.

Exercise 4.61 Use (4.53) to show that (4.54) holds if $b = q^{-k}$ and $d = aq^{N+1}$, where k and N are non-negative integers with $N \geq \max(k,n)$. Then show that, after replacing c by c/bd, both sides of (4.54) are invariant under the independent substitutions $b \mapsto bp$ and $d \mapsto dp$. Finally, use analytic continuation to deduce (4.54) in general.

Exercise 4.62 Consider the double sum

$$\sum_{0 \leq x \leq y \leq n} \frac{(a;q,p)_{x+y}(a/\lambda;q,p)_{y-x}}{(q\lambda;q,p)_{x+y}(q;q,p)_{y-x}}$$
$$\times \frac{\theta(\lambda q^{2x};p)}{\theta(\lambda;p)} \frac{(\lambda,\lambda b/a,\lambda c/a,\lambda d/a;q,p)_x}{(q,aq/b,aq/c,aq/d;q,p)_x} \left(\frac{aq}{\lambda}\right)^x$$
$$\times \frac{\theta(aq^{2y};p)}{\theta(a;p)} \frac{(q^{-n},e,f,g;q,p)_y}{(q^{n+1},aq/e,aq/f,aq/g;q,p)_y} q^y.$$

Using (4.45) to compute both the sum in x and the sum in y, deduce (4.54).

Exercise 4.63 By an argument similar to that in Exercise 4.61, deduce from Theorem 4.60 that, when $m \geq n$ are non-negative integers and $efgh = a^2 q^{1+n-m}$,

$$_{12}V_{11}(a;q^{-n},c,aq^{m+1}/c,e,f,g,h;q,p)$$
$$= (efg)^{m-n} \frac{(e/a,f/a,g/a,h/a;q,p)_{m-n}}{(a,ef/a,eg/a,fg/a;q,p)_{m-n}}$$
$$\times \, _{12}V_{11}(aq^{n-m};q^{-m},cq^{n-m},aq^{n+1}/c,e,f,g,h;q,p).$$

Exercise 4.64 In the previous exercise, take $m = n+1$ and then let $c \to q$.

Deduce the indefinite summation formula

$$\sum_{k=0}^{n} \frac{\theta(aq^{2k};p)}{\theta(a;p)} \frac{(e,f,g,h;q,p)_k}{(aq/e,aq/f,aq/g,aq/h;q,p)_k} q^k$$
$$= \frac{\theta(a/e,a/f,a/g,a/efg;p)}{\theta(a,a/ef,a/eg,a/fg;p)} \left(1 - \frac{(e,f,g,h;q,p)_{n+1}}{(a/e,a/f,a/g,a/h;q,p)_{n+1}}\right),$$

where $efgh = a^2$. Finally, prove this identity directly by induction on n.[21]

Exercise 4.65   By iterating (4.54) show that, when $bcdefg = q^{n+2}a^3$,

$${}_{12}V_{11}(a;b,c,d,e,f,g,q^{-n};q,p)$$
$$= \frac{g^n(aq,b,aq/cg,aq/dg,aq/eg,aq/fg;q,p)_n}{(aq/c,aq/d,aq/e,aq/f,aq/g,b/g;q,p)_n}$$
$$\times {}_{12}V_{11}(q^{-n}g/b;q^{-n}g/a,aq/bc,aq/bd,aq/be,aq/bf,g,q^{-n};q,p).$$

Exercise 4.66   Show that the coefficients R_k^l satisfy the "addition formula"

$$R_k^l(a,b,e,f;n) = \sum_{j=0}^{n} R_k^j(a,b,c,d;n) R_j^l(c,d,e,f;n).$$

Exercise 4.67   Show the "convolution formulas"

$$R_{k_1+k_2}^l(a,b,c,d;n_1+n_2) = \sum_{l_1+l_2=l} R_{k_1}^{l_1}(aq^{\alpha k_2}, bq^{\beta(n_2-k_2)}, c, d; n_1)$$
$$\times R_{k_2}^{l_2}(aq^{(1-\alpha)k_1}, bq^{(1-\beta)(n_1-k_1)}, cq^{l_1}, dq^{n_1-l_1}; n_2)$$

for all $\alpha, \beta \in \{0,1\}$.

4.2.6 Biorthogonal rational functions

One definition of *classical orthogonal polynomial* is a system of orthogonal polynomials whose derivatives are again orthogonal polynomials. By a classical result of Sonine, this means Jacobi, Laguerre and Hermite polynomials. During the 20th century, it was gradually realized that this definition is too restrictive, as there are further polynomials that share the main properties of Jacobi polynomials, though they are related to difference equations rather than differential equations. This development culminated in the Askey and q-Askey schemes, which together contain a large number (42 as they are counted in [18]) of polynomial systems. All these polynomials are degenerate cases of two five-parameter families: the q-Racah polynomials and the Askey–Wilson polynomials. Spiridonov and

---

[21] For generalizations to multi-basic series, see (4.57) and Exercise 4.71.

Zhedanov [44] and Spiridonov [37] obtained seven-parameter elliptic extensions of these systems. These are neither orthogonal nor polynomial; instead, they are biorthogonal rational functions.[22] In general, two systems (p_j) and (q_j) are called *biorthogonal* if, with respect to some scalar product, $\langle p_j, q_k \rangle = 0$ for $j \neq k$. For the specific systems considered here, q_j is obtained from p_j by a simple change of parameters.

A very satisfactory consequence of our construction of $_{12}V_{11}$-functions as overlap coefficients is that their discrete biorthogonality falls out immediately (we are still following [29]). Indeed, it is clear that the coefficients R_k^l satisfy

$$\delta_{k,l} = \sum_{j=0}^{n} R_k^j(a,b,c,d;n) R_j^l(c,d,a,b;n), \qquad 0 \leq k,l \leq n.$$

Let us make the substitutions

$$(a,b,c,d) \mapsto (\sqrt{c/f}, q/d\sqrt{cf}, a\sqrt{cf}, q^{-n}\sqrt{cf/a})$$

in this identity, and introduce parameters b and e such that $ab = q^{-n}$ and $abcdef = q$. We can then express the result in terms of the functions

$$r_k(X(x);a,b,c,d,e,f;q,p) = \frac{(ab,ac,ad,1/af;q,p)_k}{(aq/e;q,p)_k}$$
$$\times \, _{12}V_{11}(a/e;ax,a/x,q/be,q/ce,q/de,q^k/ef,q^{-k};q,p). \quad (4.55)$$

Here, the right-hand side is an even (in the multiplicative sense) elliptic function of x and X any generator for the field of such functions, see Proposition 4.22. Thus, r_k is a rational function of its first argument. After simplification, we find that if $ab = q^{-n}$ and $abcdef = q$ then

$$\sum_{j=0}^{n} w_j r_k\left(X(aq^j);a,b,c,d,e,f;q\right) r_l\left(X(aq^j);a,b,c,d,f,e;q\right) = C_k \delta_{k,l},$$

where

$$w_j = \frac{\theta(a^2 q^{2j};p)}{\theta(a^2;p)} \frac{(a^2,ab,ac,ad,ae,af;q,p)_j}{(q,aq/b,aq/c,aq/d,aq/e,aq/f;q,p)_j} q^j$$

and

$$C_k = \frac{(a^2 q,q/cd,q/ce,q/de;q,p)_n}{(aq/c,aq/d,aq/e,aq/cde;q,p)_n} \frac{(q,ab,ac,ad,bc,bd,cd;q,p)_k}{(1/ef;q,p)_k}$$
$$\times \frac{\theta(1/ef;p)}{\theta(q^{2k}/ef;p)} q^{-k}.$$

---

[22] In [37], even more general biorthogonal non-rational functions are considered.

This is an elliptic extension of the orthogonality relation for q-Racah polynomials; see Exercise 4.87 for an extension of Askey–Wilson polynomials.

Exercise 4.68 Using Theorem 4.54, show that the function r_k is symmetric in the parameters (a,b,c,d).

4.2.7 A quadratic summation

There are many transformation formulas between classical hypergeometric series whose variables satisfy a quadratic relation. Extensions of such results to basic hypergeometric series typically involve a mixture of q-shifted factorials $(b;q)_k$ and $(b;q^2)_k$. Accordingly, we call an elliptic hypergeometric sum *quadratic* if it combines factors of the form $(b;q,p)_k$ and $(b;q^2,p)_k$. In view of the identity

$$(b;q^2,p)_k = (\sqrt{b}, -\sqrt{b}, \sqrt{pb}, -\sqrt{pb}; q, p)_k,$$

this should be viewed more as a rule of thumb than a precise definition. We will derive just one quadratic summation formula in order to give some idea about the relevant methods. All results of this section are due to Warnaar [46].

Our starting point is a telescoping sum

$$\sum_{k=0}^{n} (A_{k+1} - A_k) = A_{n+1} - A_0.$$

Substituting $A_k = B_0 \cdots B_{k-1} C_k \cdots C_n$ gives

$$\sum_{k=0}^{n} B_0 \cdots B_{k-1} (B_k - C_k) C_{k+1} \cdots C_n = B_0 \cdots B_n - C_0 \cdots C_n.$$

In view of (4.13), it is natural to choose

$$B_k = \theta(a_k d_k^\pm, b_k c_k^\pm; p), \qquad C_k = \theta(b_k d_k^\pm, a_k c_k^\pm; p),$$

since then

$$B_k - C_k = \frac{a_k}{c_k} \theta(c_k d_k^\pm, b_k a_k^\pm; p).$$

This gives the theta function identity

$$\sum_{k=0}^{n} \frac{a_k}{c_k} \theta(c_k d_k^\pm, b_k a_k^\pm; p) \prod_{j=0}^{k-1} \theta(a_j d_j^\pm, b_j c_j^\pm; p) \prod_{j=k+1}^{n} \theta(b_j d_j^\pm, a_j c_j^\pm; p)$$

$$= \prod_{j=0}^{n} \theta(a_j d_j^\pm, b_j c_j^\pm; p) - \prod_{j=0}^{n} \theta(b_j d_j^\pm, a_j c_j^\pm; p). \quad (4.56)$$

Consider now the special case when $a_j = a$ and $b_j = b$ are independent of j, but $c_j = cq^j, d_j = dr^j$ form geometric progressions. After simplification, we obtain

$$\sum_{k=0}^{n} \frac{\theta(cdq^k r^k; p)\theta(dr^k/cq^k; p)}{\theta(cd; p)\theta(d/c; p)} \frac{(da^{\pm}; r, p)_k (cb^{\pm}; q, p)_k}{(rdb^{\pm}; r, p)_k (qca^{\pm}; q, p)_k} q^k$$

$$= \frac{\theta(ac^{\pm}, db^{\pm}; p)}{\theta(ab^{\pm}, dc^{\pm}; p)} \left(1 - \frac{(cb^{\pm}; q, p)_{n+1}(da^{\pm}; r, p)_{n+1}}{(ca^{\pm}; q, p)_{n+1}(db^{\pm}; r, p)_{n+1}} \right). \quad (4.57)$$

This is a *bibasic* sum. The case $q = r$ can be deduced from the elliptic Bailey transformation, see Exercise 4.64.

Consider (4.57) when $a = r^{-n}/d$ and $b = 1/d$. If $n > 0$, the right-hand side vanishes, whereas if $n = 0$, the zero is cancelled by the factor $\theta(a/b; p)$. We conclude that

$$\sum_{k=0}^{n} \frac{\theta(cdq^k r^k; p)\theta(dr^k/cq^k; p)}{\theta(cd; p)\theta(d/c; p)} \frac{(r^{-n}, r^n d^2; r, p)_k (cd^{\pm}; q, p)_k}{(r, rd^2; r, p)_k (qr^n cd, qr^{-n}c/d; q, p)_k} q^k$$

$$= \delta_{n0}. \quad (4.58)$$

Sums that evaluate to Kronecker's delta function are useful for deriving new summations from known ones. Indeed, suppose that $\sum_k a_{kn} = \delta_{n0}$ and that we can compute $c_k = \sum_n a_{kn} b_n$ for some sequence $\{b_n\}$. Then

$$\sum_k c_k = \sum_n \delta_{n0} b_n = b_0. \quad (4.59)$$

There are several ways to apply this idea to (4.58), but we are content with giving one example. We take $r = q^2$; that is,

$$a_{kn} = \frac{\theta(cdq^{3k}; p)\theta(dq^k/c; p)}{\theta(cd; p)\theta(d/c; p)}$$

$$\times \frac{(q^{-2n}, q^{2n}d^2; q^2, p)_k (cd^{\pm}; q, p)_k}{(q^2, q^2 d^2; q^2, p)_k (q^{2n+1}cd, q^{1-2n}c/d; q, p)_k} q^k. \quad (4.60)$$

One may expect that b_n contains the factor

$$\frac{(d^2; q^2, p)_n (d/c; q, p)_{2n}}{(q^2; q^2, p)_n (cdq; q, p)_{2n}} = \frac{(d^2, d/c, qd/c; q^2, p)_n}{(q^2, cdq, cdq^2; q^2, p)_n},$$

as it combines nicely with the n-dependent factors from (4.60) (cf. Exercise 4.47). This looks like part of a well-poised series, so we try to match the sum $\sum_n a_{kn} b_n$ with the Frenkel–Turaev sum. In the case $k = 0$ this is just $\sum_n b_n$, so we are led to take

$$b_n = \frac{\theta(d^2 q^{4n}; p)}{\theta(d^2; p)} \frac{(d^2, d/c, qd/c, e, f, q^{-2m}; q^2, p)_n}{(q^2, cdq, cdq^2, d^2 q^2/e, d^2 q^2/f, d^2 q^{2m+2}; q^2, p)_n} q^{2n},$$

with the balancing condition $c^2 d^2 q^{2m+1} = ef$. We now compute

$$\sum_{n=k}^{m} \frac{(q^{-2n}, q^{2n}d^2; q^2, p)_k}{(q^{2n+1}cd, q^{1-2n}c/d; q, p)_k} b_n$$

$$= q^{\binom{k}{2}} (dq/c)^k \frac{\theta(d^2 q^{4k}; p)}{\theta(d^2; p)} \frac{(d^2; q^2, p)_{2k}(d/c; q, p)_k(e, f, q^{-2m}; q^2, p)_k}{(cdq; q, p)_{3k}(d^2q^2/e, d^2q^2/f, d^2q^{2m+2}; q^2, p)_k}$$

$$\times \sum_{n=0}^{m-k} \frac{\theta(d^2 q^{4k+4n}; p)}{\theta(d^2 q^{4k}; p)}$$

$$\times \frac{(d^2 q^{4k}, q^k d/c, q^{k+1}d/c, eq^{2k}, fq^{2k}, q^{2k-2m}; q^2, p)_n q^{2n}}{(q^2, cdq^{1+3k}, cdq^{2+3k}, d^2q^{2+2k}/e, d^2q^{2+2k}/f, d^2q^{2m+2k+2}; q^2, p)_n}$$

$$= \frac{(c^2 q, d^2 q^2; q^2, p)_m (cdq/e; q, p)_{2m}}{(d^2 q^2/e, c^2 q/e; q^2, p)_m (cdq; q, p)_{2m}}$$

$$\times \frac{(d/c; q, p)_k(e, f, q^{-2m}; q^2, p)_k}{(c^2 q; q^2, p)_k (cdq/e, cdq/f, cdq^{1+2m}; q, p)_k},$$

where the last step is the Frenkel–Turaev sum. Thus, (4.59) takes the form

$$\sum_{k=0}^{m} \frac{\theta(cdq^{3k}; p)\theta(dq^k/c; p)}{\theta(cd; p)\theta(d/c; p)}$$

$$\times \frac{(cd, c/d, d/c; q, p)_k (e, f, q^{-2m}; q^2, p)_k}{(q^2, d^2 q^2, c^2 q; q^2, p)_k (cdq/e, cdq/f, cdq^{1+2m}; q, p)_k} q^k$$

$$= \frac{(d^2 q^2/e, c^2 q/e; q^2, p)_m (cdq; q, p)_{2m}}{(c^2 q, d^2 q^2; q^2, p)_m (cdq/e; q, p)_{2m}}.$$

After the change of variables $(c, d, e, f, m) \mapsto (\sqrt{ab}, \sqrt{a/b}, c, d, n)$, we obtain the following quadratic summation due to Warnaar.

Proposition 4.69   *For* $cd = a^2 q^{2n+1}$,

$$\sum_{k=0}^{n} \frac{\theta(aq^{3k}; p)}{\theta(a; p)} \frac{(a, b, q/b; q, p)_k (c, d, q^{-2n}; q^2, p)_k}{(q^2, aq^2/b, abq; q^2, p)_k (aq/c, aq/d, aq^{1+2n}; q, p)_k} q^k$$

$$= \frac{(aq; q, p)_{2n}(abq/c, aq^2/bc; q^2, p)_n}{(aq/c; q, p)_{2n}(abq, aq^2/b; q^2, p)_n}.$$

Exercise 4.70   Show that the case $p = 0$ of (4.56), after substituting $a_k + a_k^{-1} \mapsto a_k$ and so on, takes the form

$$\sum_{k=0}^{n} (c_k - d_k)(b_k - a_k) \prod_{j=0}^{k-1} (a_j - d_j)(b_j - c_j) \prod_{j=k+1}^{n} (b_j - d_j)(a_j - c_j)$$

$$= \prod_{j=0}^{n} (a_j - d_j)(b_j - c_j) - \prod_{j=0}^{n} (b_j - d_j)(a_j - c_j). \quad (4.61)$$

Conversely, show that substituting $a_k \mapsto X(a_k)$ and so on in (4.61), where X is as in (4.18), gives back the general case of (4.56).

Exercise 4.71 Write down the summation obtained from (4.56) by choosing all four parameter sequences as independent geometric progressions (see [14]).

Exercise 4.72 Prove Proposition 4.69 by the method of §4.2.3.

Exercise 4.73 Find a cubic summation formula by combining the Frenkel–Turaev sum with the case $r = q^3$ of (4.58) (see [46, Thm. 4.5]).

Exercise 4.74 Find a transformation formula that generalizes Proposition 4.69 (see [46, Thm. 4.2]).

Exercise 4.75 Let A and B be lower-triangular matrices (the size is irrelevant) with entries

$$A_{ij} = \frac{\prod_{k=j}^{i-1} \theta(y_j z_k^{\pm};p)}{\prod_{k=j+1}^{i} \theta(y_j y_k^{\pm};p)}, \qquad B_{ij} = \frac{y_i \theta(y_j z_j^{\pm};p) \prod_{k=j+1}^{i} \theta(y_i z_k^{\pm};p)}{y_l \theta(y_i z_i^{\pm};p) \prod_{k=j}^{i-1} \theta(y_i y_k^{\pm};p)}.$$

Show that $B = A^{-1}$. Indeed, show that the identity $AB = I$ is equivalent to (4.26) and that the identity $BA = I$ is equivalent to the case $a_j \equiv d_n, b_j \equiv d_0$ of (4.56). As any left inverse is a right inverse, this gives an alternative proof of (4.58) as a consequence of (4.26).

4.2.8 An elliptic Minton summation

Minton found the summation formula [21]

$$_{r+2}F_{r+1}\left(\begin{matrix} -n, b, c_1 + m_1, \ldots, c_r + m_r \\ b+1, c_1, \ldots, c_r \end{matrix}; 1\right) = \frac{n!(c_1 - b)_{m_1} \cdots (c_r - b)_{m_r}}{(b+1)_n (c_1)_{m_1} \cdots (c_r)_{m_r}},$$
$$(4.62)$$

where m_i and n are non-negative integers such that $m_1 + \cdots + m_r \leq n$. This was extended to non-terminating series by Karlsson, so sums with integral parameter differences are often referred to as Karlsson–Minton-type.

Following [31], we will obtain an elliptic extension of (4.62) from the elliptic partial fraction expansion (4.25). We first replace n by $n+1$ in (4.25) and rewrite it as

$$\sum_{j=0}^{n} \frac{\prod_{k=1}^{n} \theta(y_j z_k^{\pm};p)}{\theta(xy_j^{\pm};p) \prod_{k=0, k \neq j}^{n} \theta(y_j y_k^{\pm};p)} = \frac{\prod_{k=1}^{n} \theta(xz_k^{\pm};p)}{\prod_{k=0}^{n} \theta(xy_k^{\pm};p)}. \qquad (4.63)$$

We now specialize y to be a geometric progression and z to be a union of geometric progressions.[23] That is, we write $y_j = aq^j$ and

$$(z_1, \ldots, z_n) = (c_1, c_1 q, \ldots, c_1 q^{m_1}, \ldots, c_r, c_r q, \ldots, c_r q^{m_r}),$$

where m_j are non-negative integers summing to n. Then, the left-hand side of (4.63) takes the form

$$
\sum_{j=0}^{n} \frac{\prod_{l=1}^{r} \prod_{k=1}^{m_l} \theta(ac_l q^{j+k-1}, aq^{1+j-k}/c_l; p)}{\theta(xaq^j, xq^{-j}/a; p) \prod_{k=0, k \neq j}^{n} \theta(a^2 q^{j+k}, q^{j-k}; p)}
$$

$$
= \sum_{j=0}^{n} \frac{\prod_{l=1}^{r} (ac_l q^j, aq^{1+j-m_l}/c_l; q, p)_{m_l}}{\theta(xaq^j, xq^{-j}/a; p)(q, a^2 q^j; q, p)_j (q^{j-n}, a^2 q^{2j+1}; q, p)_{n-j}}
$$

$$
= \frac{\prod_{l=1}^{r} (ac_l, aq^{1-m_l}/c_l; q)_{m_l}}{\theta(xa^{\pm}; p)(q^{-n}, a^2 q; q, p)_n}
$$

$$
\times \sum_{j=0}^{n} \frac{\theta(a^2 q^{2j}; p)}{\theta(a^2; p)} \frac{(a^2, ax^{\pm}, q^{-n}; q, p)_j}{(q, aqx^{\pm}, a^2 q^{n+1}; q, p)_j} q^j \prod_{l=1}^{r} \frac{(ac_l q^{m_l}, aq/c_l; q, p)_j}{(ac_l, aq^{1-m_l}/c_l; q, p)_j},
$$

and the right-hand side is

$$
\frac{\prod_{l=1}^{r} \prod_{k=1}^{m_l} \theta(xc_l q^{k-1}, xq^{1-k}/c_l; p)}{\prod_{k=0}^{n} \theta(xaq^k, xq^{-k}/a; p)} = \frac{\prod_{l=1}^{r} (xc_l, xq^{1-m_l}/c_l; q, p)_{m_l}}{(xa, xq^{-n}/a; q, p)_{n+1}}.
$$

After the change of variables $a \mapsto \sqrt{a}, x \mapsto b/\sqrt{a}, c_l \mapsto c_l/\sqrt{a}$, we arrive at the following elliptic extension of Minton's identity.

Proposition 4.76 *If $m_1, \ldots, m_r$ are non-negative integers and $n = m_1 + \cdots + m_r$,*

$$
{}_{2r+8}V_{2r+7}(a; b, a/b, q^{-n}, c_1 q^{m_1}, \ldots, c_r q^{m_r}, aq/c_1, \ldots, aq/c_r; q, p)
$$

$$
= \sum_{j=0}^{n} \frac{\theta(aq^{2j}; p)}{\theta(a; p)} \frac{(a, b, a/b, q^{-n}; q, p)_j}{(q, bq, aq/b; q, p)_j} q^j \prod_{l=1}^{r} \frac{(c_l q^{m_l}, aq/c_l; q, p)_j}{(c_l, aq^{1-m_l}/c_l; q, p)_j}
$$

$$
= \frac{(q, aq; q, p)_n}{(aq/b, bq; q, p)_n} \prod_{l=1}^{r} \frac{(c_l/b, bc_l/a; q, p)_{m_l}}{(c_l, c_l/a; q, p)_{m_l}}.
$$

Proposition 4.76 may seem to generalize only the case $\sum_j m_j = n$ of (4.62), but it should be noted that if we start from that special case and let $c_r \to \infty$ we obtain the general case.

Exercise 4.77 Prove a transformation for Karlsson–Minton type series by starting from (4.26) and specializing y to a union of two geometric progressions (see [31, Cor. 4.5]).

Exercise 4.78 Deduce a Karlsson–Minton-type summation formula from (4.23) (the result is less attractive than Proposition 4.76; see [31, Cor. 5.3]).

---

[23] As the latter progressions may have length 1, the case of general z is included.

4.2.9 The elliptic gamma function

The classical gamma function satisfies $\Gamma(z+1) = z\Gamma(z)$, which upon iteration gives $(z)_n = \Gamma(z+n)/\Gamma(z)$. We will need an elliptic analogue $\Gamma(x;q,p)$ of the gamma function, which was introduced by Ruijsenaars [33]. It is natural to demand that

$$\Gamma(qx;q,p) = \theta(x;p)\Gamma(x;q,p), \qquad (4.64)$$

which upon iteration gives

$$(x;q,p)_n = \frac{\Gamma(q^n x;q,p)}{\Gamma(x;q,p)}. \qquad (4.65)$$

To solve (4.64), consider first in general a functional equation of the form $f(qx) = \phi(x)f(x)$. Upon iteration, it gives

$$f(x) = \frac{1}{\phi(x)\phi(qx)\cdots\phi(q^{N-1}x)} f(q^N x).$$

If $|q| < 1$ and $\phi(x) \to 1$ quickly enough as $x \to 0$, one solution will be

$$f(x) = \frac{1}{\prod_{k=0}^{\infty} \phi(q^k x)}.$$

Alternatively, we can iterate the functional equation in the other direction and obtain

$$f(x) = \phi(q^{-1}x)\phi(q^{-2}x)\cdots\phi(q^{-N}x)f(q^{-N}x).$$

In this case, if $\phi(x) \to 1$ quickly as $|x| \to \infty$, we find the solution

$$f(x) = \prod_{k=0}^{\infty} \phi(q^{-k-1}x).$$

In the case at hand, $\phi(x) = \theta(x;p)$, we can write $\phi(x) = \phi_1(x)\phi_2(x)$, where $\phi_1(x) = (x;p)_\infty$ and $\phi_2(x) = (p/x;p)_\infty$ tend to 1 as $x \to 0$ and $|x| \to \infty$, respectively. This suggests the definition

$$\Gamma(x;q,p) = \prod_{k=0}^{\infty} \frac{\phi_2(q^{-k-1}x)}{\phi_1(q^k x)} = \prod_{j,k=0}^{\infty} \frac{1 - p^{j+1}q^{k+1}/x}{1 - p^j q^k x}.$$

In a similar way as Lemma 4.3, one can then prove the following result. (In view of the symmetry between p and q, we will from now on write $\Gamma(x;p,q)$ rather than $\Gamma(x;q,p)$.)

Lemma 4.79 *The elliptic gamma function $\Gamma(x;p,q)$ is meromorphic as a function of $x \in \mathbb{C}^*$ and of p and q with $|p|, |q| < 1$, with zeroes precisely* at the points $x = p^{j+1}q^{k+1}$ and poles precisely at the points $x = p^{-j}q^{-k}$, where $j, k \in \mathbb{Z}_{\geq 0}$. Moreover, (4.64) holds.

Note the inversion formula

$$\Gamma(x;p,q)\Gamma(pq/x;p,q) = 1. \tag{4.66}$$

Just as for theta functions and elliptic shifted factorials, we will use the condensed notation

$$\Gamma(x_1,\ldots,x_m;p,q) = \Gamma(x_1;p,q)\cdots\Gamma(x_m;p,q),$$
$$\Gamma(ax^{\pm};p,q) = \Gamma(ax;p,q)\Gamma(a/x;p,q).$$

One would expect that the elliptic gamma function degenerates to the classical gamma function if we first take the trigonometric limit $p \to 0$ and then the rational limit $q \to 1$. Indeed, we have $\Gamma(x;q,0) = 1/(x;q)_\infty$ and [13, §1.10]

$$\lim_{q\to 1}(1-q)^{1-x}\frac{(q;q)_\infty}{(q^x;q)_\infty} = \Gamma(x).$$

Exercise 4.80   Prove that the meromorphic solutions to $f(qx) = \theta(x;p)f(x)$ are precisely the functions $f(x) = \Gamma(x;q,p)g(x)$, where g is an arbitrary multiplicatively elliptic function with period q.

Exercise 4.81   Give an analogue of the reflection formula

$$\Gamma(z)\Gamma(1-z) = \frac{\pi}{\sin(\pi z)}$$

for $\Gamma(x,q/x;p,q)$.

Exercise 4.82   Give an analogue of the duplication formula

$$\sqrt{\pi}\,\Gamma(2z) = 2^{2z-1}\Gamma(z)\Gamma(z+1/2)$$

for the function $\Gamma(x^2;p,q)$.

4.2.10 Elliptic hypergeometric integrals

As was mentioned in §4.2.1, if we want to consider elliptic analogues of infinite hypergeometric series, it is often better to define them by integrals. The study of such integrals was initiated by Spiridonov, see e.g., [35, 37].

A model result for converting series to integrals is Barnes's integral representation

$$_2F_1\left(\begin{matrix}a,b\\c\end{matrix};z\right) = \frac{\Gamma(c)}{\Gamma(a)\Gamma(b)}\int_{-i\infty}^{i\infty}\frac{\Gamma(a+s)\Gamma(b+s)\Gamma(-s)}{\Gamma(c+s)}(-z)^s\,\frac{ds}{2\pi i}. \tag{4.67}$$

This holds for $z \notin \mathbb{R}_{\geq 0}$ and $a, b \notin \mathbb{Z}_{\leq 0}$. The integrand has poles at $s = -a - n$, $s = -b - n$ and $s = n$, where $n \in \mathbb{Z}_{\geq 0}$. The contour of integration should pass to the right of the first two sequences of poles but to the left of the third sequence.

To prove (4.67) one needs to know that the residue of the gamma function at $-n$ is $(-1)^n / n!$. Consequently, the residue of the integrand at n is

$$\frac{1}{2\pi i} \frac{\Gamma(c)}{\Gamma(a)\Gamma(b)} \frac{\Gamma(a+n)\Gamma(b+n)}{\Gamma(c+n)n!} z^n = \frac{1}{2\pi i} \frac{(a)_n (b)_n}{(c)_n n!} z^n.$$

Thus, (4.67) simply means that the integral is $2\pi i$ times the sum of all residues to the right of the contour. This follows from Cauchy's residue theorem together with an estimate of the integrand that we will not go into.

In analogy with Barnes's integral, we will now consider a class of integrals closely related to very well-poised elliptic hypergeometric series [37]. They have the form

$$I(t_0, \ldots, t_m; p, q) = \int \frac{\Gamma(t_0 x^\pm, \ldots, t_m x^\pm; p, q)}{\Gamma(x^{\pm 2}; p, q)} \frac{dx}{2\pi i x}. \tag{4.68}$$

Note that the factor

$$\frac{1}{\Gamma(x^{\pm 2}; p, q)} = -\frac{\theta(x^2; p)\theta(x^2; q)}{x^2}$$

is analytic for $x \neq 0$. Thus, the only poles of the integrand are at

$$x = p^j q^k t_l, \qquad j, k \in \mathbb{Z}_{\geq 0}, \quad l = 0, \ldots, m \tag{4.69}$$

and at the reciprocal of these points. The integration is over a closed positively oriented contour such that the poles (4.69) are inside the contour and their reciprocals are outside. Such a contour exists if $t_j t_k \notin p^{\mathbb{Z}_{\leq 0}} q^{\mathbb{Z}_{\leq 0}}$ for $1 \leq j, k \leq m$. For instance, if $|t_j| < 1$ for all j we may integrate over the unit circle.

Let $f(x)$ denote the integrand in (4.68). To explain the connection to elliptic hypergeometric series, we first use (4.65) and Exercise 4.47 to show that

$$\frac{f(q^k x)}{f(x)} = C^k q^{\binom{k}{2}(m-3)} x^{k(m-3)} \frac{\theta(q^{2k} x^2; p)}{\theta(x^2; p)} \prod_{j=0}^{m} \frac{(t_j x; q, p)_k}{(qx/t_j; q, p)_k},$$

where $C = (-1)^{m+1} q^{m-3} / t_0 \cdots t_m$. This clearly has a very well-poised structure, but we dislike the quadratic exponent of q. To get rid of it, we substitute $t_j = p^{l_j} u_j$, where l_j are integers. Using again Exercise 4.47, we find that

$$\frac{f(q^k x)}{f(x)} = \left(Cp^{|l|}q^{-|l|}\right)^k \left(q^{\binom{k}{2}}x^k\right)^{m-3-2|l|} \frac{\theta(q^{2k}x^2;p)}{\theta(x^2;p)} \prod_{j=0}^{m} \frac{(u_j x;q,p)_k}{(qx/u_j;q,p)_k}.$$

Thus, it is natural to take m odd and $|l| = (m-3)/2$. If we in addition assume that

$$t_0 \cdots t_m = (pq)^{(m-3)/2}, \tag{4.70}$$

then $C = p^{-|l|}q^{|l|}$ and

$$\frac{f(q^k x)}{f(x)} = \frac{\theta(q^{2k}x^2;p)}{\theta(x^2;p)} \prod_{j=0}^{m} \frac{(u_j x;q,p)_k}{(qx/u_j;q,p)_k}.$$

This gives in turn

$$\frac{\operatorname{Res}_{x=u_0 q^k} f(x)}{\operatorname{Res}_{x=u_0} f(x)} = \frac{\lim_{x \to u_0 q^k}(x-u_0 q^k)f(x)}{\lim_{x \to u_0}(x-u_0)f(x)} = q^k \lim_{x \to u_0} \frac{f(q^k x)}{f(x)}$$

$$= \frac{\theta(q^{2k}u_0^2;p)}{\theta(u_0^2;p)} \frac{(u_0^2, u_0 u_1 \cdots, u_0 u_m;q,p)_k}{(q, qu_0/u_1, \ldots, qu_0/u_m;q,p)_k} q^k. \tag{4.71}$$

Thus, the (typically divergent) sum of residues at the points $x = u_0 q^k, k \geq 0$, is a constant times

$$_{m+5}V_{m+4}(u_0^2; u_0 u_1, \ldots, u_0 u_m; q, p).$$

In contrast to (4.67), we are not claiming that the integral is equal to this sum. However, we can still think of (4.68) as a substitute for the series $_{m+5}V_{m+4}$, at least when m is odd and (4.70) holds.

Exercise 4.83 In view of the definition of elliptic hypergeometric series, it is natural to call an integral

$$\int f(x)\,dx$$

elliptic hypergeometric if $f(qx)/f(x)$ is multiplicatively elliptic with period p. Show that the integral (4.68) is elliptic hypergeometric if and only if $t_0^2 \cdots t_m^2 = (pq)^{m-3}$.

4.2.11 Spiridonov's elliptic beta integral

In view of the Frenkel–Turaev formula, one may hope that (4.68) can be computed in closed form for $m = 5$ and $t_0 \cdots t_5 = pq$. Indeed, we have the following beautiful integral evaluation due to Spiridonov. The limit case when $p \to 0$ (with $t_0, \ldots, t_4$ fixed) is due to Nasrallah and Rahman [22]; the subsequent limit $t_4 \to 0$ is the famous Askey–Wilson integral [3].

Theorem 4.84 *Assume that $|p|, |q| < 1$ and that $t_0, \ldots, t_5$ are parameters such that*

$$t_j t_k \notin p^{\mathbb{Z}_{\le 0}} q^{\mathbb{Z}_{\le 0}}, \qquad 0 \le j, k \le 5, \tag{4.72}$$

$$t_0 \cdots t_5 = pq. \tag{4.73}$$

Then,

$$I(t_0, \ldots, t_5; p, q) = \frac{2 \prod_{0 \le i < j \le 5} \Gamma(t_i t_j; p, q)}{(p; p)_\infty (q; q)_\infty}. \tag{4.74}$$

We will give an elementary proof of Theorem 4.84 found by Spiridonov [38] a few years after the original proof in [35]. Consider first $I(t_0, \ldots, t_5; p, q)$ as a function of p. For fixed values of the parameters t_j, there are only finitely many values of p such that $|p| < 1$ and (4.72) is violated. Outside these points, the integral is analytic in p. Thus, by analytic continuation, we may assume $0 < p < 1$. By symmetry, we may assume that $0 < q < 1$ and, again by analytic continuation, (4.34). Consider now the integral as a function of t_0, where $t_1, \ldots, t_4$ are fixed and t_5 is determined from (4.73). It is analytic as long as t_0 avoids the points forbidden by (4.72). Since p and q are real, these forbidden values are on a finite number of rays starting at the origin. We will assume that t_0 avoids these rays; that is,

$$t_j t_k \notin \mathbb{R}_{>0}, \qquad j = 0, 5, \quad 0 \le k \le 5. \tag{4.75}$$

We proceed to show that, under this condition, the quotient

$$F(t_0) = \frac{I(t_0, \ldots, t_5; p, q)}{\prod_{0 \le i < j \le 5} \Gamma(t_i t_j; p, q)}$$

satisfies $F(q t_0) = F(t_0)$.

Let

$$f(t_0; x) = \frac{\prod_{j=0}^5 \Gamma(t_j x^\pm; p, q)}{\prod_{0 \le i < j \le 5} \Gamma(t_i t_j; p, q) \cdot \Gamma(x^{\pm 2}; p, q)}$$

denote the integrand of $F(t_0)$ (apart from $1/2\pi ix$). We use (4.64) to compute the difference

$$f(t_0;x) - f(qt_0;x)$$

$$= \frac{\Gamma(t_0 x^{\pm}, \ldots, t_4 x^{\pm}, q^{-1} t_5 x^{\pm}; p, q)}{\Gamma(x^{\pm 2}, t_0 t_5, q t_0 t_1, \ldots, q t_0 t_4; p, q) \prod_{1 \le i < j \le 5} \Gamma(t_i t_j; p, q)}$$

$$\times \left\{ \theta(t_0 t_1, \ldots, t_0 t_4, q^{-1} t_5 x^{\pm}; p) - \theta(q^{-1} t_1 t_5, \ldots, q^{-1} t_4 t_5, t_0 x^{\pm}; p) \right\}.$$

We now apply the case $n = 2$ of (4.27), where we substitute

$$(x, a_1, a_2, b_1, \ldots, b_4) \mapsto (\lambda t_0, \lambda x, \lambda/x, t_1/\lambda, \ldots, t_4/\lambda), \qquad \lambda^2 = \frac{pq}{t_0 t_5}.$$

After simplification, we find that the factor in brackets equals

$$\frac{t_0 \theta(t_5/qt_0; p)}{x^{-1} \theta(x^2; p)} \left\{ x^{-2} \theta(t_0 x, \ldots, t_4 x, t_5 x/q; p) - x^2 \theta(t_0/x, \ldots, t_4/x, t_5/qx; p) \right\}.$$

It follows that

$$f(t_0;x) - f(qt_0;x) = g(t_0;x) - g(t_0;qx), \tag{4.76}$$

where

$$g(t_0;x) =$$

$$\frac{t_0 x \theta(t_5/qt_0; p) \theta(x^2; q) \Gamma(t_0 x, \ldots, t_4 x, t_5 x/q, t_0 q/x, \ldots, t_4 q/x, t_5/x; p, q)}{\Gamma(t_0 t_5; p, q) \prod_{j=1}^4 \Gamma(q t_0 t_j; p, q) \prod_{1 \le i < j \le 5} \Gamma(t_i t_j; p, q)}.$$

Integrating (4.76) over a contour $\mathscr{C}$ gives

$$\int_{\mathscr{C}} f(t_0;x) \frac{dx}{2\pi i x} - \int_{\mathscr{C}} f(qt_0;x) \frac{dx}{2\pi i x} = \int_{\mathscr{C}} g(t_0;x) \frac{dx}{2\pi i x} - \int_{q\mathscr{C}} g(t_0;x) \frac{dx}{2\pi i x}. \tag{4.77}$$

We choose $\mathscr{C}$ so that the points (4.69) (with $m = 5$) are inside $\mathscr{C}$ and their reciprocals outside. Then, the first integral is equal to $F(t_0)$. The second integral equals $F(qt_0)$, provided that these conditions still hold when t_0, t_5 are replaced by $t_0 q, t_5/q$. This gives the additional requirement that the points $x = p^j q^{-1} t_5$ are inside $\mathscr{C}$ for $j \in \mathbb{Z}_{\ge 0}$ and their reciprocals outside $\mathscr{C}$. We can choose $\mathscr{C}$ in this way, provided that none of the points $p^j q^{-1} t_5$ is equal to the reciprocal of one of the points (4.69). This follows from our assumption (4.75).

The function $g(t_0;x)$ has poles at

$$x = t_l p^j q^{k+1}, \quad x = t_5 p^j q^k, \qquad 0 \le l \le 4, \quad j,k \in \mathbb{Z}_{\ge 0}$$

and at

$$x = t_l^{-1} p^{-j} q^{-k}, \quad x = t_5^{-1} p^{-j} q^{1-k}, \qquad 0 \le l \le 4, \quad j,k \in \mathbb{Z}_{\ge 0}.$$

The first set of poles are inside both the contours $q\mathscr{C}$ and $\mathscr{C}$, whereas the second set of poles are outside both contours. Thus, we can deform $\mathscr{C}$ to $q\mathscr{C}$ without crossing any poles of g. It follows that the right-hand side of (4.77) vanishes. This completes our proof that $F(qt_0) = F(t_0)$. By symmetry, $F(pt_0) = F(t_0)$. Since p and q are real, we may iterate these equations without violating (4.75). Thus, $F(p^k q^l t_0) = F(t_0)$ for $k, l \in \mathbb{Z}$. Since we assume (4.34), the points $p^k q^l t_0$ have a limit point in the open set defined by (4.75). By analytic continuation, $F(t_0) = F$ is a constant.

To compute the constant F, we consider the limit $t_0 \to t_1^{-1}$. The obstruction from letting $t_0 = t_1^{-1}$ in the definition of I comes from the condition that $x = t_0$ and $x = t_1$ are inside $\mathscr{C}$, whereas $x = t_0^{-1}$ and $x = t_1^{-1}$ are outside. To resolve this problem, we write

$$F = \int_{\mathscr{C}} f(t_0;x) \frac{dx}{2\pi i x} = \operatorname*{Res}_{x=t_0} \frac{f(t_0;x)}{x} - \operatorname*{Res}_{x=t_0^{-1}} \frac{f(t_0;x)}{x} + \int_{\mathscr{C}'} f(t_0;x) \frac{dx}{2\pi i x},$$

(4.78)

where $\mathscr{C}'$ is a modification of $\mathscr{C}$ running outside $x = t_0$ and inside $x = t_0^{-1}$. As f vanishes in the limit $t_0 \to t_1^{-1}$, so does the integral over $\mathscr{C}'$. Moreover, since $f(t_0;x) = f(t_0;x^{-1})$, the first two terms can be combined and we obtain

$$F = 2 \lim_{t_0 \to t_1^{-1}} \operatorname*{Res}_{x=t_0} \frac{f(t_0;x)}{x}.$$

We compute

$$\operatorname*{Res}_{x=t_0} \Gamma(t_0/x; p, q)$$

$$= \lim_{x \to t_0} \frac{x - t_0}{(1 - t_0/x) \prod_{j=0}^{\infty} (1 - p^j t_0/x)(1 - q^j t_0/x)} \prod_{j,k=0}^{\infty} \frac{1 - p^{j+1} q^{k+1} x/t_0}{1 - p^{j+1} q^{k+1} t_0/x}$$

$$= \frac{t_0}{(p;p)_\infty (q;q)_\infty}$$

and consequently

$$\operatorname*{Res}_{x=t_0} \frac{f(t_0;x)}{x} = \frac{\prod_{j=1}^{5} \Gamma(t_j/t_0; p, q)}{(p;p)_\infty (q;q)_\infty \Gamma(t_0^{-2}; p, q) \prod_{1 \le i < j \le 5} \Gamma(t_i t_j; p, q)}.$$

If $t_0 \to t_1^{-1}$, this becomes

$$\frac{1}{(p;p)_\infty (q;q)_\infty \prod_{2 \le i < j \le 5} \Gamma(t_i t_j; p, q)} = \frac{1}{(p;p)_\infty (q;q)_\infty},$$

where we used (4.66) and the fact that $t_2 t_3 t_4 t_5 = pq$ in the limit. In conclusion, $F = 2/(p;p)_\infty (q;q)_\infty$, which is Spiridonov's integral evaluation.

Exercise 4.85 By considering the integral

$$\int \frac{\Gamma(cz^{\pm}w^{\pm};p,q)\prod_{j=1}^{4}\Gamma(a_jz^{\pm},b_jw^{\pm};p,q)}{\Gamma(z^{\pm2},w^{\pm2};p,q)}\,\frac{dz}{2\pi iz}\frac{dw}{2\pi iw},$$

prove that

$$I(t_1,\dots,t_8;p,q) =$$
$$\prod_{1\le j<k\le 4}\Gamma(t_jt_k,t_{j+4}t_{k+4};p,q)I(t_1/\lambda,\dots,t_4/\lambda,t_5\lambda,\dots,t_8\lambda),$$

where $\lambda^2 = t_1t_2t_3t_4/pq = pq/t_5t_6t_7t_8$. By iterating this identity, obtain several further integral transformations (see [39, §5.1]).

Exercise 4.86 Consider the limit $t_0t_1 \to q^{-n}$ of (4.74). Generalizing the splitting (4.78) and using (4.71), deduce the Frenkel–Turaev summation.

Exercise 4.87 Show, under appropriate assumptions on the parameters and the contour of integration, the one-parameter family of biorthogonality relations

$$\int \frac{\Gamma(t_0x^{\pm},\dots,t_5x^{\pm};p,q)\theta(\lambda x^{\pm};q)}{\Gamma(x^{\pm2};p,q)}$$

$$\times r_k(X(x);t_0,t_1,t_2,t_3,t_4,t_5;q,p)\,r_l(X(x);t_0,t_1,t_2,t_3,t_5,t_4;q,p)\,\frac{dx}{2\pi ix} = 0,$$

where $k \neq l$, $t_0\cdots t_5 = q$, the rational functions r_k are defined in (4.55) and λ is a free parameter. This gives an elliptic analogue of Rahman's biorthogonal rational functions [24], which generalize the Askey–Wilson polynomials [3].[24]

4.3 Solvable lattice models

4.3.1 Solid-on-solid models

We will now explain how elliptic hypergeometric series first appeared, as fused Boltzmann weights for Baxter's elliptic solid-on-solid model. The main reference for this section is [8]. The reader who wants more background on exactly solvable models in statistical mechanics is referred to the first few chapters of [16] for a brief introduction and to the standard textbook [5] for more details.

The goal of statistical mechanics is to predict the large-scale behaviour

---

[24] Hint: Use the symmetry of Exercise 4.68 to expand r_k as a sum with numerator parameters $t_0x^{\pm}$ and r_l as a sum with numerator parameters $t_1x^{\pm}$. Then use that, since $\theta(\lambda x^{\pm};q)$ is in the linear span of $\theta(t_2x^{\pm};q)$ and $\theta(t_3x^{\pm};q)$, it is sufficient to take $\lambda = t_3$. One can give more general two-index biorthogonality relations for functions of the form $r_{k_1}(\cdots;p,q)r_{k_2}(\cdots;q,p)$, see [37].

of a system described by local rules. In solid-on-solid (SOS) models, the system can be viewed as a random surface. Let us first consider a model whose *states* are rectangular arrays of fixed size with real entries. We may think of the entries as the height of a discrete surface over the rectangle. To get a statistical model, we need to associate a weight to each state, which is proportional to the probability that the state is assumed.[25] The weight is the product of local *Boltzmann weights*, associated to 2×2-blocks of adjacent entries in the array. If our array is $(M+1) \times (N+1)$, the 2×2-blocks naturally form an $(M \times N)$ array. Giving the blocks coordinates (i, j) in a standard way, suppose that the block with coordinates (i, j) is $\left[\begin{smallmatrix} a & b \\ c & d \end{smallmatrix}\right]$. We then assign to this block the Boltzmann weight

$$ W \left(\begin{matrix} a & b \\ c & d \end{matrix} \middle| u_i, v_j \right), $$

where, for the moment, we think of W as an arbitrary function of six variables, and where $u_1, \ldots, u_M, v_1, \ldots, v_N$ are parameters associated to the vertical and horizontal lines separating the heights. These parameters are known as *rapidities* (or spectral parameters) and the lines are called *rapidity lines*. As an example, the state

$$ \begin{matrix} 1 & 2 & 3 \\ 4 & 5 & 6 \\ 7 & 8 & 9 \end{matrix} $$

has weight

$$ W \left(\begin{matrix} 1 & 2 \\ 4 & 5 \end{matrix} \middle| u_1, v_1 \right) \cdot W \left(\begin{matrix} 2 & 3 \\ 5 & 6 \end{matrix} \middle| u_1, v_2 \right) $$

$$ \times W \left(\begin{matrix} 4 & 5 \\ 7 & 8 \end{matrix} \middle| u_2, v_1 \right) \cdot W \left(\begin{matrix} 5 & 6 \\ 8 & 9 \end{matrix} \middle| u_2, v_2 \right). $$

In the models that we will consider, only finitely many states are allowed. For instance, for Baxter's elliptic SOS model described in §4.3.4, one rule is that the height of adjacent squares differ by exactly one.[26] Since we may shift all heights by an arbitrary real number we need further restrictions, which we will take to be boundary conditions. For instance, we may fix the whole boundary or just the height at a corner. Assuming in addition that W is positive, we can then define the probability of a state to

---

[25] Physically, the weight is $e^{-E/kT}$, where $k > 0$ is Boltzmann's constant, E the energy of the state and T the temperature. Thus, high energy states are less likely than low energy states, but become more likely as temperature increases.

[26] This condition is natural for body-centered cubic crystals such as iron at room temperature.

be its weight divided by the *partition function*

$$\sum_{\text{states}} \text{weight,}$$

where we sum over all states satisfying our boundary conditions.

We now observe that SOS models make sense on more general geometries. The rapidity lines, separating regions with constant height, could be quite arbitrary oriented curves in a portion of the plane. The main restriction is that only two curves may cross at any point. Then, each crossing looks like

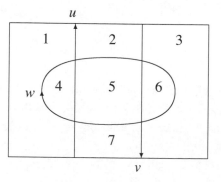

to which we assign the weight

$$W\begin{pmatrix} a & b \\ c & d \end{pmatrix} u, v \Big).$$

Note that the orientation determines how the heights and rapidities should be inserted in W. To give an example, the state

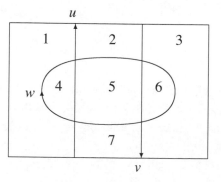

has weight

$$W\begin{pmatrix} 1 & 2 \\ 4 & 5 \end{pmatrix} w, u \Big) \cdot W\begin{pmatrix} 3 & 6 \\ 2 & 5 \end{pmatrix} v, w \Big) \cdot W\begin{pmatrix} 1 & 4 \\ 7 & 5 \end{pmatrix} u, w \Big) \cdot W\begin{pmatrix} 3 & 7 \\ 6 & 5 \end{pmatrix} w, v \Big).$$

4.3.2 The Yang–Baxter equation

In the models of interest to us, the weights satisfy the *Yang–Baxter equation*, which can be viewed as an integrability criterion.[27] Roughly speaking, the Yang–Baxter equation gives a natural way to make sense of triple crossings. More precisely, we want to allow the type of crossing to the left in the following picture, but not the one to the right.

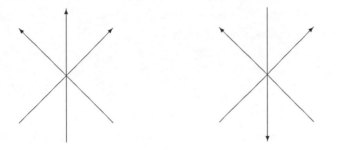

Imagine that our triple crossing is composed of three single crossings viewed from a distance. This can happen in two ways, as illustrated in the following picture. We have also introduced symbols for the adjacent heights and the rapidities.

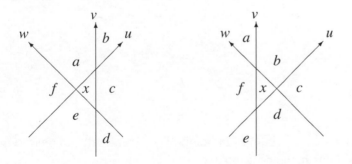

We now postulate that the corresponding two systems have the same partition function. We allow $a, \ldots, f$ to be fixed, for instance by boundary conditions, but sum over all possibilities for x (as before, we assume that the resulting sums are finite). This gives the Yang–Baxter equation in the

---

[27] To be slightly more precise, Baxter's elliptic SOS model is closely related to a one-dimensional quantum mechanical model known as the XYZ spin chain. Using the Yang–Baxter equation, one can find an infinite family of operators commuting with the Hamiltonian of the spin chain. This is a quantum analogue of Liouville integrability, where there exists a maximal set of invariants that Poisson commute with the classical Hamiltonian.

form

$$\sum_x W\begin{pmatrix} a & b \\ x & c \end{pmatrix} u,v \end{pmatrix} W\begin{pmatrix} f & a \\ e & x \end{pmatrix} u,w \end{pmatrix} W\begin{pmatrix} e & x \\ d & c \end{pmatrix} v,w \end{pmatrix}$$

$$= \sum_x W\begin{pmatrix} f & a \\ x & b \end{pmatrix} v,w \end{pmatrix} W\begin{pmatrix} x & b \\ d & c \end{pmatrix} u,w \end{pmatrix} W\begin{pmatrix} f & x \\ e & d \end{pmatrix} u,v \end{pmatrix}. \quad (4.79)$$

There is another natural relation for Boltzmann weights known as the unitarity relation. Pictorially, it means that two consecutive crossings of rapidity lines cancel; that is, that the systems

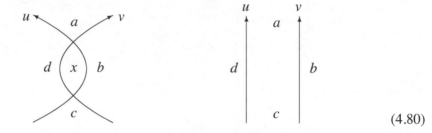

(4.80)

have the same partition function. Note that, if $a \neq c$, the system on the right does not exist (heights may only change as we cross a rapidity line) so we set its weight to zero. If $a = c$, there are no crossings, so the weight is the empty product 1. On the left, we should as before sum over x. This leads to the condition

$$\sum_x W\begin{pmatrix} d & x \\ c & b \end{pmatrix} u,v \end{pmatrix} W\begin{pmatrix} d & a \\ x & b \end{pmatrix} v,u \end{pmatrix} = \delta_{ac}. \quad (4.81)$$

4.3.3 The R-operator

Focusing on the relative changes in height, it is often useful to rewrite the Boltzmann weights in the notation

$$W\begin{pmatrix} a & b \\ c & d \end{pmatrix} u,v \end{pmatrix} = R_{d-b,b-a}^{c-a,d-c}(a|u,v),$$

or equivalently

$$R_{kl}^{mn}(\lambda|u,v) = W\begin{pmatrix} \lambda & \lambda+l \\ \lambda+m & \lambda+N \end{pmatrix} u,v \end{pmatrix}, \qquad k+l = m+n = N.$$

We want to view R_{kl}^{mn} as matrix elements of an operator, which we can think of as representing a crossing of two rapidity lines. To this end, let

Λ be the set of allowed differences between adjacent heights, V a vector space with basis $(e_\lambda)_{\lambda \in \Lambda}$ and $R(\lambda|u,v)$ the operator on $V \otimes V$ defined by

$$R(\lambda|u,v)(e_k \otimes e_l) = \sum_{m+n=k+l} R^{kl}_{mn}(\lambda|u,v)(e_m \otimes e_n).$$

In (4.79), let

$$i = e - f, \quad j = d - e, \quad k = c - d, \quad l = c - b, \quad m = b - a, \quad n = a - f$$

denote the various height differences encountered when travelling around the triple crossing. We replace x by $c - x$ on the left-hand side and $x + f$ on the right and finally let $f = \lambda$. This gives

$$\sum_x R^{l+m-x,x}_{lm}(\lambda + n|u,v)R^{i,j+k-x}_{l+m-x,n}(\lambda|u,w)R^{jk}_{x,j+k-x}(\lambda + i|v,w)$$

$$= \sum_x R^{x,m+n-x}_{mn}(\lambda|v,w)R^{i+j-x,k}_{l,m+n-x}(\lambda + x|u,w)R^{ij}_{i+j-x,x}(\lambda|u,v). \quad (4.82)$$

We consider the two sides as matrix elements for operators on $V \otimes V \otimes V$, where we act on $e_i \otimes e_j \otimes e_k$ and pick out the coefficient of $e_l \otimes e_m \otimes e_n$. The resulting coordinate-free form of (4.79) was first given by Felder [11]. It can be written as

$$R^{12}(\lambda + h^3|u,v)R^{13}(\lambda|u,w)R^{23}(\lambda + h^1|v,w)$$

$$= R^{23}(\lambda|v,w)R^{13}(\lambda + h^2|u,w)R^{12}(\lambda|u,v).$$

Here, the upper indices determine the spaces in the tensor product where the operators are acting, and h is the grading operator $he_j = je_j$. For instance,

$$R^{12}(\lambda|u,v)(e_i \otimes e_j \otimes e_k) = (R(\lambda|u,v)(e_i \otimes e_j)) \otimes e_k$$

$$= \sum_x R^{ij}_{i+j-x,x}(\lambda|u,v)\,e_{i+j-x} \otimes e_x \otimes e_k,$$

$$R^{23}(\lambda + h^1|v,w)(e_i \otimes e_j \otimes e_k) = e_i \otimes (R(\lambda + i|v,w)(e_j \otimes e_k))$$

$$= \sum_x R^{jk}_{x,j+k-x}(\lambda + i|v,w)\,e_i \otimes e_x \otimes e_{j+k-x}.$$

We mention that there are several versions of the Yang–Baxter equation. The one encountered here is sometimes called the quantum dynamical Yang–Baxter equation. Important special cases include the quantum Yang–Baxter equation, when $R(\lambda|u,v)$ is independent of λ, and the hexagon identity for $6j$-symbols (Racah and q-Racah polynomials), when $R(\lambda|u,v)$ is independent of u and v.

Exercise 4.88 Show that the unitarity relation (4.81) can be expressed as

$$R^{12}(\lambda\,|u,v)R^{21}(\lambda\,|v,u) = \text{Id}.$$

4.3.4 The elliptic SOS model

The elliptic SOS model (also called eight-vertex-solid-on-solid model) was introduced by Baxter [4] in his solution of a related model known as the eight-vertex model. In the elliptic SOS model, neighbouring heights differ by exactly 1. In particular, if one height is a then necessarily all heights are in $a + \mathbb{Z}$. The Boltzmann weights only depend on the quotient of the two rapidities at a crossing. For this reason, we write

$$W\begin{pmatrix} a & b \\ c & d \end{pmatrix}u,v\end{pmatrix} = W\begin{pmatrix} a & b \\ c & d \end{pmatrix}u/v\end{pmatrix}, \qquad R^{mn}_{kl}(\lambda\,|u,v) = R^{mn}_{kl}(\lambda\,|u/v).$$

$$(4.83)$$

The indices in (4.83) satisfy $k, l, m, n \in \{\pm 1\}$ and $k+l = m+n$, which has six solutions. Writing $\pm$ instead of ± 1, the Boltzmann weights are given by

$$R^{++}_{++}(\lambda\,|u) = R^{--}_{--}(\lambda\,|u) = 1, \tag{4.84a}$$

$$R^{+-}_{+-}(\lambda\,|u) = \frac{\theta(q^{1-\lambda},u;p)}{\theta(q^{-\lambda},uq;p)}, \qquad R^{-+}_{-+}(\lambda\,|u) = \frac{\theta(q^{\lambda+1},u;p)}{\theta(q^{\lambda},uq;p)}, \tag{4.84b}$$

$$R^{+-}_{-+}(\lambda\,|u) = \frac{\theta(q,q^{-\lambda}u;p)}{\theta(q^{-\lambda},uq;p)}, \qquad R^{-+}_{+-}(\lambda\,|u) = \frac{\theta(q,q^{\lambda}u;p)}{\theta(q^{\lambda},uq;p)}. \tag{4.84c}$$

Here, p and q are fixed parameters with $|p| < 1$. We will assume (4.34), though the case when q is a root of unity is in fact of special interest. To make physical sense, one should choose the parameters so that the Boltzmann weights are positive, but that will not be a concern for us. It will be useful to note the symmetries

$$R^{mn}_{kl}(\lambda\,|u) = R^{-m,-n}_{-k,-l}(-\lambda\,|u) = R^{nm}_{lk}(-\lambda - k - l\,|u) \tag{4.85}$$

or, equivalently,

$$W\begin{pmatrix} a & b \\ c & d \end{pmatrix}u\end{pmatrix} = W\begin{pmatrix} -a & -b \\ -c & -d \end{pmatrix}u\end{pmatrix} = W\begin{pmatrix} d & b \\ c & a \end{pmatrix}u\end{pmatrix}. \tag{4.86}$$

Theorem 4.89 *The Boltzmann weights (4.84) satisfy the Yang–Baxter equation (4.82).*

Unfortunately, we cannot present an elegant proof of Theorem 4.89, so we resort to brute force verification. A priori, we need to check (4.82) for each choice of $i, j, k, l, m, n \in \{\pm 1\}$ with $i + j + k = l + m + n$. This leads to 20 equations. Fortunately, the number can be reduced by exploiting symmetries of the Boltzmann weights. Indeed, applying the first equation in (4.85) to (4.82) and replacing λ by $-\lambda$ gives back (4.82) with $(i, j, k, l, m, n) \mapsto (-i, -j, -k, -l, -m, -n)$. Thus, we may assume $i + j + k > 0$, leaving us with 10 equations. Similarly, applying the second equation in (4.85) to (4.82) and replacing

$$(\lambda, u, v, w) \mapsto (-\lambda - i - j - k, w^{-1}, v^{-1}, u^{-1})$$

leads to (4.82) after the permutations $i \leftrightarrow k$ and $l \leftrightarrow n$. Thus, we may assume $i \geq k$ and if $i = k$ we may in addition assume $l \geq n$. Now, we are down to the six equations given in the following table, where we also write x_L and x_R for the admissible values of x at the left-hand and right-hand side of (4.82).

i	j	k	l	m	n	x_L	x_R
+	+	+	+	+	+	+	+
+	+	-	+	+	-	+	+
+	+	-	+	-	+	$\pm$	+
+	+	-	-	+	+	$\pm$	+
+	-	+	+	+	-	+	$\pm$
+	-	+	+	-	+	$\pm$	$\pm$

In particular, the number of terms in these identities are, respectively, 2, 2, 3, 3, 3 and 4. The two-term identities are trivial. It is easy to check that the three-term identities are all equivalent to Weierstrass's identity (4.13). Finally, the four-term identity has the form

$$R_{+-}^{-+}(\lambda + 1 | u/v) R_{-+}^{+-}(\lambda | u/w) R_{+-}^{-+}(\lambda + 1 | v/w)$$
$$+ R_{+-}^{+-}(\lambda + 1 | u/v) R_{++}^{++}(\lambda | u/w) R_{-+}^{-+}(\lambda + 1 | v/w)$$
$$= R_{-+}^{+-}(\lambda | v/w) R_{+-}^{-+}(\lambda + 1 | u/w) R_{-+}^{+-}(\lambda | u/v)$$
$$+ R_{-+}^{-+}(\lambda | v/w) R_{++}^{++}(\lambda - 1 | u/w) R_{+-}^{+-}(\lambda | u/v)$$

or, equivalently,

$$\frac{\theta(q;p)^3\,\theta(q^{\lambda+1}u/v,q^{-\lambda}u/w,q^{\lambda+1}v/w;p)}{\theta(q^{\lambda+1},uq/v,q^{-\lambda},uq/w,q^{\lambda+1},vq/w;p)}$$

$$+\frac{\theta(q^{-\lambda},u/v,q^{\lambda+2},v/w;p)}{\theta(q^{-\lambda-1},uq/v,q^{\lambda+1},vq/w;p)}$$

$$=\frac{\theta(q;p)^3\,\theta(q^{-\lambda}v/w,q^{\lambda+1}u/w,q^{-\lambda}u/v;p)}{\theta(q^{-\lambda},vq/w,q^{\lambda+1},uq/w,q^{-\lambda},uq/v;p)}$$

$$+\frac{\theta(q^{\lambda+1},v/w,q^{1-\lambda},u/v;p)}{\theta(q^{\lambda},vq/w,q^{-\lambda},uq/v;p)}. \tag{4.87}$$

We prove this by interpolation. We first consider both sides as functions of v. By Proposition 4.27, it is enough to verify the identity for two values, $v = v_1$ and $v = v_2$, such that neither v_1/v_2 nor v_1v_2/uw is in $p^{\mathbb{Z}}$. We choose $v_1 = u$, which gives a trivial identity, and $v_2 = q^{\lambda}w$, which after simplification gives

$$-q^{\lambda-1}\,\theta(q;p)^3\,\theta(q^{2\lambda+1};p)+\theta(q^{\lambda};p)^3\,\theta(q^{\lambda+2};p)$$
$$=\theta(q^{\lambda+1};p)^3\,\theta(q^{\lambda-1};p).$$

This is a special case of Weierstrass's identity (4.13).

Exercise 4.90 Give an alternative proof of (4.87) by rearranging the terms and using (4.13).

Exercise 4.91 Show directly that the unitarity relation (4.81) holds in the elliptic SOS model. (We will see another way to do this in Exercise 4.95.)

4.3.5 Fusion and elliptic hypergeometry

For Baxter's model, adjacent heights differ by exactly 1. One can obtain less restrictive models by applying a so called fusion procedure. Date et al. [8] found that the Boltzmann weights of these fused models are given by $_{12}V_{11}$-sums. We will give a new approach to this result, by relating it to our construction of $_{12}V_{11}$-sums as connection coefficients described in §4.2.5.

As a starting point, the following result gives a description of Baxter's Boltzmann weights as connection coefficients in a two-dimensional space of theta functions.

Proposition 4.92 *For $c = a \pm 1$, let $\phi(a,c|u)$ denote the function*

$$\phi(a,c|u)(x) = \frac{\theta(q^{a(c-a)/2}\sqrt{u}x^{\pm};p)}{q^{a(c-a)/2}\sqrt{u}}.$$

Then, the Boltzmann weights of the elliptic SOS model are determined by the expansion

$$\phi(b,d|u) = \sum_c W\left(\begin{matrix} a & b \\ c & d \end{matrix}\middle| u\right)\phi(a,c|uq). \tag{4.88}$$

Proof Fixing a, there are four possible choices for the pair (b,d), corresponding to the identities

$$\phi(a+1,a|u) = \sum_{j=\pm1} W\left(\begin{matrix} a & a+1 \\ a+j & a \end{matrix}\middle| u\right)\phi(a,a+j|uq),$$

$$\phi(a-1,a|u) = \sum_{j=\pm1} W\left(\begin{matrix} a & a-1 \\ a+j & a \end{matrix}\middle| u\right)\phi(a,a+j|uq),$$

$$\phi(a+1,a+2|u) = W\left(\begin{matrix} a & a+1 \\ a+1 & a+2 \end{matrix}\middle| u\right)\phi(a,a+1|uq),$$

$$\phi(a-1,a-2|u) = W\left(\begin{matrix} a & a-1 \\ a-1 & a-2 \end{matrix}\middle| u\right)\phi(a,a-1|uq)$$

or, equivalently,

$$\theta(\sqrt{uq}^{-\frac{\lambda+1}{2}}x^\pm;p) = q^{-\lambda-1}R_{-+}^{+-}(\lambda|u)\theta(\sqrt{uq}^{\frac{\lambda+1}{2}}x^\pm;p)$$
$$+q^{-1}R_{-+}^{-+}(\lambda|u)\theta(\sqrt{uq}^{\frac{1-\lambda}{2}}x^\pm;p),$$

$$\theta(\sqrt{uq}^{\frac{\lambda-1}{2}}x^\pm;p) = q^{-1}R_{+-}^{+-}(\lambda|u)\theta(\sqrt{uq}^{\frac{\lambda+1}{2}}x^\pm;p)$$
$$+q^{\lambda-1}R_{+-}^{-+}(\lambda|u)\theta(\sqrt{uq}^{\frac{1-\lambda}{2}}x^\pm;p),$$

$$\theta(\sqrt{uq}^{\frac{\lambda+1}{2}}x^\pm;p) = R_{++}^{++}(\lambda|u)\theta(\sqrt{uq}^{\frac{\lambda+1}{2}}x^\pm;p),$$

$$\theta(\sqrt{uq}^{\frac{1-\lambda}{2}}x^\pm;p) = R_{--}^{--}(\lambda|u)\theta(\sqrt{uq}^{\frac{1-\lambda}{2}}x^\pm;p).$$

The first two expansions are of the form (4.12), so the coefficients are given by Weierstrass's identity (4.13). The second two identities are trivial. This leads to the explicit expressions for Boltzmann weights given in (4.84). $\qquad\square$

Fusion of the SOS model corresponds to iterating (4.88). We have

$$\phi(b,d|u) = \sum_f W\left(\begin{matrix} e & b \\ f & d \end{matrix}\middle| u\right)\phi(e,f|uq)$$

$$= \sum_{fc} W\left(\begin{matrix} a & e \\ c & f \end{matrix}\middle| uq\right)W\left(\begin{matrix} e & b \\ f & d \end{matrix}\middle| u\right)\phi(a,c|uq^2).$$

As the two functions $\phi(a,a\pm1|uq^2)$ are generically linearly independent, it follows that

$$\sum_f W\left(\begin{matrix} a & e \\ c & f \end{matrix}\middle| uq\right) W\left(\begin{matrix} e & b \\ f & d \end{matrix}\middle| u\right) \tag{4.89}$$

is independent of e, as long as $|a-e| = |b-e| = 1$. Equivalently, the partition function for the system

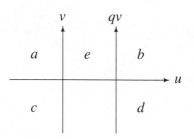

where we sum over the admissible heights of the empty slot, is independent of e. Thus, we can forget about e, and think of the vertical lines as coalescing. We may view this sum as a Boltzmann weight for a *fused* SOS-model, for which $a-c$ and $b-d$ are in $\{\pm 1\}$, whereas $b-a$ and $d-c$ are in $\{-2,0,2\}$.

Next, we observe that fusion works also in the vertical direction. Indeed, by (4.86), the partition function for

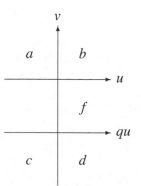

is independent of f. Iterating fusion in both directions, we find that the

partition function for

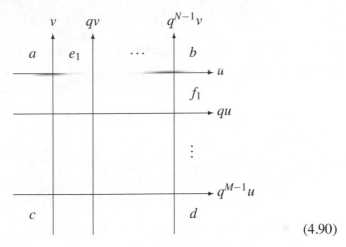

$$(4.90)$$

where we fix the corners together with the top and right boundary and sum over the admissible heights of all other squares, is independent of the interior boundary heights $e_1,\ldots,e_{N-1}, f_1,\ldots,f_{M-1}$, provided that adjacent boundary heights differ by 1. We denote the resulting quantity

$$W_{MN}\left(\begin{matrix} a & b \\ c & d \end{matrix}\,\middle|\, u/v\right).$$

Here, $a-c$ and $b-d$ are in $\{-M, 2-M,\ldots,M\}$ whereas $a-b$ and $c-d$ are in $\{-N, 2-N,\ldots,N\}$. We view W_{MN} as Boltzmann weights for a model, where each rapidity line is labelled by a multiplicity that gives the maximal allowed height difference across that line.

In order to identify the fused Boltzmann weights W_{MN} with elliptic hypergeometric sums, we will need the following fact.

Lemma 4.93 *If $|b-a| = |c-b| = 1$, the product $\phi(a,b|uq^{-1})\phi(b,c|u)$ is independent of b.*

Proof If $|a-c| = 2$ there is only one admissible value for b and there is nothing to prove. Else, $a = c$ and the claim is that

$$\phi(a,a+1|uq^{-1})\phi(a+1,a|u) = \phi(a,a-1|uq^{-1})\phi(a-1,a|u).$$

This is trivial to verify. $\qquad\square$

We will write

$$\phi_M(a,c|u) = \phi(a,b_1|uq^{1-M})\phi(b_1,b_2|uq^{2-M})\cdots\phi(b_{M-1},c|u),$$

where $a - c \in \{-M, 2-M, \ldots, M\}$ and

$$b_1 - a, \; b_2 - b_1, \ldots, \; c - b_{M-1} \in \{\pm 1\}. \qquad (4.91)$$

By Lemma 4.93, ϕ_M is independent of the parameters b_j. Thus, we may take the first $(M+c-a)/2$ of the numbers (4.91) as 1 and the remaining $(M+a-c)/2$ as -1. After simplification, this gives the explicit formula

$$\phi_M(a,c|u) = \frac{q^{\frac{1}{4}(c^2-a^2-M^2)}}{(\sqrt{u})^M}$$
$$\times (q^{(1-M+a)/2}\sqrt{u}x^{\pm};q,p)_{(M+c-a)/2}(q^{(1-M-a)/2}\sqrt{u}x^{\pm};q,p)_{(M+a-c)/2}.$$

We can now generalize Proposition 4.92 to fused models.

Theorem 4.94 *The Boltzmann weights of the fused elliptic SOS models satisfy*

$$\phi_M(b,d|u) = \sum_c W_{MN}\begin{pmatrix} a & b \\ c & d \end{pmatrix} u\; \phi_M(a,c|uq^N). \qquad (4.92)$$

Proof For simplicity we give the proof for $M = N = 2$, but it should be clear that the same argument works in general. By definition,

$$W_{22}\begin{pmatrix} a & b \\ c & d \end{pmatrix} u = \sum_{ghk} W\begin{pmatrix} a & e \\ g & h \end{pmatrix} u\; W\begin{pmatrix} e & b \\ h & f \end{pmatrix} q^{-1}u$$
$$\times W\begin{pmatrix} g & h \\ c & k \end{pmatrix} qu\; W\begin{pmatrix} h & f \\ k & d \end{pmatrix} u,$$

where e and f are arbitrary admissible heights. Since $|a-g| = |c-g| = 1$ we may write

$$\phi_2(a,c|uq^2) = \phi(a,g|uq)\phi(g,c|uq^2)$$

independently of g. Thus, the right-hand side of (4.92) is

$$\sum_{cghk} W\begin{pmatrix} a & e \\ g & h \end{pmatrix} u\; W\begin{pmatrix} e & b \\ h & f \end{pmatrix} q^{-1}u$$
$$\times W\begin{pmatrix} g & h \\ c & k \end{pmatrix} qu\; W\begin{pmatrix} h & f \\ k & d \end{pmatrix} u\; \phi(a,g|qu)\phi(g,c|q^2u).$$

Computing this sum by repeated application of (4.88), first for the sum over c, then for g and k and finally for h, we eventually arrive at

$$\phi(b,f|q^{-1}u)\phi(f,d|u) = \phi_2(b,d|u). \qquad \square$$

We now observe that (4.92) is a special case of the expansion (4.51). Explicitly,

$$W_{MN}\begin{pmatrix} a & b \\ c & d \end{pmatrix} u) = q^{\frac{1}{4}(a^2+d^2-b^2-c^2-2MN)}$$

$$\times R_{(M+d-b)/2}^{(M+c-u)/2}(q^{\frac{1-M+b}{2}}\sqrt{u}, q^{\frac{1-M-b}{2}}\sqrt{u}, q^{\frac{1-M+N+a}{2}}\sqrt{u}, q^{\frac{1-M+N-a}{2}}\sqrt{u}; M).$$

Thus, it follows from (4.52) that W_{MN} can be written as an elliptic hypergeometric sum.

Finally, we mention that the fused Boltzmann weights satisfy the Yang–Baxter equation in the form

$$\sum_x W_{MN}\begin{pmatrix} a & b \\ x & c \end{pmatrix} u/v) W_{MP}\begin{pmatrix} f & a \\ e & x \end{pmatrix} u/w) W_{NP}\begin{pmatrix} e & x \\ d & c \end{pmatrix} v/w)$$

$$= \sum_x W_{NP}\begin{pmatrix} f & a \\ x & b \end{pmatrix} v/w) W_{MP}\begin{pmatrix} x & b \\ d & c \end{pmatrix} u/w) W_{MN}\begin{pmatrix} f & x \\ e & d \end{pmatrix} u/v), \quad (4.93)$$

which is quite non-trivial when viewed as a hypergeometric identity. We will prove (4.93) for $M = N = P = 2$, but it will be clear that the argument works in general. To this end, we start with the partition functions defined by the pictures

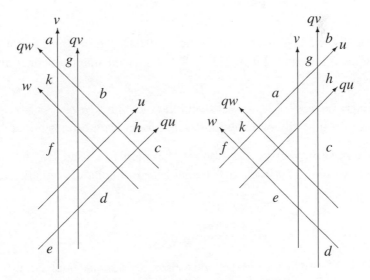

As usual, we sum over all admissible heights for the empty slots. We claim that these two partition functions are equal. Indeed, using the Yang–Baxter equation we can pull the lines labelled v and qv through the crossings of the other four lines, thus passing between the two pictures without changing the partition function. Note that this one-sentence pictorial (though rigorous) argument corresponds to an eight-fold application of the identity (4.82) and would thus look quite daunting if written out with explicit formulas. Let us now write the partition function on the left as

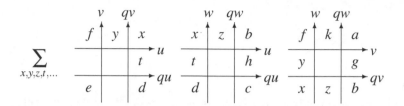

Here, we only indicate the summation variables that are shared by two factors; each empty slot carries an additional independent summation variable. For fixed t, x and y, the first factor is of the form (4.90). Thus, after summing over the empty slots, it is independent of y and t and equal to

$$W_{22}\begin{pmatrix} f & x \\ e & d \end{pmatrix} u/v \Big).$$

As t now only appears in the second factor, we can sum over the empty slots together with t and obtain

$$W_{22}\begin{pmatrix} x & b \\ d & c \end{pmatrix} u/w \Big).$$

Finally, we sum over y, z and the empty slot in the third factor and arrive at the right-hand side of (4.93). The left-hand side is obtained in the same way.

Exercise 4.95 Using (4.86) and the fact that $\phi(a,c|u) = \phi(c,a|q/u)$, derive the unitarity of the elliptic SOS model from Proposition 4.92.

Exercise 4.96 Prove a unitarity relation for fused Boltzmann weights and verify that it leads to a special case of the biorthogonality relations of §4.2.6.

References

[1] G.E. Andrews, R. Askey and R. Roy, *Special Functions*, Cambridge University Press, 1999.
[2] T.M. Apostol, *Modular Functions and Dirichlet Series in Number Theory.* Springer-Verlag, 1976.
[3] R. Askey and J. Wilson, *Some basic hypergeometric orthogonal polynomials that generalize Jacobi polynomials*, Mem. Amer. Math. Soc. 54 (1985), no. 319.

[4] R.J. Baxter, *Eight-vertex model in lattice statistics and one-dimensional anisotropic Heisenberg chain II. Equivalence to a generalized ice-type model*, Ann. Phys. 76 (1973), 25–47.

[5] R.J. Baxter, *Exactly Solved Models in Statistical Mechanics*. Academic Press, 1982.

[6] V V Bazhanov and S.M. Sergeev, *A master solution of the quantum Yang-Baxter equation and classical discrete integrable equations*, Adv. Theor. Math. Phys. 16 (2012), 65–95.

[7] D. Betea, *Elliptically distributed lozenge tilings of a hexagon*, SIGMA 14 (2018), paper 032.

[8] E. Date, M. Jimbo, A. Kuniba, T. Miwa and M. Okado, *Exactly solvable SOS models. II. Proof of the star-triangle relation and combinatorial identities*, in Conformal Field Theory and Solvable Lattice Models, Academic Press, 1988, pp. 17–122.

[9] S.E. Derkachov and V. P. Spiridonov, *Finite dimensional representations of the elliptic modular double*, Theor. Math. Phys. 183 (2015), 597–618.

[10] F.A. Dolan and H. Osborn, *Applications of the superconformal index for protected operators and q-hypergeometric identities to $N = 1$ dual theories*, Nucl. Phys. B 818 (2009), 137–178.

[11] G. Felder, *Conformal field theory and integrable systems associated to elliptic curves*. In *Proceedings of the International Congress of Mathematicians (Zürich, 1994)*, Birkhäuser, 1995, pp. 1247–1255.

[12] I.B. Frenkel and V.G. Turaev, *Elliptic solutions of the Yang-Baxter equation and modular hypergeometric functions*. In *The Arnold–Gelfand Mathematical Seminars*, Birkhäuser, 1997, pp. 171–204.

[13] G. Gasper and M. Rahman, *Basic Hypergeometric Series*, Second edition, Cambridge University Press, 2004.

[14] G. Gasper and M.J. Schlosser, *Summation, transformation, and expansion formulas for multibasic theta hypergeometric series*, Adv. Stud. Contemp. Math. (Kyungshang) 11 (2005), 67–84.

[15] M.E.H. Ismail, *A review of q-series*, this volume.

[16] M. Jimbo and T. Miwa, *Algebraic Analysis of Solvable Lattice Models*, Amer. Math. Soc., 1995.

[17] Y. Kajihara and M. Noumi, *Multiple elliptic hypergeometric series. An approach from the Cauchy determinant*, Indag. Math. 14 (2003), 395–421.

[18] R. Koekoek and R. F. Swarttouw, The Askey scheme of hypergeometric orthogonal polynomials and its q-analogue, Delft University of Technology, 1998, homepage.tudelft.nl/11r49/askey/.

[19] E. Koelink, Y. van Norden, and H. Rosengren, *Elliptic $U(2)$ quantum group and elliptic hypergeometric series*, Comm. Math. Phys. 245 (2004), 519–537.

[20] T.H. Koornwinder, *On the equivalence of two fundamental theta identities*, Anal. Appl. (Singap.) 12 (2014), 711–725.

[21] B.M. Minton, *Generalized hypergeometric function of unit argument*, J. Math. Phys. 11 (1970), 1375–1376.

[22] B. Nassrallah and M. Rahman, *Projection formulas, a reproducing kernel and a generating function for q-Wilson polynomials*, SIAM J. Math. Anal. 16 (1985), 186–197.

[23] H. Rademacher, Topics in Analytic Number Theory, Springer-Verlag, 1973.

[24] M. Rahman, *An integral representation of a $_{10}\phi_9$ and continuous biorthogonal $_{10}\phi_9$ rational functions*, Canad. J. Math. 38 (1986), 605–618.

[25] E.M. Rains, *BC_n-symmetric abelian functions*, Duke Math. J. 135 (2006), 99–180.

[26] E.M. Rains, *Transformations of elliptic hypergeometric integrals*, Ann. Math. 171 (2010), 169–243.

[27] H. Rosengren, *Elliptic hypergeometric series on root systems*, Adv. Math. 181 (2004), 417–447.

[28] H. Rosengren, *Sklyanin invariant integration*, Int. Math. Res. Not. 2004 (2004), 3207–3232.

[29] H. Rosengren, *An elementary approach to 6j-symbols (classical, quantum, rational, trigonometric, and elliptic)*, Ramanujan J. 13 (2007), 133–168.

[30] H. Rosengren, Bibliography of Elliptic Hypergeometric Functions, www.math.chalmers.se/~hjalmar/bibliography.html.

[31] H. Rosengren and M. Schlosser, *On Warnaar's elliptic matrix inversion and Karlsson–Minton-type elliptic hypergeometric series*, J. Comput. Appl. Math. 178 (2005), 377–391.

[32] H. Rosengren and S.O. Warnaar, *Elliptic hypergeometric functions associated with root systems*. To appear in *Multivariable Special Functions*, T.H. Koornwinder and J.V. Stokman (eds.), Cambridge University Press. arXiv:1704.08406.

[33] S.N.M. Ruijsenaars, *First order analytic difference equations and integrable quantum systems*, J. Math. Phys. 38 (1997), 1069–1146.

[34] M.J. Schlosser, *Elliptic enumeration of nonintersecting lattice paths*, J. Combin. Theory A 114 (2007), 505–521.

[35] V.P. Spiridonov, *An elliptic beta integral*. In *New Trends in Difference Equations*, Taylor & Francis, 2002, pp. 273-282.

[36] V.P. Spiridonov, *Theta hypergeometric series*. In *Asymptotic Combinatorics with Applications to Mathematical Physics*, Kluwer, 2002, pp. 307–327.

[37] V.P. Spiridonov, *Theta hypergeometric integrals*, St. Petersburg Math. J. 15 (2004), 929–967.

[38] V.P. Spiridonov, *Short proofs of the elliptic beta integrals*, Ramanujan J. 13 (2007), 265–283.

[39] V.P. Spiridonov, *Essays on the theory of elliptic hypergeometric functions*, Russian Math. Surveys 63 (2008), 405–472.

[40] V.P. Spiridonov, *Elliptic beta integrals and solvable models of statistical mechanics*. In *Algebraic Aspects of Darboux Transformations, Quantum Integrable Systems and Supersymmetric Quantum Mechanics*, Contemp. Math. 563, Amer. Math. Soc., 2012, pp. 345–353.

[41] V. P. Spiridonov, *Aspects of elliptic hypergeometric functions*. In *The Legacy of Srinivasa Ramanujan*, Ramanujan Math. Soc., 2013, pp. 347–361.

[42] V.P. Spiridonov and G.S. Vartanov, *Elliptic hypergeometry of supersymmetric dualities*, Comm. Math. Phys. 304 (2011), 797–874.

[43] V.P. Spiridonov and G.S. Vartanov, *Elliptic hypergeometry of supersymmetric dualities II. Orthogonal groups, knots, and vortices*, Comm. Math. Phys. 325 (2014), 421–486.

[44] V.P. Spiridonov and A.S. Zhedanov, *Spectral transformation chains and some new biorthogonal rational functions*, Comm. Math. Phys. 210 (2000), 49–83.

[45] J. Tannery and J. Molk, *Éléments de la Théorie des Fonctions Elliptiques, Tome III: Calcul Intégral*, Gauthier-Villars, 1898.

[46] S.O. Warnaar, *Summation and transformation formulas for elliptic hypergeometric series*, Constr. Approx. 18 (2002), 479–502.

[47] E.T. Whittaker and G.N. Watson, *A Course of Modern Analysis*, Cambridge University Press, 1927.

5

Combinatorics of Orthogonal Polynomials and their Moments

Jiang Zeng

Abstract: The aim of this chapter is to introduce the formal theory of general orthogonal polynomials and present the two dual combinatorial approaches due to Foata for the special function aspects of the orthogonal polynomials, and to Flajolet and Viennot for the lattice paths models used for the moments and general orthogonal polynomials. After reviewing the standard interplay between orthogonal polynomials and combinatorics, influenced by their pioneering works, we will report on some recent topics developed in this cross-cutting field of these two branches of mathematics.

5.1 Introduction

There are many interactions between special functions/orthogonal polynomials and combinatorics. In this lecture we will give an introduction to the basic theory of orthogonal polynomials of one variable, combinatorial theory of generating functions and the combinatorial approaches to orthogonal polynomials, which were initiated by Dominique Foata, Philippe Flajolet and G. Xavier Viennot in the 1980s. We will also survey some recent results influenced or inspired by their combinatorial views.

In the late 1970s using various combinatorial models, such as permutations, injections, set partitions, etc., Foata and his coauthors provided combinatorial or geometrical proofs of numerous important identities in the domain of special functions and orthogonal polynomials, which were originally established analytically. An illustrative example of this approach is Foata's combinatorial proof [18] of Mehler's formula for Hermite polynomials $\tilde{H}_n(x)$ (see (5.38))

$$\sum_{n \geq 0} \tilde{H}_n(x)\tilde{H}_n(y)\frac{t^n}{n!} = \frac{1}{\sqrt{1-t^2}} \exp\left[\frac{xyt - (x^2+y^2)t^2/2}{1-t^2}\right]. \qquad (5.1)$$

Basically, this approach deals with the special function aspects of the orthogonal polynomials, i.e., one starts by interpreting a family of polynomials as generating functions of certain combinatorial structures and then looks for combinatorial structures underlying classical formulae. Once the combinatorial structure is identified, it is then possible to evaluate its generating function using the tools of combinatorial theory of generating functions, whose fundamental ingredient is the *exponential formula* of labeled structures. At about the same time Flajolet [16] was developing his combinatorial theory of continued fractions of Jacobi type or J-continued fractions

$$\sum_{n\geq0}\mu_n t^n = \cfrac{1}{1 - b_0 t - \cfrac{\lambda_1 t^2}{1 - b_1 t - \cfrac{\lambda_2 t^2}{\ddots \cfrac{}{1 - b_k t - \cfrac{\lambda_{k+1} t^2}{\ddots}}}}}. \tag{5.2}$$

His starting observation was that the coefficient μ_n in the above formal power series expansion is the generating polynomial for Motzkin paths of length n under the weight function assigning weight λ_k (resp. b_k, 1) on southeast (resp. east, northeast) steps starting at level k. By a classical result in the theory of orthogonal polynomials, the coefficients μ_n in the formal power series expansion in (5.2) are the moments of the orthogonal polynomial sequence (OPS) defined by $p_{n+1}(x) = (x - b_n)p_n(x) - \lambda_n p_{n-1}(x)$, $p_0(x) = 1$. Thus Flajolet's result gives the first combinatorial description of their moments μ_n.

The results are extremely nice when one deals with Sheffer polynomials using the above combinatorial approach. Recall that Sheffer polynomials are generated by

$$f(t)\exp(xg(t)) = \sum_{n\geq0} p_n(x)\frac{t^n}{n!},$$

where $f(0) = 1$, $g(0) = 0$, $g'(0) \neq 0$, and f and g are either analytic in a neighborhood of 0 or have formal power series there. Meixner determined all Sheffer polynomials which are orthogonal. They include Hermite and Laguerre polynomials, discrete analogues named after Charlier and Meixner, and Meixner–Pollaczek polynomials with an absolutely continuous weight function. In view of Flajolet's combinatorial theory of continued fractions, the bijection found by Françon and Viennot [26] can be considered as a combinatorial proof of the simple fact that the n-th moment for the Laguerre polynomials is $n!$, the number of permutations of $\{1,\ldots,n\}$. Based on the above two papers, Viennot [57] wrote the seminal memoir in which he rebuilt combinatorially the formal theory of orthogonal polynomials. In the same vein, Gessel and Viennot [27] applied non-intersecting lattice paths to problems in enumerative combinatorics. It is remarkable that the moment sequences of the above five Sheffer OPS correspond to the classical enumerative sequences of perfect matchings, partitions and permutations with suitable weights. Furthermore, the two dual approaches of Foata, Flajolet and Viennot have finally merged to a unified treatment of the linearization coefficients for the five families of orthogonal Sheffer polynomials; see [3, 5, 15, 24, 64].

When we enter the q-world of orthogonal polynomials, the main difficulty is the choice of a good q-analogue, also the generating function approach to the Sheffer polynomials does not work. The Askey–Wilson polynomials are the most general classical orthogonal polynomials. Their orthogonality relies on the evaluation of the Askey–Wilson q-beta integral

$$\frac{(q;q)_\infty}{2\pi} \int_0^\pi \frac{h(\cos 2\theta;1)}{h(\cos\theta;t_1,t_2,t_3,t_4)}\,d\theta = \frac{(t_1 t_2 t_3 t_4;q)_\infty}{\prod_{1\leq r<s\leq 4}(t_r t_s;q)_\infty}, \tag{5.3}$$

where $\max\{|t_1|,|t_2|,|t_3|,|t_4|\} < 1$ and

$$h(\cos\theta;t_1,\ldots,t_r) = \prod_{j=1}^r (t_j e^{i\theta}, t_j e^{-i\theta};q)_\infty.$$

Ismail, Stanton and Viennot's combinatorial evaluation of Askey–Wilson integral was a spectacular achievement of this combinatorial approach [35]. However, after that, no other combinatorial q-versions of the linearization coefficients formula for the Sheffer orthogonal polynomials appeared until Anshelevich [1] came up with a right one in 2005 using stochastic processes. Shortly later, a combinatorial proof à la Viennot of Anshelevich's result was given in [41] based on the combinatorics of moments and polynomials. The latter approach led finally to a new q-version of the linearization coefficients of Laguerre polynomials [42].

We assume that the reader is familiar with the elementary theory of OPS [9] and the basic enumerative techniques [59]. For completeness we first review the formal theory of the OPS in §5.2.1, and then introduce the Flajolet–Viennot path model for the moments and orthogonal polynomials in §5.2.2. In §5.3.1 we summarize the combinatorial theory of exponential generating functions for counting labeled structures and give some examples of applications for this theory. In particular, we show how to prove, by using only the *exponential formula*, MacMahon's Master Theorem and its β-analogue due to Foata–Zeilberger [24], and a recent Mehler-type formula for 2D-Hermite polynomials [37]. In Section 5.4 we review various combinatorial aspects of the orthogonal Sheffer polynomials such as the moments, the polynomials themselves and the linearization coefficients [64]. In Section 5.5, recalling some basic facts about the Al-Salam–Chihara polynomials in §5.5.1, we review the moments and linearization coefficients of the continuous q-Hermite polynomials [35], q-Charlier polynomials [41] and q-Laguerre polynomials [42] in §5.5.2 and §5.5.3. In §5.5.4 we examine a class of non-orthogonal q-Hermite polynomials [10], which are inverses of continuous q-Hermite polynomials; they are also connected to the normal ordering problem of two non-commutative variables [56]. In §5.5.5 we show how to apply the combinatorics of continued fractions to derive the γ-positivity of certain q-Eulerian polynomials [53]. Finally, we state some open problems related to the topics developed in these lectures.

5.2 General and combinatorial theories of formal OPS

5.2.1 Formal theory of orthogonal polynomials

Let $\mathbb{K}$ be a *commutative ring with zero characteristic*. In the following $\mathbb{K}$ is usually a polynomial ring of some indeterminates with coefficients in $\mathbb{Q}, \mathbb{R}$ or $\mathbb{C}$.

Definition 5.1 Let $\mathscr{F} : \mathbb{K}[x] \to \mathbb{K}$ be a linear functional. A sequence of polynomials $\{p_n(x)\}_{n \geq 0}$ in $\mathbb{K}[x]$ is said to be *orthogonal* with respect to the linear functional $\mathscr{F}$ if:

(i) $p_n(x)$ is of degree n;
(ii) $\mathscr{F}(p_n(x) p_m(x)) = K_n \delta_{n,m}, K_n \neq 0$.

The sequence $\{\mu_n\}_{n \geq 0}$ such that $\mu_n := \mathscr{F}(x^n)$ for $n \geq 0$ is called the *moment sequence associated* with $\mathscr{F}$.

Sometimes the polynomials $\{p_n(x)\}$ are also said to be orthogonal with respect to the sequence of moments $\{\mu_n\}_{n \geq 0}$.

Theorem 5.2 *If $\{p_n(x)\}$ and $\{q_n(x)\}$ are two OPS with respect to $\mathscr{F}$, there are non-zero constants $c_n \neq 0$ such that*

$$q_n(x) = c_n p_n(x), \quad n = 0, 1, 2, \ldots.$$

Proof By definition, we can write $q_n(x) = \sum_{k=0}^{n} c_{n,k} p_k(x)$. It follows that

$\mathscr{F}(q_n(x)p_k(x)) = c_{n,k}\mathscr{F}(p_k^2(x))$ for $0 \le k \le n$. Now, for $0 \le k \le n$ writing $p_k(x) = \sum_{l=0}^k \beta_{k,l}q_l(x)$ we see that

$$\mathscr{F}(q_n(x)p_k(x)) = \sum_{l=0}^k \beta_{k,l}\mathscr{F}(q_n(x)q_l(x)) = \beta_{n,n}\delta_{n,k}.$$

Since $\beta_{n,n} \ne 0$, this proves $c_{n,k} = 0$ for $k = 0,\dots,n-1$. $\qquad\square$

Definition 5.3   An OPS $\{p_n(x)\}$ is *monic* if $p_n(x) = x^n + \cdots$, and *orthonormal* if $\mathscr{F}(p_n(x)p_m(x)) = \delta_{n,m}$.

Not every sequence can be a *moment sequence*. For example, if $\mu_0 = \mu_1 = \mu_2 = 1$ and suppose that $\{p_n(x)\}$ is the corresponding monic OPS. Then $P_0(x) = 1$ and $P_1(x) = x - b_0$. But $\mathscr{F}(P_0(x)P_1(x)) = \mathscr{F}(x) - b_0 = 0$, so $b_0 = 1$ and $\mathscr{F}(P_1^2(x)) = \mathscr{F}((x-1)^2) = 1 - 2 + 1 = 0$.

We give a characterization of the moment sequence using Hankel determinants. Given a sequence $\{\mu_n\}$, its Hankel determinant is defined by

$$\Delta_n = \det(\mu_{i+j})_{i,j=0}^n = \begin{vmatrix} \mu_0 & \mu_1 & \cdots & \mu_n \\ \mu_1 & \mu_2 & \cdots & \mu_{n+1} \\ \vdots & \vdots & \ddots & \vdots \\ \mu_{n-1} & \mu_n & \cdots & \mu_{2n-1} \\ \mu_n & \mu_{n+1} & \cdots & \mu_{2n} \end{vmatrix}. \qquad (5.4)$$

Theorem 5.4   *Let $\mathscr{F}$ be the linear functional associated with the moment sequence $\{\mu_n\}$. There is an OPS with respect to $\mathscr{F}$ if and only if $\Delta_n \ne 0$ for all $n \in \mathbb{N}$.*

Proof ($\implies$) Let $p_n(x) = \sum_{k=0}^n c_{n,k}x^k$ $(n \ge 0)$ be the monic OPS with respect to the moment sequence $\{\mu_n\}$. Then $\mathscr{F}(x^m p_n(x)) = \sum_{k=0}^n c_{n,k}\mu_{k+m} = k_n\delta_{m,n}$ for $0 \le m \le n$. This can be written as

$$\begin{pmatrix} \mu_0 & \mu_1 & \cdots & \mu_n \\ \mu_1 & \mu_2 & \cdots & \mu_{n+1} \\ \vdots & \vdots & \ddots & \vdots \\ \mu_n & \mu_{n+1} & \cdots & \mu_{2n} \end{pmatrix} \cdot \begin{pmatrix} c_{n,0} \\ c_{n,1} \\ \vdots \\ c_{n,n} \end{pmatrix} = \begin{pmatrix} 0 \\ 0 \\ \vdots \\ k_n \end{pmatrix}.$$

Since $c_{n,n} = 1$, for $i = 1,2,\dots,n$, multiplying the ith column of Δ_n by $c_{n,i-1}$ and adding them to the last column we get $\Delta_n = k_n\Delta_{n-1} = k_nk_{n-1}\cdots k_1 \ne 0$.

($\Longleftarrow$) If $\Delta_n \neq 0$ for all n, we verify readily that

$$p_n(x) = \frac{1}{\Delta_{n-1}} \begin{vmatrix} \mu_0 & \mu_1 & \cdots & \mu_n \\ \vdots & \vdots & \ddots & \vdots \\ \mu_{n-1} & \mu_n & \cdots & \mu_{2n-1} \\ 1 & x & \cdots & x^n \end{vmatrix} \tag{5.5}$$

is the corresponding monic OPS.   □

Theorem 5.5 (Favard's theorem) *A sequence of polynomials $\{p_n(x)\}_{n \geq 0}$ in $\mathbb{K}[x]$ is a monic OPS if and only if there is a sequence $\{b_n\}_{n \geq 0}$ and a non-zero sequence $\{\lambda_n\}_{n \geq 0}$ such that $P_0(x) = 1$, $P_1(x) = x - b_0$ and*

$$p_{n+1}(x) = (x - b_n)p_n(x) - \lambda_n p_{n-1}(x) \quad \text{for} \quad n \geq 1. \tag{5.6}$$

Proof ($\Longrightarrow$) As the polynomial $xp_n(x)$ is of degree $n+1$ we have

$$xp_n(x) = p_{n+1}(x) + b_n p_n(x) + \lambda_n p_{n-1}(x) + \sum_{k=0}^{n-2} \alpha_{n,k} p_k(x).$$

Suppose that $\{p_n(x)\}_{n \geq 0}$ is orthogonal with respect to a linear functional $\mathscr{F}$ on $\mathbb{K}[x]$. By the orthogonality we get

$$\alpha_{n,k} = \frac{\mathscr{F}(xp_n(x)p_k(x))}{\mathscr{F}(p_k^2(x))}.$$

It follows that $\alpha_{n,k} = 0$ for $0 \leq k \leq n-2$, because $xp_k(x)$ is of degree $\leq n-1$ and $\lambda_n = \mathscr{F}(p_n^2(x))/\mathscr{F}(p_{n-1}^2(x)) \neq 0$.
  ($\Longleftarrow$) We define a linear functional $\mathscr{F}$ on $\mathbb{K}[x]$ step by step by

$$\mathscr{F}(x^0) = \lambda_0 \neq 0, \quad \mathscr{F}(p_n(x)) = 0, \quad n \geq 1. \tag{5.7}$$

We begin by showing by induction on $n+k$ that this functional satisfies $\mathscr{F}(x^k p_n(x)) = 0$ for $0 \leq k \leq n-1$. First, if $n+k = 1$ then the result is obvious for $n = 1$ and $k = 0$. Suppose that it is true for $(n,k) \in \mathbb{N}^2$ such that $k < n$ and $n+k < \ell$ for some $\ell \geq 1$. Then for $n+k = \ell$ and $k < n$, we have

$$\mathscr{F}(x^k p_n(x)) = \mathscr{F}(x^{k-1}p_{n+1}(x)) + b_n \mathscr{F}(x^{k-1}p_n(x)) + \lambda_n \mathscr{F}(x^{k-1}p_{n-1}(x)). \tag{5.8}$$

So, by induction $\mathscr{F}(x^k p_n(x)) = \mathscr{F}(x^{k-1}p_{n+1}(x)) = \cdots = \mathscr{F}(p_{n+k}(x)) = 0$. Now, it follows from (5.8) that

$$\mathscr{F}(x^n p_n(x)) = \lambda_n \mathscr{F}(x^{n-1}p_{n-1}(x)) = \lambda_n \cdots \lambda_1 \lambda_0 \neq 0. \quad \square$$

Remark 5.6 If we fix $\mu_0 := \lambda_0 \neq 0$, then the functional $\mathscr{F}$ is unique. Suppose that the polynomials $(p_n(x))$ satisfy (5.6) and $Q_n(x) = a^{-n}p_n(ax + b)\,(a \neq 0)$. Then

$$Q_{n+1}(x) = \left(x - \frac{b_n - b}{a}\right)Q_n(x) - \frac{\lambda_n}{a^2}Q_{n-1}(x), \tag{5.9}$$

and if $(p_n(x))$ are the OPS with respect to the moments (μ_n), then $(Q_n(x))$ are the OPS with respect to the moments m_n given by

$$m_n = a^{-n}\sum_{k=0}^{n}\binom{n}{k}(-b)^{n-k}\mu_k. \tag{5.10}$$

Corollary 5.7 *If $\{p_n(x)\}$ is a monic OPS satisfying (5.6) and $\mathscr{F}$ the corresponding linear functional such that $\mathscr{F}(x^0) = \lambda_0$, then*

$$\mathscr{F}(x^n p_n(x)) = \lambda_n\lambda_{n-1}\cdots\lambda_1\lambda_0, \tag{5.11}$$

$$\mathscr{F}(x^{n+1}p_n(x)) = \lambda_n\lambda_{n-1}\cdots\lambda_1\lambda_0(b_0 + b_1 + \cdots b_n). \tag{5.12}$$

Proof It follows from (5.6) that

$$\mathscr{F}(x^n p_n(x)) = \lambda_n\mathscr{F}(x^{n-1}p_{n-1}(x)) = \lambda_n\lambda_{n-1}\cdots\lambda_1\lambda_0,$$

and

$$\begin{aligned}\mathscr{F}(x^{n+1}p_n(x)) &= b_n\mathscr{F}(x^n p_n(x)) + \lambda_n\mathscr{F}(x^n p_{n-1}(x)) \\ &= b_n\mathscr{F}(x^n p_n(x)) + b_{n-1}\lambda_n\mathscr{F}(x^{n-1}p_{n-1}(x)) \\ &\quad + \lambda_n\lambda_{n-1}\mathscr{F}(x^{n-1}p_{n-2}(x)).\end{aligned}$$

The proof is completed by induction. $\qquad\qquad\qquad\square$

Let

$$\chi_n = \begin{vmatrix} \mu_0 & \mu_1 & \cdots & \mu_{n-1} & \mu_{n+1} \\ \mu_1 & \mu_2 & \cdots & \mu_n & \mu_{n+2} \\ \vdots & \vdots & \ddots & \vdots & \vdots \\ \mu_{n-1} & \mu_n & \cdots & \mu_{2n-2} & \mu_{2n} \\ \mu_n & \mu_{n+1} & \cdots & \mu_{2n-1} & \mu_{2n+1} \end{vmatrix}. \tag{5.13}$$

By (5.5) we can derive the following Hankel determinant formulae.

Corollary 5.8 *We have*

$$\begin{aligned}\Delta_n &= \lambda_0^{n+1}\lambda_1^n\lambda_2^{n-1}\cdots\lambda_{n-1}^2\lambda_n, \\ \chi_n &= (b_0 + b_1 + \cdots + b_n)\Delta_n.\end{aligned}$$

Remark 5.9 We can derive formulae for b_n and λ_n in terms of Hankel determinants of μ_n. There are many interesting interrelationships between orthogonal polynomials and (Hankel) determinant formulae; see [31] for a recent quadratic formula for basic hypergeometric series motivated by a determinant formula for Askey–Wilson polynomials.

Note that the above proof does not give any explicit relationship between the moments $\{\mu_n\}_{n\geq 0}$ and the coefficients $\{b_n\}_{n\geq 0}$, $\{\lambda_n\}_{n\geq 0}$ in the three-term recurrence of the corresponding orthogonal polynomials. To express their interdependence we need to introduce the notion of continued fractions. Consider the (finite) continued fraction:

$$J_n(t) = \cfrac{1}{1 - b_0 t - \cfrac{\lambda_1 t^2}{1 - b_1 t - \cfrac{\lambda_2 t^2}{\ddots \atop 1 - b_{n-1}t}}} = \frac{N_n(t)}{D_n(t)}. \qquad (5.14)$$

Proposition 5.10 *For $n \geq 1$ there hold*

$$N_n(t) = (1 - b_{n-1}t)N_{n-1}(t) - \lambda_{n-1}t^2 N_{n-2}(t), \qquad (5.15)$$
$$D_n(t) = (1 - b_{n-1}t)D_{n-1}(t) - \lambda_{n-1}t^2 D_{n-2}(t), \qquad (5.16)$$

where $N_0(t) = 0$, $N_1(t) = 1$, $D_0(t) = 1$ and $D_1(t) = 1 - b_0 t$.

Proof Note that

$$\frac{N_1(t)}{D_1(t)} = \frac{1}{1 - b_0 t}, \quad \frac{N_2(t)}{D_2(t)} = \frac{1 - b_1 t}{1 - (b_0 + b_1)t + (b_0 b_1 - \lambda_1)t^2}.$$

Clearly, the polynomials $N_n(t)$ and $D_n(t)$ depend only on $b_0, \ldots, b_{n-1}$ and $\lambda_1, \ldots, \lambda_{n-1}$. Suppose the result is true for n. Then

$$J_{n+1}(t) = \frac{(1 - (b_{n-1} + \frac{\lambda_n t}{1 - b_n t}t))N_{n-1}(t) - \lambda_{n-1}t^2 N_{n-2}(t)}{(1 - (b_{n-1} + \frac{\lambda_n t}{1 - b_n t}t))D_{n-1}(t) - \lambda_{n-1}t^2 D_{n-2}(t)}$$

$$= \frac{N_n(t) - \frac{\lambda_n t^2}{1 - b_n t}N_{n-1}(t)}{D_n(t) - \frac{\lambda_n t^2}{1 - b_n t}D_{n-1}(t)}$$

$$= \frac{(1 - b_n t)N_n(t) - \lambda_n t^2 N_{n-1}(t)}{(1 - b_n t)D_n(t) - \lambda_n t^2 D_{n-1}(t)}.$$

This completes the proof. □

Note that

$$N_n(t)D_{n-1}(t) - N_{n-1}(t)D_n(t) = \lambda_{n-1}t^2 \left[N_{n-1}(t)D_{n-2}(t) - N_{n-2}(t)D_{n-1}(t) \right]$$
$$= \lambda_{n-1} \cdots \lambda_1 t^{2(n-1)}.$$

Therefore

$$\frac{N_n(t)}{D_n(t)} - \frac{N_{n-1}(t)}{D_{n-1}(t)} = \frac{N_n(t)D_{n-1}(t) - N_{n-1}(t)D_n(t)}{D_n(t)D_{n-1}(t)}$$
$$= \frac{\lambda_{n-1} \cdots \lambda_1 t^{2n}}{D_n(t)D_{n-1}(t)} \in t^{2n}\mathbb{K}[[t]].$$

In other words, when $m \geq n-1$ all the fractions $N_m(t)/D_m(t)$ have the same coefficients of t^k for $0 \leq k \leq 2n-1$ in their power series expansions. This leads to the definition of the infinite continued fraction $\lim_{n\to\infty} J_n(t) := \sum_{p\geq 0} c_p t^p$ in $\mathbb{K}[[t]]$.

Definition 5.11   The formal power series $\lim_{n\to\infty} J_n(t)$ is denoted by the continued fraction (5.2), which is called a J-*fraction*, and an S-*fraction* if $b_n = 0$ for $n \geq 0$.

For example, we have $c_0 = 1, c_1 = b_0$ and $c_2 = b_0^2 + \lambda_1$. Note that $N_n(t)$ and $D_n(t)$ are polynomials in t of degree $n-1$ and n respectively. For $n \geq 1$ set

$$q_n(t) = t^{n-1}N_n(1/t), \quad p_n(t) = t^n D_n(1/t) \qquad (n \geq 0). \qquad (5.17)$$

Then the recurrences (5.15) and (5.16) imply that for $n \geq 2$ we have

$$q_n(t) = (t - b_{n-1})q_{n-1}(t) - \lambda_{n-1}q_{n-2}(t), \qquad (5.18)$$
$$p_n(t) = (t - b_{n-1})p_{n-1}(t) - \lambda_{n-1}p_{n-2}(t), \qquad (5.19)$$

where $q_0(t) = 0$, $q_1(t) = 1$, $p_0(t) = 1$ and $p_1(t) = t - b_0$. Hence, the sequence $\{p_n(t)\}$ is the same monic OPS defined by (5.6).

If $(p_n(x))_{n\geq 0}$ is the monic OPS in $\mathbb{K}[x]$, then it forms a basis of $\mathbb{K}[x]$. Thus there are entries $a_{n,k}$ ($0 \leq k \leq n$) in $\mathbb{K}$ satisfying

$$x^n = \sum_{k=0}^{n} a_{n,k}p_k(x). \qquad (5.20)$$

The infinite matrix $(a_{n,k})$ ($0 \leq k \leq n$) is usually called a *Stieltjes tableau*.

Proposition 5.12   *Let $(p_n(x))_{n\geq 0}$ be a monic OPS satisfying (5.6). Then the array $(a_{n,k})$ is characterized by the recurrence relation:*

$$\begin{cases} a_{0,k} = \delta_{k,0}, \quad a_{n,0} = b_0 a_{n-1,0} + \lambda_1 a_{n-1,1}, \\ a_{n,k} = a_{n-1,k-1} + b_k a_{n-1,k} + \lambda_{k+1}a_{n-1,k+1}. \end{cases} \qquad (5.21)$$

Proof   Let φ be the linear functional as defined in (5.7). Then

$$\varphi(x^n p_k(x)) = a_{n,k}\varphi(p_k(x)^2) \tag{5.22}$$

with $\varphi(p_k(x)^2) = \lambda_0 \cdots \lambda_k$. Multiplying the equation in (5.21) by $\lambda_0 \cdots \lambda_k$ the latter is equivalent to

$$\varphi(x^n p_k(x)) = \lambda_k \varphi(x^{n-1}p_{k-1}(x)) + b_k\varphi(x^{n-1}p_k(x)) + \varphi(x^{n-1}p_{k+1}(x)).$$

The last identity is obvious in view of Favard's recurrence relation (5.6).

□

The coefficients $a_{n,k}$ are then polynomials in b_k and λ_k with non-negative integers. As the moments $\mu_n := \mathscr{F}(x^n)$ are equal to $a_{n,0}$, we can compute the moment sequence of the OPS satisfying (5.6) in Favard's theorem by (5.20) starting from the sequences (b_k) and (λ_k). Actually, we have the following more convenient formula.

Theorem 5.13   *The generating function of the moments $\{\mathscr{F}(x^n)\}$ has the continued fraction expansion:*

$$\sum_{n\geq 0}\mathscr{F}(x^n)z^n = \cfrac{\lambda_0}{1 - b_0 z - \cfrac{\lambda_1 z^2}{1 - b_1 z - \cfrac{\lambda_2 z^2}{\ddots}}}. \tag{5.23}$$

Proof   Let $h_k(z) = \sum_{n\geq k}a_{n,k}z^n$. Then (5.21) is equivalent to

$$h_0(z) = \lambda_0 + b_0 z h_0(z) + \lambda_1 z h_1(z);$$
$$h_i(z) = z h_{i-1}(z) + b_i z h_i(z) + \lambda_{i+1}z h_{i+1}(z) \qquad (i \geq 1).$$

It follows that

$$h_0(z) = \cfrac{\lambda_0}{1 - b_0 z - \lambda_1 z \dfrac{h_1(z)}{h_0(z)}},$$

and for $i \geq 1$,

$$\frac{h_i(z)}{h_{i-1}(z)} = \cfrac{z}{1 - b_i z - \lambda_{i+1}z\dfrac{h_{i+1}(z)}{h_i(z)}}.$$

The result follows then by iteration of the last identity.

□

The following contraction formulae of Stieltjes continued fractions are well-known in the analytical theory of continued fractions [58, p. 21–22] and very useful in manipulating continued fractions and computation of moments [53]. Combining with the idea of symmetric moment functionals [9, p. 40] for OPS, the corresponding formulae are called the *odd-even trick* in [12]. Here we sketch a proof in the ring of formal power series.

Proposition 5.14 (Contraction formula)　*For any sequence $\{c_n\}_{n\geq 0}$ in $\mathbb{K}$ we have*

$$\cfrac{c_0}{1-\cfrac{c_1 x}{1-\cfrac{c_2 x}{\ddots}}} = c_0 + \cfrac{c_0 c_1 x}{1-(c_1+c_2)x-\cfrac{c_2 c_3 x^2}{1-(c_3+c_4)x-\cfrac{c_4 c_5 x^2}{\ddots}}} \tag{5.24}$$

$$= \cfrac{c_0}{1-c_1 x-\cfrac{c_1 c_2 x^2}{1-(c_2+c_3)x-\cfrac{c_3 c_4 x^2}{\ddots}}}. \tag{5.25}$$

Proof　For any formal power series $A(x) = 1 + \sum_{n\geq 1} a_n x^n$ in $\mathbb{K}[[x]]$ we have

$$\cfrac{1}{1+\cfrac{ax}{A(x)}} = 1 - \cfrac{ax}{ax+A(x)}.$$

By iterating the above formula starting from the first and the second row, respectively, we obtain (5.24) and (5.25). □

For example, contracting Euler's formula

$$\sum_{n\geq 0} n! x^n = \cfrac{1}{1-\cfrac{1x}{1-\cfrac{1x}{1-\cfrac{2x}{1-\cfrac{2x}{\ddots}}}}}, \tag{5.26}$$

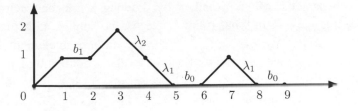

Figure 5.1 A Motzkin path $\gamma \in \mathcal{M}_{0,0}(9)$ with weight $b_0^2 b_1 \lambda_1^2 \lambda_2$.

we obtain the following two equivalent formulae

$$\sum_{n=0}^{\infty} n! x^n = \cfrac{1}{1 - x - \cfrac{1^2 x^2}{1 - 3x - \cfrac{2^2 x^2}{\ddots}}}, \tag{5.27}$$

$$\sum_{n=0}^{\infty} (n+1)! x^n = \cfrac{1}{1 - 2x - \cfrac{1 \cdot 2 x^2}{1 - 4x - \cfrac{2 \cdot 3 x^2}{\ddots}}}. \tag{5.28}$$

Remark 5.15 We will give a proof of Euler's formula (5.27) as an illustration of the *addition formula of Rogers and Stieltjes* in the proof of Theorem 5.30.

5.2.2 The Flajolet–Viennot combinatorial approach

A lattice path ω in the plane $\mathbb{Z} \times \mathbb{Z}$ is a sequence $\omega = (s_0, s_1, \ldots, s_n)$ of points $s_i = (x_i, y_i)$ of $\mathbb{Z} \times \mathbb{Z}$. Each pair of consecutive points (s_i, s_{i+1}) is called a *step* of the path ω. In particular, the step (s_i, s_{i+1}) is called

- *East* if $s_i = (x, y)$ and $s_{i+1} = (x+1, y)$;
- *North-East* if $s_i = (x, y)$ and $s_{i+1} = (x+1, y+1)$;
- *South-East* if $s_i = (x, y)$ and $s_{i+1} = (x+1, y-1)$.

The y_i is called *level* of the step (s_i, s_{i+1}) and n the *length of the path* ω.

Definition 5.16 A *Motzkin path* is a lattice path $\omega = (s_0, s_1, \ldots, s_n)$ in $\mathbb{N} \times \mathbb{N}$ satisfying the following two conditions : (i) the elementary steps are of three types : North-East, East, South-East; (ii) $s_0 = (0,0)$ and $s_n = (n,0)$. In particular, a Motzkin path without East step is called a *Dyck path*.

We can depict a Motzkin path γ by drawing a line between γ_i and γ_{i+1}, see Figure 5.1. In what follows, we fix the ring $\mathbb{K}$ as being $\mathbb{K} := \mathbf{Z}[b_0, b_1, \ldots, b_k, \ldots; \lambda_1, \lambda_2, \ldots, \lambda_k, \ldots]$. The weight of each elementary step North-East (resp. East resp. South-East) of level k is defined by 1 (resp. b_k, resp. λ_k) and the *weight* $w(\omega)$ of a Motzkin path ω is defined to be the product of the weights of its steps: $w(\omega) = \prod_{i=1}^{n} w(s_{i-1}, s_i)$. Let $\mathcal{M}_{k,l}(n)$ be the set of Motzkin paths from $(0, k)$ to (n, l) for $k, l \geq 0$. There is an explicit polynomial formula in the variables (λ_i) and (b_i) with non-negative integral coefficients for the moments μ_n. An equivalent version of (5.29) is proved in [30, p. 306]. The original version of (5.29) given in [16, Proposition 3A] is inaccurate. We shall sketch a proof of this formula using the idea in the latter paper.

Theorem 5.17 (Jacobi–Rogers polynomials)   *If*

$$\sum_{n \geq 0} \mu_n z^n = \cfrac{1}{1 - b_0 z - \cfrac{\lambda_1 z^2}{1 - b_1 z - \cfrac{\lambda_2 z^2}{\ddots}}},$$

then

$$\mu_n = b_0^n + \sum_{h \geq 0} \sum_{\substack{n_0, \ldots, n_{h-1} \geq 1 \\ m_0, \ldots, m_h \geq 0}} b_0^{m_0} \cdots b_{h+1}^{m_h} \lambda_1^{n_0} \cdots \lambda_{h+1}^{n_h} \cdot \rho(\mathbf{n}, \mathbf{m}), \qquad (5.29)$$

where $2(n_0 + \cdots + n_h) + (m_0 + \cdots + m_{h+1}) = n$ *and*

$$\rho(\mathbf{n}, \mathbf{m}) = \prod_{j=0}^{h+1} \binom{n_j + n_{j-1} - 1}{n_{j-1} - 1} \binom{m_j + n_j + n_{j-1} - 1}{m_j}$$

with the convention $n_{-1} = 1$ *and* $n_{h+1} = 0$.

Sketch of proof   We first consider the case of Dyck paths. Hence $b_j = 0$ ($j \geq 0$) and the formula reduces to $\mu_{2n+1} = 0$ and

$$\mu_{2n} = \sum_{h \geq 0} \sum_{\substack{n_0, \ldots, n_h \geq 1 \\ n_0 + \cdots + n_h = n}} (\lambda_1^{n_0} \cdots \lambda_{h+1}^{n_h}) \prod_{j=1}^{h} \binom{n_j + n_{j-1} - 1}{n_{j-1} - 1} \qquad (n \geq 1). \quad (5.30)$$

A Dyck path of length $2n$ with n_j up-steps from height j to $j+1$ ($0 \leq j \leq h$) is completely determined by associating the n_j up-steps with the n_{j-1} down-steps from height j to $j-1$ ($1 \leq j \leq h$) in a way consistant with the rules defining paths; clearly this number is equal to the number of non-negative integral solutions of $x_1 + \cdots + x_{n_{j-1}} = n_j$, namely $\binom{n_j + n_{j-1} - 1}{n_{j-1} - 1}$. This proves (5.30).

Next, we consider Motzkin paths of length n with n_j up-steps from height j to $j+1$ $(j = 0, \ldots, h)$ and m_j level-steps at height j $(j = 0, \ldots, h+1)$. Hence, each of these Motzkin paths is bounded by $h+1$. If $m_0 = n$, then the path has n level-steps at height 0 and the weight is clearly b_0^n. Assume that $n_0, \ldots, n_h \geq 1$. We can first construct a Dyck path of length $2(n_0 + \cdots + n_h)$ and then insert m_j level-steps at each height $j (0 \leq j \leq h+1)$. By (5.30) there are $\prod_{h=0}^{h+1} \binom{n_j + n_{j-1} - 1}{n_{j-1} - 1}$ such Dyck paths, and at each height j there are $\binom{n_j + n_{j-1} + m_j - 1}{m_j}$ ways to associate m_j level-steps with n_j up-steps and n_{j-1} down-steps in a way consistant with the rules defining paths. Summarizing we get (5.29). $\qquad\square$

Remark 5.18 Formula (5.29) has been applied recently to deduce the so-called γ-positivity of Eulerian polynomials [53]; see §5.5.5.

Definition 5.19 Let Seg_n be the segment graph $G = (V, E)$, where $V = \{1, \ldots, n\}$ and $E = \{(1,2), (2,3), \ldots, (n-1, n)\}$. A *tiling* of the segment Seg_n is a collection of disjoint singletons (called *monominos*) and doubletons (called *dominos*). The elements of a domino must be consecutive integers, and vertices not in any of the sets of singletons and doubletons are called *isolated points*. The weight of an isolated point i is x, a monomino $\{i\}$ is $-b_{i-1}$, a domino $\{i, i+1\}$ is $-\lambda_i$.

For example, for the tiling $\alpha = \{\{2,3\}, \{4\}, \{6\}, \{8,9\}\}$ of Seg_9 depicted as follows:

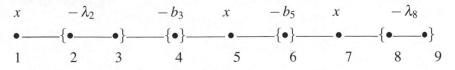

the weight is $v(\alpha) = b_3 b_5 \lambda_2 \lambda_8 x^3$. There is a general model for any OPS based on Favard's theorem.

Theorem 5.20 *The enumerating polynomial $P_n(x)$ of tilings of Seg_n satisfies $P_1(x) = x - b_0$ and*

$$P_{n+1}(x) = (x - b_n) P_n(x) - \lambda_n P_{n-1}(x).$$

Proof We partition the tilings of $\{1, \ldots, n+1\}$ into three subsets:

- $n+1$ is an isolated point: $x P_n(x)$,
- $\{n+1\}$ is a monomino: $-b_n P_n(x)$,
- $\{n, n+1\}$ is a domino: $-\lambda_n P_{n-1}(x)$.

Summing up all the three cases we obtain $P_{n+1}(x)$.   □

Using the tiling model for polynomials and the lattice path model for the moments, Viennot [57] gave a bijective proof of the following result which generalizes the orthogonality of OPS and the Stieltjes tableau (5.20).

Theorem 5.21 (Viennot)   *Let φ be the linear functional associated to the OPS defined by (5.6). For all integers $k, l \geq 0$, the following formula holds*

$$\varphi(x^n P_k(x) P_l(x)) = \lambda_0 \lambda_1 \cdots \lambda_l \sum_{\gamma \in \mathcal{M}_{k,l}(n)} w(\gamma).$$

Proof   We proceed by induction on n using Favard's theorem. By (5.11) the formula is true for $n = 0$. Suppose that the formula is true for $n \geq 0$. Now, enumerating the paths $\gamma \in \mathcal{M}_{k,l}(n+1)$ according to the type of the first step, i.e., North-East, East or South-East we obtain the following decomposition

$$\sum_{\gamma \in \mathcal{M}_{k,l}(n+1)} w(\gamma) = \sum_{\gamma \in \mathcal{M}_{k+1,l}(n)} w(\gamma) + b_k \sum_{\gamma \in \mathcal{M}_{k,l}(n)} w(\gamma) + \lambda_k \sum_{\gamma \in \mathcal{M}_{k-1,l}(n)} w(\gamma).$$

Multiplying by $\lambda_0 \lambda_1 \cdots \lambda_l$ and applying the induction hypothesis, we have

$$\lambda_0 \cdots \lambda_l \sum_{\gamma \in \mathcal{M}_{k,l}(n+1)} w(\gamma) = \varphi(x^n p_{k+1}(x) p_l(x)) + b_k \varphi(x^n p_k(x) p_l(x))$$

$$+ \lambda_k \varphi(x^n p_{k-1}(x) p_l(x))$$
$$= \varphi(x^n (p_{k+1}(x) + b_k p_k(x) + \lambda_k p_{k-1}(x)) p_l(x))),$$

which is equal to $\varphi(x^{n+1} p_k(x) p_l(x))$ by (5.6).   □

In view of (5.22) we immediately obtain the following combinatorial interpretation for the coefficients in Stieltjes tableau (5.21).

Corollary 5.22   *For $n, k \geq 0$ we have*

$$a_{n,k} = \sum_{\gamma \in \mathcal{M}_{0,k}(n)} w(\gamma).$$

In particular, the nth moment $\varphi(x^n)$ is the enumerating polynomial of $\mathcal{M}_{0,0}(n)$.

5.3 Combinatorics of generating functions

5.3.1 Exponential formula and Foata's approach

An accessible exposition on the combinatorial theory of generating functions can be found in the books [59, 17]. Many objects of classical combinatorics present themselves naturally as labeled structures where each "atom" of a structure bears a distinctive integer label. For instance, the cycle decomposition of a permutation represents the permutation as an unordered collection of cyclic graphs whose nodes are labeled by integers. In order to count labeled objects, we appeal to exponential generating functions. The exponential generating function (*EGF*) for a sequence $(A_n)_{n \geq 0}$ is the formal power series

$$A(z) = \sum_{n=0}^{\infty} A_n \frac{z^n}{n!}.$$

Let $\mathscr{A}$ be a class of structures and $\mathscr{A}_n$ be the set of structures on a set of n elements. Some examples of structures on a finite set are: permutations, set partitions, labeled graphs and labeled trees, functional graphs that are associated with mappings of a finite set into itself, as well as structures related to occupancy problems.

The exponential generating function (EGF) of a class of structures $\mathscr{A}$ is the exponential generating function for the numbers $A_n = \text{card}(\mathscr{A}_n)$. Equivalently, the EGF of class $\mathscr{A}$ is

$$A(z) = \sum_{n=0}^{\infty} A_n \frac{z^n}{n!} = \sum_{\alpha \in \mathscr{A}} \frac{z^{|\alpha|}}{|\alpha|!}.$$

A labeled structure (*LS*) is formed of "atoms", nodes in the case of graphs of trees, that are labeled in some way by distinct integers. The size of a structure is the number of its atoms, which is thus also the number of its labels. An LS is said to be *well-labeled* (or *canonically labeled*) if its labels form an initial segment of $\mathbb{N}^*$. The empty(null) structure ε of size 0 bears no label; it is considered as a special case of a labeled structure. An LS may be relabeled. We only consider relabelings that preserve the order relations between labels.

- Reduction: For a non-canonically LS of size n, this operation reduces its labels to the standard interval $[1 \ldots n]$ while preserving the relative order of labels. For instance, the sequence $\langle 7, 3, 9, 2 \rangle$ reduces to $\langle 3, 2, 4, 1 \rangle$. We note $\rho_0(\alpha)$, the reduction of the structure α.
- Expansion: This operation is defined by means of a relabeling function $[1 \ldots n] \to \mathbb{N}^*$ assumed to be strictly increasing. Let $e(\alpha)$ denote the result of relabeling α by e.

Given two LS $\alpha \in \mathscr{A}$ and $\beta \in \mathscr{B}$, the product $\alpha \star \beta$ is a finite collection of structures that are ordered pairs (α', β') of relabeled copies of (α, β),

$$\alpha \star \beta = \{(\alpha', \beta') \mid \rho_0(\alpha') = \alpha, \rho_0(\beta') = \beta, (\alpha', \beta') \text{ is well-labeled}\}.$$

An equivalent form is via expansion of labels:

$$\alpha \star \beta = \{(e(\alpha), f(\beta)) \mid \Im(e) \cap \Im(f) = \emptyset, \Im(e) \cup \Im(f) = [1..|\alpha| + |\beta|]\}.$$

If $\mathscr{A}$ and $\mathscr{B}$ are two classes of combinatorial structures, the product $\mathscr{C} = \mathscr{A} \star \mathscr{B}$ is defined by the usual extension of operations to sets:

$$\mathscr{A} \star \mathscr{B} = \bigcup_{\substack{\alpha \in \mathscr{A} \\ \beta \in \mathscr{B}}} (\alpha \star \beta).$$

The main constructions of union, product, sequence, set, and cycle for labeled structures together with their translation into exponential generating functions are summarized in the following table.

Construction	EGF
Union $\mathscr{A} = \mathscr{B} + \mathscr{C}$	$A(x) = B(x) + C(x)$
Product $\mathscr{A} = \mathscr{B} \star \mathscr{C}$	$A(x) = B(x) \cdot C(x)$
Sequence $\mathscr{A} = \mathfrak{S}(\mathscr{B})$	$A(x) = \frac{1}{1 - B(x)}$
Set $\mathscr{A} = \mathfrak{B}(\mathscr{B})$	$A(x) = \exp(B(x))$
Cycle $\mathscr{A} = \mathfrak{C}(\mathscr{B})$	$A(x) = \log \frac{1}{1 - B(x)}$

Roughly speaking, if, for each $n \geq 0$, Y_n is some structure on $[1 \ldots n]$ and Y_n^+ the assemblage (set) of these structures on $[1 \ldots n]$, and $g_n = \mu(Y_n)$, $f_n = \mu(Y_n^+)$, then the following *exponential formula* holds:

$$\sum_{n \geq 0} f_n \frac{t^n}{n!} = \exp\left(\sum_{n \geq 1} g_n \frac{t^n}{n!}\right). \tag{5.31}$$

Since the classical orthogonal polynomials can be considered as enumerators of finite structures with respect to some statistics, we can use the "combinatorial models" to first combinatorially explain known identities of special functions in the literature and then, guided by combinatorial structures, discover new identities. Foata and his collaborators [18, 19, 21, 23] have worked out many significant examples. We shall give some examples of such applications in the next two subsections.

5.3.2 *Models of orthogonal Sheffer polynomials*

Recall that a sequence of polynomials $\{p_n(x) : n = 0, 1, 2, \ldots\}$ is said to be of *Sheffer type* if the exponential generating function has the form

$$\sum_{n \geq 0} p_n(x) \frac{t^n}{n!} = f(t) \exp(xg(t)), \tag{5.32}$$

where $f(t)$, $g(t)$ are formal power series with $f(0) = 1$, $g(0) = 0$ and $g'(0) \neq 0$. The interested reader may consult [50] for a modern treatment of Sheffer polynomials from the umbral calculus point of view. There are five OPS of Sheffer type, viz., the polynomials of Hermite, Laguerre, Charlier, Meixner and Meixner–Pollaczek [9]. The shifted factorials are defined by

$$(x)_0 = 1; \quad (x)_n = x(x+1) \cdots (x+n-1), \quad \text{for} \quad n \geq 1.$$

(a) Hermite polynomials $(H_n(x))_{n \geq 0}$:

$$\sum_{n \geq 0} H_n(x) \frac{t^n}{n!} = \exp(2xt - t^2), \tag{5.33}$$

orthogonal on $\mathbf{R}$ with respect to the weight function $x \mapsto e^{-x^2/2}$.

(b) Charlier polynomials $(C_n(x; a))_{n \geq 0}$:

$$\sum_{n \geq 0} C_n(x; a) \frac{t^n}{n!} = e^t \left(1 - t/a\right)^x, \tag{5.34}$$

orthogonal on $\mathbf{Z}_+$ with respect to the weight function $x \mapsto a^x/x!$.

(c) Laguerre polynomials $(L_n^{(\alpha)}(x))_{n \geq 0}$:

$$\sum_{n \geq 0} L_n^{(\alpha)}(x) t^n = (1 - t)^{-\alpha - 1} \exp\left(\frac{xt}{t - 1}\right), \tag{5.35}$$

orthogonal on $\mathbf{R}_+$ with respect to the weight function $x \mapsto x^\alpha e^{-x}$.

(d) Meixner polynomials $(M_n(x; \beta, c))_{n \geq 0}$:

$$\sum_{n \geq 0} M_n(x; \beta, c) \frac{t^n}{n!} = \left(1 - t/c\right)^x (1 - t)^{-x - \beta}, \tag{5.36}$$

orthogonal on $\mathbf{Z}_+$ with respect to the weight function $x \mapsto (\beta)_x c^x/x!$.

(e) Meixner–Pollaczek polynomials $(P_n(x; \delta, \eta))_{n \geq 0}$:

$$\sum_{n \geq 0} P_n(x; \delta, \eta) \frac{t^n}{n!} = \left[(1 + \delta t)^2 + t^2\right]^{-\eta/2} \exp\left[x \arctan\left(\frac{t}{1 + \delta t}\right)\right], \tag{5.37}$$

orthogonal on $\mathbf{R}$ with respect to the weight function

$$x \mapsto [\Gamma(\eta/2)]^{-2} \left|\Gamma((\eta + ix)/2)\right|^2 \exp(x \arctan \delta).$$

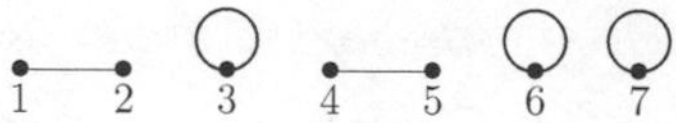

Figure 5.2 An involution on $[1..7]$ with weight $x^3(-1)^2$.

Using the combinatorial theory of the previous section we can readily find combinatorial structures counted by the five Sheffer OPS. First we consider the rescaled Hermite polynomials $\tilde{H}(x) = 2^{-n/2} H_n(x/\sqrt{2})$, whose EGF is then

$$\sum_{n\geq 0} \tilde{H}(x) \frac{t^n}{n!} = \exp(xt - t^2/2). \tag{5.38}$$

An involution on $[1\ldots n]$ is a permutation, whose cycles have lengths 1 or 2 only. We can visualize such a structure on $[1\ldots n]$ by drawing an edge between i and j if i and j are in a cycle and a loop around $\{i\}$ if i is in a cycle of length 1; see Figure 5.2. Thus $Y_1 = \{\bigcirc\!\!-\!1\}$ with $\mu(Y_1) = x$, $Y_2 = \{1\bullet\!\!-\!\!\!-\!\!\bullet 2\}$ with $\mu(Y_2) = -1$, and $Y_n = \emptyset$ $(n \geq 3)$ with $\mu(Y_n) = 0$ $(n \geq 3)$. Then

$$\sum_{n\geq 0} \mu(Y_n^+) \frac{t^n}{n!} = \exp(xt - t^2/2).$$

It follows that

$$\tilde{H}_n(x) = \sum_{\sigma \in \mathscr{I}_n} x^{\text{fix}\,\sigma}(-1)^{\text{cyc}_2\,\sigma}, \tag{5.39}$$

where $\mathscr{I}_n := Y_n^+$ is the set of involutions on $[1\ldots n]$, the numbers of fixed points and edges of σ are denoted by $\text{fix}(\sigma)$ and $\text{cyc}_2(\sigma)$, respectively. Using this model and the exponential formula Foata [18] gave the combinatorial proof of the Mehler formula (5.1).

Formula (5.35) is the exponential generating function for the normalized Laguerre polynomials $n! L_n^{(\alpha)}(x)$. The right-hand side of (5.35) is the product of the following two series:

$$F(t) := (1-t)^{-(\alpha+1)} = \exp\big(-(\alpha+1)\log(1-t)\big)$$

$$= \exp\left((\alpha+1)\sum_{n=1}^{\infty}(n-1)!\frac{t^n}{n!}\right),$$

$$G(t) := \exp\left(\frac{-xt}{1-t}\right) = \exp\left(-x\sum_{n=1}^{\infty}n!\frac{t^n}{n!}\right).$$

Thus F enumerates the sets of cycles (i.e., permutations) with weight $\alpha+1$ for each cycle and G enumerates the sets of paths (i.e., linear orderings) with weight $-x$ for each path. Hence, the Laguerre polynomial $n!L_n^{(\alpha)}(x)$ enumerates all the quadruples $(U,V;f,g)$ on $[1\ldots n]$, where $\{U,V\}$ is a partition of $[1\ldots n]$ with f (resp. g) a set of cycles (resp. paths) of U (resp. V). Let B be the set of of end points of the paths in g and $A = [1\ldots n]\setminus B$. Then we can recast the latter quadruples as pairs (A,f), where A is a subset of $[1\ldots n]$ and $f:A\to[1\ldots n]$ an injection, i.e.,

$$n!L_n^{(\alpha)}(x) = \sum_{(A,f)} (\alpha+1)^{\mathrm{cyc}\,f}(-x)^{n-|A|}, \qquad (5.40)$$

where $A\subset[1\ldots n]$ and $f:A\to[1\ldots n]$ is an injection, and cyc f is the number of cycles of f. The above combinatorial model for Laguerre polynomials was first given by Foata and Strehl [23]; see also [54].

5.3.3 MacMahon's Master Theorem and a Mehler-type formula

Theorem 5.23 (MacMahon's Master Theorem) *Let V_m be the determinant $\det(\delta_{i,j} - b(i,j)x_j)$ $(1\le i,\,j\le m)$. Then the coefficient of $x_1^{n_1}\cdots x_m^{n_m}$ in the expansion of V_m^{-1} is equal to the coefficient of $x_1^{n_1}\cdots x_m^{n_m}$ in the product*

$$(b(1,1)x_1+\cdots+b(1,m)x_m)^{n_1}\cdots(b(m,1)x_1+\cdots+b(m,m)x_m)^{n_m}. \quad (5.41)$$

For each $\mathbf{n}=(n_1,\ldots,n_m)\in\mathbb{N}^m$ let $[\mathbf{n}]=\bigcup_{i=1}^m\{i\}\times[1..n_i]$. It is convenient to consider the first term i of the ordered pair $a=(i,j)$ the color of a and write $i=c(a)$. Let $\mathfrak{S}_{\mathbf{n}}$ be the set of permutations of $[\mathbf{n}]$ and define the weight of each $\pi\in\mathfrak{S}_{\mathbf{n}}$ by

$$v(\pi) = \prod_{a\in[\mathbf{n}]} b(c(a),\,c(\pi(a))).$$

The following result is due to Foata–Zeilberger [24].

Theorem 5.24 (β-extension of MacMahon's Master Theorem) *We have*

$$\sum_{\mathbf{n}} \frac{x_1^{n_1}}{n_1!}\cdots\frac{x_m^{n_m}}{n_m!}\sum_{\pi\in\mathfrak{S}_{\mathbf{n}}}\beta^{\mathrm{cyc}\,\pi}v(\pi) = V_m^{-\beta}. \qquad (5.42)$$

Proof Let $\Phi(\beta)$ be the left-hand side of (5.42). By the exponential formula we have

$$\Phi(\beta) = \exp\left(\beta\sum_{\mathbf{n}}\frac{x_1^{n_1}}{n_1!}\cdots\frac{x_m^{n_m}}{n_m!}\sum_{\pi\in\mathscr{C}_{\mathbf{n}}}v(\pi)\right) = (\Phi(-1))^{-\beta},$$

where $\mathscr{C}_{\mathbf{n}}$ is the set of cyclic permutations of $[\mathbf{n}]$. As $\varepsilon(\sigma) = (-1)^{n-\mathrm{cyc}\,\sigma}$

for any $\sigma \in \mathfrak{S}_n$, we have

$$\Phi(-1) = \sum_{\mathbf{n}} \frac{x_1^{n_1}}{n_1!} \cdots \frac{x_m^{n_m}}{n_m!} (-1)^{n_1 + \cdots + n_m} \sum_{\pi \in \mathfrak{S}_n} \varepsilon(\sigma) \prod_{a \in [n]} b(c(a), c(\pi(a))).$$

Since $\sum_{\pi \in \mathfrak{S}_n} \varepsilon(\sigma) \prod_{a \in [n]} b(c(a), c(\pi(a)))$ is a determinant with at least two identical rows if one of the n_i's is at least equal to 2, we can reduce the above formula to

$$\Phi(-1) = \sum_{\mathbf{n} \in \{0,1\}^m} (-x_1)^{n_1} \cdots (-x_m)^{n_m} \sum_{\pi \in \mathfrak{S}_n} \varepsilon(\sigma) \prod_{a \in [n]} b(c(a), c(\pi(a))),$$

which is equal to the determinant $V_m = \det(\delta_{i,j} - b(i,j)x_j)$, for $1 \le i, j \le m$. $\qquad\square$

Remark 5.25 The idea of the above proof was originally used in [63] to derive a β-extension of the multivariable Lagrange inversion formula. The Foata–Zeilberger proof uses the exponential formula to reduce (5.42) to the $\beta = 1$ case, i.e., MacMahon's Master Theorem. Our proof shows that MacMahon's Master Theorem itself follows from the exponential formula and the trivial identity $\exp(x) = (\exp(-x))^{-1}$.

The *2D-Hermite* (or *complex Hermite*) polynomials $\{H_{m,n}(z_1, z_2)\}_{m,n=0}^{\infty}$ are defined by

$$H_{m,n}(z_1, z_2) = \sum_{k=0}^{m \wedge n} (-1)^k k! \binom{m}{k} \binom{n}{k} z_1^{m-k} z_2^{n-k}, \qquad (5.43)$$

where $m \wedge n := \min\{m, n\}$. They may be also defined by the generating function

$$\sum_{m,n \ge 0} \frac{u^m v^n}{m! \, n!} H_{m,n}(z_1, z_2) = \exp(uz_1 + vz_2 - uv). \qquad (5.44)$$

They have many applications in physical problems; see [37] and the references therein. They satisfy the orthogonality relation

$$\frac{1}{\pi} \int_{\mathbb{R}^2} H_{m,n}(x+iy, x-iy) \overline{H_{p,q}(x+iy, x-iy)} e^{-x^2 - y^2} \, dx \, dy = m! \, n! \, \delta_{m,p} \delta_{n,q}.$$

Carlitz [7] (rediscovered by Wünsche [60]) proved the following Mehler-type formula

$$\sum_{m,n \ge 0} \frac{u^m v^n}{m! \, n!} H_{m,n}(z_1, \bar{z}_1) H_{n,m}(z_2, \bar{z}_2)$$

$$= (1 - uv)^{-1} \exp \left(\frac{-uvz_1\bar{z}_1 + uz_1\bar{z}_2 + v\bar{z}_1 z_2 - uvz_2\bar{z}_2}{1 - uv} \right). \qquad (5.45)$$

This shows the positivity of the Poisson kernel

$$\sum_{m,n\geq 0} \frac{u^m\,\bar{u}^n}{m!\,n!} H_{m,n}(z_1,\bar{z}_1) H_{n,m}(z_2,\bar{z}_2)$$

$$= (1-u\bar{u})^{-1}\exp\left(\frac{-u\bar{u}z_1\bar{z}_1 + uz_1\bar{z}_2 + \bar{u}\bar{z}_1 z_2 - u\bar{u}z_2\bar{z}_2}{1-u\bar{u}}\right),\ |u|<1. \quad (5.46)$$

Below we reproduce the Foata style proof of (5.45) in [37], where a multi-variable generalization of (5.45) is proved in the same vein.

A *complete bipartite graph* is a graph whose vertices can be partitioned into two subsets V_1 and V_2 such that no edge has both endpoints in the same subset, and every possible edge that could connect vertices in different subsets is part of the graph. That is, it is a bipartite graph (V_1, V_2, E) such that for every two vertices $v_1 \in V_1$ and $v_2 \in V_2$, $v_1 v_2$ is an edge in E. A complete bipartite graph (V_1, V_2, E) with $|V_1| = m$ and $|V_2| = n$ is denoted $K_{m,n}$. A *matching* of $K_{m,n}$ is a set of edges without common vertices. If α is such a matching, we denote by $\mathrm{ed}(\alpha)$ the number of edges in α, $\mathrm{iso}_1(\alpha)$ (resp. $\mathrm{iso}_2(\alpha)$) the number of isolated vertices in V_1 (resp. V_2). For convenience, we will draw a matching of $K_{m,n}$ by adding a loop at each isolated vertex. For example, two matchings are given in Figure 5.3.

It is clear from (5.43) that

$$H_{m,n}(z_1, z_2) = \sum_{\alpha \in \mathscr{M}(K_{m,n})} (-1)^{\mathrm{ed}(\alpha)} z_1^{\mathrm{iso}_1(\alpha)} z_2^{\mathrm{iso}_2(\alpha)}, \quad (5.47)$$

where $\mathscr{M}(K_{m,n})$ denotes the set of matchings of $K_{m,n}$. It follows from (5.47) that we have the following combinatorial interpretation

$$H_{m,n}(z_1, \bar{z}_1) H_{n,m}(z_2, \bar{z}_2)$$

$$= \sum_{(\sigma_1,\sigma_2) \in \mathscr{M}(K_{m,n}) \times \mathscr{M}(K_{n,m})} \prod_{j=1}^{2} (-1)^{\mathrm{ed}(\sigma_j)} z_j^{\mathrm{iso}_1(\sigma_j)} \bar{z}_j^{\mathrm{iso}_2(\sigma_j)}. \quad (5.48)$$

For the two examples in Figure 5.3 we have the corresponding weights

$$w(\alpha) = (-1)^2 z_1^3 \bar{z}_1^2, \qquad w(\beta) = (-1)^3 z_2 \bar{z}_2^2.$$

We now reformulate a pair of matchings (σ_1, σ_2) on $V_1 + V_2$ by a bi-colored graph.

Definition 5.26 A bi-involutionary ℓ-graph is a bipartite graph (V_1, V_2, E) with m vertices labeled $1, 2, \ldots, \ell$ with edges and loops bi-colored with 1 or 2 in such a way that

(i) each vertex has valency 2 (i.e., each vertex is incident to two edges, one edge and one loop, or two loops);

(ii) each vertex is incident to two different colors;

(iii) each edge matches a vertex in V_1 with a vertex in V_2.

Let $G = (V_1, V_2, E)$ be such a graph. Denote by $\mathrm{ed}_j(G)$ (resp. $\mathrm{iso}_{i,j}(G)$) the number of edges (resp. loops in V_i) of color j ($i, j = 1, 2$) and define the weight of G by

$$\mu(G) = u^{|V_1|} v^{|V_2|} (-1)^{\mathrm{ed}_1(G) + \mathrm{ed}_2(G)} z_1^{\mathrm{iso}_{1,1}(G)} \bar{z}_1^{\mathrm{iso}_{2,1}(G)} z_2^{\mathrm{iso}_{1,2}(G)} \bar{z}_2^{\mathrm{iso}_{2,2}(G)}.$$

(5.49)

Note that $|V_i| = \mathrm{ed}_j(G) + \mathrm{iso}_{i,j}(G)$ for $i, j = 1, 2$. In view of (5.48) and (5.49), the left-hand side of (5.45) can be written as

$$\sum_{\ell \geq 0} \frac{1}{\ell!} \sum_G \mu(G),$$

(5.50)

the second sum being over all the bi-involutionary ℓ-graphs $G = (V_1, V_2, E)$.

It remains to find the connected components of a generic bi-involutionary ℓ-graph $G = (V_1, V_2, E)$. There are five types of connected components.

I–II. Paths starting from a vertex from V_1 (resp. V_2) and ending to a vertex in V_1 (resp. V_2) and edges are colored alternatively with 1 and 2, see Figure 5.3. The generating functions of these paths are respectively given by

$$\exp\left(uz_1\bar{z}_2 \sum_{k \geq 0} u^k v^k\right) = \exp\left(\frac{uz_1\bar{z}_2}{1 - uv}\right)$$

(5.51)

and

$$\exp\left(v\bar{z}_1 z_2 \sum_{k \geq 0} u^k v^k\right) = \exp\left(\frac{v\bar{z}_1 z_2}{1 - uv}\right).$$

(5.52)

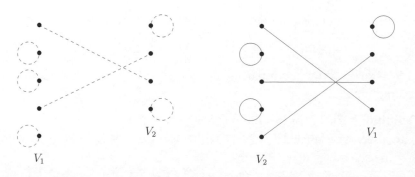

Figure 5.3 Two matchings $\alpha \in \mathcal{M}(K_{5,4})$ and $\beta \in \mathcal{M}(K_{4,5})$.

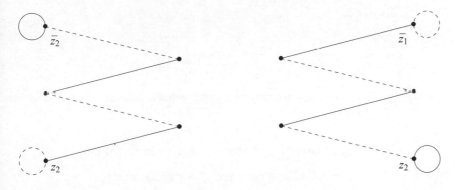

Figure 5.4 Paths of types I and II.

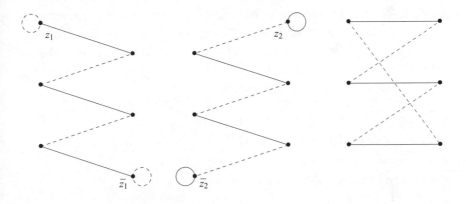

Figure 5.5 Paths of types III, VI and V (cycles).

III–IV. Paths starting from a vertex in V_1 (resp. V_2) and ending in a vertex in V_2 (resp. V_1) and edges are colored alternatively with 1 and 2; see Figure 5.5. The generating functions of the paths of type III and IV are respectively given by

$$\exp\left(-uvz_1\bar{z}_1 \sum_{k\geq 0} u^k v^k\right) = \exp\left(\frac{-uvz_1\bar{z}_1}{1-uv}\right) \qquad (5.53)$$

and

$$\exp\left(-uvz_2\bar{z}_2 \sum_{k\geq 0} u^k v^k\right) = \exp\left(\frac{-uvz_2\bar{z}_2}{1-uv}\right). \qquad (5.54)$$

V. The paths are cycles and edges are colored alternatively with 1 and 2; see Figure 5.5. The generating function of the paths of type V is given by

$$\frac{1}{1-uv} = \exp\left(\sum_{k\geq 1} \frac{u^k v^k}{k}\right). \tag{5.55}$$

Multiplying (5.51)–(5.55) we obtain the right-hand side of (5.45).

5.4 Moments of orthogonal Sheffer polynomials

5.4.1 Combinatorics of the moments

We consider a generalization of Eulerian polynomials by computing the generating function of permutations with respect to five statistics. The continued fraction expansion of its ordinary generating function is then derived by the methods of Stieltjes and Rogers, which in particular implies the combinatorial interpretation of the moments of the orthogonal Sheffer polynomials. Given a permutation σ of $[1\ldots n]$, each value $x = \sigma(i)$, $1 \leq i \leq n$, belongs to the following five cases

- a *cycle peak*, if $\sigma^{-1}(x) < x > \sigma(x)$;
- a *cycle valley*, if $\sigma^{-1}(x) > x < \sigma(x)$;
- a *cycle double rise*, if $\sigma^{-1}(x) < x < \sigma(x)$;
- a *cycle double fall*, if $\sigma^{-1}(x) > x > \sigma(x)$;
- a *fixed point*, if $\sigma(i) = i$.

The numbers of cycle peaks, cycle valleys, cycle double rises, cycle double falls and fixed points of σ are denoted by $\mathrm{cpeak}\,\sigma$, $\mathrm{cvalley}\,\sigma$, $\mathrm{cdrise}\,\sigma$, $\mathrm{cdfall}\,\sigma$ and $\mathrm{fix}\,\sigma$, respectively. Let $\mathrm{cyc}\,\sigma$ be the number of cycles of σ. We are interested in the following generating polynomial:

$$\mu_n = \sum_{\sigma \in \mathfrak{S}_n} u_1^{\mathrm{cvalley}\,\sigma} u_2^{\mathrm{cpeak}\,\sigma} u_3^{\mathrm{cdrise}\,\sigma} u_4^{\mathrm{cdfall}\,\sigma} \alpha^{\mathrm{fix}\,\sigma} \beta^{\mathrm{cyc}\,\sigma}. \tag{5.56}$$

Note that $\mathrm{cpeak}\,\sigma = \mathrm{cvalley}\,\sigma$ for any $\sigma \in \mathfrak{S}_n$. For example, if

$$\sigma = (1, 5, 2, 3, 9)\,(4, 8, 10, 7)\,(6)\,(11) \in \mathfrak{S}_{11},$$

where we write a cycle containing a as $(a, \sigma(a), \ldots, \sigma^k(a))$ with $\sigma^{k+1}(a) = a$ for some integer $k \geq 0$, then the corresponding statistics are as follows:

$$\mathrm{cpeak}\,\sigma = |\{5, 9, 10\}| = 3, \quad \mathrm{cvalley}\,\sigma = |\{1, 2, 4\}| = 3,$$
$$\mathrm{cdrise}\,\sigma = |\{3, 8\}| = 2, \quad \mathrm{cdfall}\,\sigma = |\{7\}| = 1,$$
$$\mathrm{fix}\,\sigma = |\{6, 11\}| = 2, \quad \mathrm{cyc}\,\sigma = 4.$$

The first three μ_n are given by

$$\mu_1 = \alpha\beta,$$
$$\mu_2 = \alpha^2\beta^2 + u_1 u_2 \beta,$$
$$\mu_3 = \alpha^3\beta^3 + 3u_1 u_2 \alpha\beta^2 + u_1 u_2 (u_3 + u_4)\beta.$$

Special cases of the polynomial (5.56) or its variants have been studied by many people; see [30]. The following three theorems are established in [65].

Theorem 5.27 *We have*

$$1 + \sum_{n \geq 1} \mu_n \frac{x^n}{n!} = e^{\alpha\beta x} \left(\frac{\alpha_1 - \alpha_2}{\alpha_1 e^{\alpha_2 x} - \alpha_2 e^{\alpha_1 x}} \right)^\beta, \tag{5.57}$$

where $\alpha_1 \alpha_2 = u_1 u_2$ and $\alpha_1 + \alpha_2 = u_3 + u_4$.

Note that

$$\frac{a - b}{a e^{bx} - b e^{ax}} = \left(1 - ab \sum_{n \geq 2} \frac{a^{n-1} - b^{n-1}}{a - b} \frac{x^n}{n!} \right)^{-1}.$$

Hence, when expanded as a power series in $\mathbb{K}[[x]]$, the coefficient of $x^n/n!$ is a polynomial in ab and $a + b$ with integral coefficients. In particular, when $u_1 = u_3 = u$ and $u_2 = u_4 = 1$, we can take $\alpha_1 = 1$ and $\alpha_2 = u$, as $\mathrm{exc} = \mathrm{cdrise} + \mathrm{cvalley}$, the polynomial in (5.56) becomes

$$P_n(u, \alpha, \beta) := \sum_{\sigma \in \mathfrak{S}_n} u^{\mathrm{exc}\,\sigma} \alpha^{\mathrm{fix}\,\sigma} \beta^{\mathrm{cyc}\,\sigma},$$

and the generating function (5.57) reduces to

$$1 + \sum_{n \geq 1} P_n(u, \alpha, \beta) \frac{x^n}{n!} = \left(\sum_{n \geq 0} \frac{(u - \alpha)^n - u(1 - \alpha)^n}{1 - u} \frac{x^n}{n!} \right)^{-\beta}. \tag{5.58}$$

A combinatorial proof of (5.58) was given in [46].

The *formal Laplace transformation* is a linear operator $\mathscr{L}$ on $K[[x]]$ defined by

$$\mathscr{L}(t^n, x) = n! x^{n+1} = \int_0^\infty t^n e^{-t/x} dt \quad \text{for} \quad n \geq 0.$$

The integral expression should be understood as a symbolic notation. Indeed, for any $f(t) = \sum_{n \geq 0} a_n \frac{t^n}{n!} \in K[[x]]$ we have

$$\mathscr{L}(f(t), x) = a_0 x + x\mathscr{L}(f'(t), x), \quad \mathscr{L}(f(-t), -x) = -\mathscr{L}(f(t), x).$$

Applying the Laplace transformation to (5.57) yields the following formula for the ordinary generating function.

Theorem 5.28   *We have*

$$1 + \sum_{n \geq 1} \mu_n x^n = \sum_{n \geq 0} \frac{\alpha_2^n(\beta)_n x^n}{\prod_{k=0}^n [1 - ((\alpha - \alpha_2)\beta + k(\alpha_1 - \alpha_2))x]}, \qquad (5.59)$$

where $\alpha_1 \alpha_2 = u_1 u_2$ and $\alpha_1 + \alpha_2 = u_3 + u_4$.

To derive the continued fraction expansion of the ordinary generating function of $\{\mu_n\}$ from (5.57) we need the Stieltjes–Rogers addition formula [30, p. 295].

Theorem 5.29 (Addition formula)   *The formal power series $f(x) = \sum_{i \geq 0} a_i x^i / i!$ has the property that*

$$f(x+y) = \sum_{m \geq 0} \alpha_m f_m(x) f_m(y),$$

where α_m is independent of x and y and

$$f_m(x) = \frac{x^m}{m!} + \beta_m \frac{x^{m+1}}{(m+1)!} + O(x^{m+2}),$$

if and only if the formal power series $\hat{f}(x) = \sum_{i \geq 0} a_i x^i$ has the J-continued fraction expansion with the parameters

$$b_m = \beta_{m+1} - \beta_m \quad \text{and} \quad \lambda_m = \alpha_{m+1}/\alpha_m, \quad m \geq 0.$$

Theorem 5.30   *We have*

$$1 + \sum_{n \geq 1} \mu_n x^n =$$

$$\cfrac{1}{1 - \alpha\beta x - \cfrac{\beta u_1 u_2 x^2}{1 - (\alpha\beta + u_3 + u_4)x - \cfrac{2(\beta+1)u_1 u_2 x^2}{\ddots \cfrac{}{1 - b_n x - \cfrac{\lambda_{n+1} x^2}{\ddots}}}}}, \qquad (5.60)$$

where $b_n = (\alpha\beta + n(u_3 + u_4))$ and $\lambda_{n+1} = (n+1)(\beta+n)u_1 u_2$ for $n \geq 0$.

Proof   (When $u_1 = u_2 = u_3 = \alpha = \beta = 1$). In this case we have

$$f(x) = \sum_{n \geq 0} n! \frac{x^n}{n!} = \frac{1}{1-x}.$$

It follows that

$$f(x+y) = \frac{1}{(1-x)(1-y)-xy}$$

$$= \sum_{m \geq 0} (m!)^2 \frac{x^m(1-x)^{-m-1}}{m!} \frac{y^m(1-y)^{-m-1}}{m!}.$$

Hence,

$$f_m(x) = \frac{x^m(1-x)^{-m-1}}{m!} = \frac{x^m}{m!} + (m+1)^2 \frac{x^{m+1}}{(m+1)!} + O(x^{m+1}),$$

and $\alpha_m = (m!)^2$ and $\beta_{m+1} = (m+1)^2$. Thus we get the parameters in the addition formula:

$$b_m = \beta_{m+1} - \beta_m = 2m+1 \quad \text{and} \quad \lambda_{m+1} = \frac{\alpha_{m+1}}{\alpha_m} = (m+1)^2.$$

This completes the proof. $\qquad\square$

The moments of the Hermite, Charlier, Laguerre, Meixner and Meixner–Pollaczeck polynomials have the following integral representations [9, Chapter VI]:

$$\nu(x^n) = \frac{1}{\sqrt{2\pi}} \int_{-\infty}^{+\infty} x^n e^{-x^2/2} dx; \tag{5.61}$$

$$\mu(x^n) = e^{-a} \sum_{x=0}^{\infty} x^n \frac{a^x}{x!}; \tag{5.62}$$

$$\psi(x^n) = \frac{1}{\Gamma(\alpha+1)} \int_0^{+\infty} x^n x^\alpha e^{-x} dx; \tag{5.63}$$

$$\rho(x^n) = (1-c)^\beta \sum_{x=0}^{\infty} x^n \frac{c^x(\beta)_x}{x!}; \tag{5.64}$$

$$\varphi(x^n) = \frac{1}{\int_{-\infty}^{+\infty} w(x)dx} \int_{-\infty}^{+\infty} x^n w(x) dx, \tag{5.65}$$

where $w(x) = [\Gamma(\eta/2)]^{-2} |\Gamma((\eta+ix)/2)|^2 \exp(-x \arctan \delta)$.

Proposition 5.31 ([64]) *The five moment sequences* $(\nu(x^n))$, $(\mu(x^n))$, $(\psi(x^n))$, $(\rho(x^n))$ *and* $(\varphi(x^n))$ *have, respectively, the EGF:*

$$\exp(t^2/2); \ \exp(a(e^t - 1)); \ (1-t)^{-\alpha-1}; \ \left(\frac{1-c}{1-ce^t}\right)^\beta; \ (\cos t - \delta \sin t)^{-\eta}.$$

$$\tag{5.66}$$

Recall that a permutation $\sigma \in \mathfrak{S}_n$ has an *excedance* at $i \in [1\ldots n]$ if $\sigma(i) > i$, a *drop* at $i \in [1\ldots n]$ if $\sigma(i) < i$, and a *maximum* (or cycle peak) at

$i \in [1 \ldots n]$ if $\sigma^{-1}(i) < i > \sigma(i)$. Let exc σ (resp. drop σ and max σ) be the number of excedances (resp. drops and maxima) of σ. Furthermore, call σ a *derangement* if $\sigma(i) \neq i$ for all $i \in [1 \ldots n]$. Let $\mathscr{D}_n$ be the set of derangements of $[1 \ldots n]$. Finally, call σ an *involution* if $\sigma^2 = id.$ and denote by $\mathscr{I}_n$ the set of *fixed point free* involutions on $[1 \ldots n]$. For a finite set S, we denote by $|S|$ the *cardinality* of S. A *partition* of S is an assemblage of the disjoint subsets (or *blocs*) $\pi = \{E_1, E_2, \ldots, E_l\}$ such that $E_1 \cup \cdots \cup E_l = S$. Let bloc π be the number of blocks of π and $\mathscr{J}_n$ the set of partitions of $[1 \ldots n]$.

Comparing the coefficients (b_n) and (λ_n) of the continued fraction (5.60) with the coefficients in the three-term recurrence of Sheffer orthogonal polynomials (see [9]) we obtain the combinatorial interpretations for their moments. Of course, we can also derive these formulae by comparing the EGF of the moments (5.66) with the EGF (5.57).

Theorem 5.32 ([64])   *The following formulae hold true:*

$$\nu(x^n) = |\mathscr{I}_n|; \tag{5.67}$$

$$\mu(x^n) = \sum_{\pi \in \mathscr{J}_n} a^{\text{bloc}\,\pi}; \tag{5.68}$$

$$\psi(x^n) = \sum_{\sigma \in \mathfrak{S}_n} (\alpha+1)^{\text{cyc}\,\sigma} = (\alpha+1)_n; \tag{5.69}$$

$$\rho(x^n) = (1-c)^{-n} \sum_{\sigma \in \mathfrak{S}_n} c^{\text{drop}\,\sigma} \beta^{\text{cyc}\,\sigma}; \tag{5.70}$$

$$\varphi(x^n) = \delta^n \sum_{\sigma \in \mathfrak{S}_n} (1+1/\delta^2)^{\max\,\sigma} \eta^{\text{cyc}\,\sigma}. \tag{5.71}$$

Remark 5.33   For the moments of the orthogonal Sheffer polynomials, Viennot [57] worked out explicit bijections between the Motzkin path models and certain classes of permutations using the bijection of Françon–Viennot [26]. Note that Viennot's original interpretation of $\rho(x^n)$ (resp. $\varphi(x^n)$) uses the number of *descents* (resp. *pics*) and *saillants*; these statistics have the same distribution as $(\text{drop}, \text{cyc})$ (resp. $(\max, \text{cyc})$) through the *transformation fondamentale* [22].

The following result is crucial to figure out the weight function in the combinatorial interpretation of the linearization coefficients for Meixner–Pollaczek polynomials.

Corollary 5.34   *If $i^2 = -1$, then*

$$\sum_{\sigma \in \mathscr{D}_n} (\delta+i)^{\text{drop}\,\sigma} (\delta-i)^{\text{exc}\,\sigma} \eta^{\text{cyc}\,\sigma} = \delta^n \sum_{\sigma \in \mathscr{D}_n} (1+1/\delta^2)^{\max\,\sigma} \eta^{\text{cyc}\,\sigma}.$$

Remark 5.35 This identity is not valid on $\mathfrak{S}_n$.

5.4.2 Linearization coefficients of Sheffer polynomials

Let $\{p_n(x) : n = 0, 1, 2, \ldots\}$ be an OPS with respect to a positive measure $d\alpha(x)$. The expansion

$$p_m(x)p_n(x) = \sum_{k=0}^{n+m} a(m,n,k)\, p_k(x) \tag{5.72}$$

is equivalent to

$$\int_{-\infty}^{+\infty} p_n(x)p_m(x)p_k(x)d\alpha(x) = a(m,n,k) \int_{-\infty}^{+\infty} p_k(x)p_k(x)d\alpha(x). \tag{5.73}$$

This leads to consideration of the integral $\int_{-\infty}^{+\infty} \prod_{i=1}^{m} p_{n_i}(x)\alpha(x)$ for $m \geq 1$. For a fixed integer m, we define $\mathbf{n} = (n_1, \ldots, n_m)$ and $\mathbf{n}! = n_1! \cdots n_m!$, $\mathbf{x}^{\mathbf{n}} = x_1^{n_1} \cdots x_m^{n_m}$. The linearization coefficients of Sheffer polynomials are then defined with appropriate normalization as follows:

$$\mathcal{H}(\mathbf{n}) = \frac{1}{\sqrt{2\pi}} \int_{-\infty}^{+\infty} \left(\prod_{i=1}^{m} 2^{-n/2} H_{n_i}(x/\sqrt{2}) \right) e^{-x^2/2} dx; \tag{5.74}$$

$$\mathcal{C}(\mathbf{n};a) = e^{-a} \sum_{x=0}^{\infty} \left(\prod_{i=1}^{m} (-a)^{n_i} C_{n_i}(x;a) \right) \frac{a^x}{x!}; \tag{5.75}$$

$$\mathcal{L}(\mathbf{n};\alpha) = \frac{1}{\Gamma(\alpha+1)} \int_{0}^{+\infty} \left(\prod_{i=1}^{m} (-1)^{n_i} n_i! L_{n_i}^{(\alpha)}(x) \right) x^\alpha e^{-x} dx; \tag{5.76}$$

$$\mathcal{M}(\mathbf{n};\beta,c) = (1-c)^\beta \sum_{x=0}^{\infty} \left(\prod_{i=1}^{m} (-c)^{n_i} M_{n_i}(x;\beta,c) \right) \frac{c^x (\beta)_x}{x!}; \tag{5.77}$$

$$\mathcal{P}(\mathbf{n};\delta,\eta) = \frac{1}{\int_{-\infty}^{+\infty} w(x)dx} \int_{-\infty}^{+\infty} \left(\prod_{i=1}^{m} P_{n_i}(x;\delta,\eta) \right) w(x)dx; \tag{5.78}$$

here

$$w(x) = [\Gamma(\eta/2)]^{-2} \left| \Gamma((\eta+ix)/2) \right|^2 \exp(-x \arctan \delta).$$

We are now in position to compute the generating function for the linearization coefficients of all the Sheffer polynomials and express them in terms of the elementary symmetric functions of $x_1, \ldots, x_m$ with the help of the generating functions of the moments and the polynomials. In what follows, we shall denote by e_k for the kth $(0 \leq k \leq m)$ elementary symmetric polynomial of $x_1, \ldots, x_m$; that is,

$$\sum_{j=0}^{m} e_j t^j = \prod_{k=1}^{m}(1 + x_k t).$$

By convention, we set $e_k = 0$ if $k > m$.

Theorem 5.36 ([64]) *The linearization coefficients of the five Sheffer OPS have the following EGF:*

$$\sum_{n \geq 0} \mathscr{H}(\mathbf{n}) \frac{\mathbf{x}^{\mathbf{n}}}{\mathbf{n}!} = e^{e_2}; \tag{5.79}$$

$$\sum_{n \geq 0} \mathscr{C}(\mathbf{n}; a) \frac{\mathbf{x}^{\mathbf{n}}}{\mathbf{n}!} = e^{ae_2 + \cdots + ae_m}; \tag{5.80}$$

$$\sum_{n \geq 0} \mathscr{L}(\mathbf{n}; \alpha) \frac{\mathbf{x}^{\mathbf{n}}}{\mathbf{n}!} = (1 - e_2 - 2e_3 - \cdots - (m-1)e_m)^{-(\alpha+1)}; \tag{5.81}$$

$$\sum_{n \geq 0} \mathscr{M}(\mathbf{n}; \beta, c) \frac{\mathbf{x}^{\mathbf{n}}}{\mathbf{n}!} = \left\{ 1 - \sum_{k=2}^{m}(c + \cdots + c^{k-1})e_k \right\}^{-\beta}; \tag{5.82}$$

$$\sum_{n \geq 0} \mathscr{P}(\mathbf{n}; \delta, \eta) \frac{\mathbf{x}^{\mathbf{n}}}{\mathbf{n}!} = \left\{ 1 - \sum_{k=2}^{m}(\delta^k + \delta^{k-2}) \sum_{l \geq 0} \binom{k-1}{2l+1} \left(\frac{-1}{\delta^2}\right)^l e_k \right\}^{-\eta}. \tag{5.83}$$

Remark 5.37 Formulae (5.79) and (5.80) are quite easy. The generating functions (5.81) and (5.82) are due to Askey and Ismail [3] while (5.83) was first proved in [64].

In order to interpret combinatorially these coefficients we have to introduce some multi-analogues of the notions and notations for permutations of $[1..n]$. For each $i \in [1..m]$, define the colored sets $\mathbb{N}_i = \{i\} \times \mathbb{N}$ and $[n]_i = \{i\} \times [1 \ldots n] = \{(i, 1), \ldots, (i, n)\}$ since one can imagine that i is the color of $(i, j) \in \mathbb{N}_i$ and simply write $c(i, j) = i$. Let $\mathbb{N}_* = [1..m] \times \mathbb{N}$ be the disjoint union $\mathbb{N}_1 \cup \cdots \cup \mathbb{N}_m$ and let $[\mathbf{n}] = [1..n_1]_1 \cup \cdots \cup [1..n_m]_m$. Note that the lexicographic order is a *total order* on the set $\mathbb{N}_*$.

Let $\mathfrak{S}(\mathbf{n})$ be the set of permutations of $[\mathbf{n}]$. If π is a permutation of $[\mathbf{n}]$, an element $a \in [\mathbf{n}]$ is said to be a *c-fixed point* of π if $c(a) = c(\pi(a))$. A permutation without c-fixed point is said to be a *c-derangement* on $[\mathbf{n}]$. Let $\mathscr{D}(\mathbf{n})$ and $\mathscr{M}(\mathbf{n})$ respectively denote the sets of c-derangements and of c-fixed point free involutions (or matchings) on $[\mathbf{n}]$. Similarly, we say that π has a *c-excedance* (resp. *c-drop*) at $a \in [\mathbf{n}]$ if $c(\pi(a)) > c(a)$

(resp. $c(\pi(a)) < c(a)$) and denote by exc π (resp. drop π) the number of c-exceedances (resp. c-drops) of π.

Example 5.38 If $\mathbf{n} = (3,1,1)$ and

$$\pi = \left(\begin{array}{ccccc} (1,1) & (1,2) & (1,3) & (2,1) & (3,1) \\ (1,3) & (2,1) & (3,1) & (1,2) & (1,1) \end{array} \right),$$

then exc $\pi = 2$, drop $\pi = 2$ and cyc $\pi = 2$.

A color (or box) partition of $[\mathbf{n}]$ is a partition of $[\mathbf{n}]$ such that no two elements of each block have the same color. Let $\Pi(\mathbf{n})$ be the set of color partitions without singletons of $[\mathbf{n}]$. The linearization coefficients of the orthogonal Sheffer polynomials have the following combinatorial interpretations.

Theorem 5.39 *The following identities hold :*

$$\mathscr{H}(\mathbf{n}) = |\mathscr{M}(\mathbf{n})|; \tag{5.84}$$

$$\mathscr{C}(\mathbf{n};a) = \sum_{\pi \in \Pi(\mathbf{n})} a^{\mathrm{bloc}\,\pi}; \tag{5.85}$$

$$\mathscr{L}(\mathbf{n};\alpha) = \sum_{\pi \in \mathscr{D}(\mathbf{n})} (1+\alpha)^{\mathrm{cyc}\,\pi}; \tag{5.86}$$

$$\mathscr{M}(\mathbf{n};\beta,c) = \sum_{\pi \in \mathscr{D}(\mathbf{n})} c^{\mathrm{drop}\,\pi} \beta^{\mathrm{cyc}\,\pi}; \tag{5.87}$$

$$\mathscr{P}(\mathbf{n};\delta,\eta) = \sum_{\pi \in \mathscr{D}(\mathbf{n})} (\delta+i)^{\mathrm{drop}\,\pi} (\delta-i)^{\mathrm{exc}\,\pi} \eta^{\mathrm{cyc}\,\pi}. \tag{5.88}$$

Proof A quick proof of the above identities is to show that the two sides of these identities have the same generating function. The first two equations can be derived from the *multivariable exponential formula* and the others from the β-extension of the MacMahon Master Theorem. We sketch the proof of (5.87). Set $b(i,i) = 0$, $b(i,j) = c$ if $i > j$ and $b(i,j) = 1$ if $i < j$. Then, for each $\pi \in \mathfrak{S}(\mathbf{n})$, we have

$$v(\pi) = \prod_{i=1}^{m} \prod_{j=1}^{n_i} b(i, c(\pi(i,j))) = \left\{ \begin{array}{ll} c^{\mathrm{drop}\,\pi} & \text{if } \pi \text{ is a c-derangement;} \\ 0 & \text{otherwise.} \end{array} \right.$$

Hence

$$\sum_{\pi \in \mathscr{P}(\mathbf{n})} \beta^{\mathrm{cyc}\,\pi} v(\pi) = \sum_{\pi \in \mathscr{D}(\mathbf{n})} c^{\mathrm{drop}\,\pi} \beta^{\mathrm{cyc}\,\pi}.$$

On the other hand, the determinant of "Master Theorem" is equal to

$$
V_m = (-1)^m x_1 \cdots x_m
\begin{vmatrix}
-x_1^{-1} & 1 & \cdots & 1 \\
c & -x_2^{-1} & \cdots & 1 \\
\vdots & \vdots & \ddots & \vdots \\
c & c & \cdots & -x_m^{-1}
\end{vmatrix}
$$

$$
= 1 - \sum_{k=2}^{m} (c + \cdots + c^{k-1}) e_k.
$$

The result follows then from the β-extension of MacMahon's Master Theorem. $\qquad\square$

Remark 5.40 The combinatorial interpretations for Hermite polynomials (5.84) and Laguerre polynomials (5.86) are due to Azor et al. [5] and Foata–Zeilberger [24], respectively. The interpretations (5.85) and (5.87) were given in [62] while (5.88) in [64].

When $\alpha = 0$ and $\mathbf{n} = (1, \ldots, 1)$, identity (5.86) reduces to the well-known formula for the number of derangements of $[1 \ldots n]$:

$$
\mathscr{L}(\mathbf{n}; 0) = (-1)^m \int_0^\infty (x-1)^m e^{-x} dx = m! \sum_{k=0}^{m} \frac{(-1)^k}{k!}.
$$

If $m = 3$, we can derive explicit formulae for the linearization coefficients from the above theorem or directly from their combinatorial interpretations [64, 65]. Since the q-versions for the Hermite and Charlier polynomials are given in (5.118) and (5.120) we shall only list the formulae for Laguerre, Meixner and Meixner–Pollaczek polynomials. Note that from (5.35) or (5.40) we derive immediately the explicit formula of Laguerre polynomials

$$
L_n^{(\alpha)}(x) = \frac{1}{n!} \sum_{k=0}^{n} \binom{n}{k} (\alpha + k + 1)_{n-k} (-x)^k. \tag{5.89}
$$

Corollary 5.41 ([64]) *Let $\mathbf{n} = (n_1, n_2, n_3)$; we have the explicit formulae:*

$$
\mathscr{L}(\mathbf{n}; \alpha) = \sum_s \frac{n_1! n_2! n_3! 2^{n_1+n_2+n_3-2s} (\alpha+1)_s}{(s-n_1)!(s-n_2)!(s-n_3)!(n_1+n_2+n_3-2s)!}; \tag{5.90}
$$

$$
\mathscr{M}(\mathbf{n}; \beta, c) = \sum_s \frac{n_1! n_2! n_3! (1+c)^{n_1+n_2+n_3-2s} c^s (\beta)_s}{(s-n_1)!(s-n_2)!(s-n_3)!(n_1+n_2+n_3-2s)!}; \tag{5.91}
$$

$$
\mathscr{P}(\mathbf{n}; \delta, \eta) = \sum_s \frac{n_1! n_2! n_3! (\eta)_s (1+1/\delta^2)^s \delta^{n_1+n_2+n_3} 2^{n_1+n_2+n_3-2s}}{(s-n_1)!(s-n_2)!(s-n_3)!(n_1+n_2+n_3-2s)!},
$$

$$
\tag{5.92}
$$

where each sum is over s such that $\max(n_1, n_2, n_3) \leq s \leq \lfloor \frac{n_1+n_2+n_3}{2} \rfloor$.

We conclude this section with the following remarks.

(i) From (5.90) we have $\mathscr{L}(n, m; \alpha) = \mathscr{L}(n, m, 0; \alpha) = n!(\alpha+1)_n \delta_{n,m}$. In view of (5.73) we can rewrite the formula (5.90) as the linearization formula for Laguerre polynomials

$$L_m^{(\alpha)}(x) L_n^{(\alpha)}(x) = \sum_{k=|n-m|}^{m+n} (-1)^{n+m+k} a(m,n,k;\alpha) L_k^{(\alpha)}(x), \quad (5.93)$$

where

$$a(m,n,k;\alpha) =$$

$$k! \sum_{s=\max(0,m-k,n-k)}^{\lfloor \frac{n+m-k}{2} \rfloor} \frac{2^{n+m-k-2s}(\alpha+1+k)_s}{(s+k-n)!(s+k-m)!(n+m-k-2s)!s!}.$$

An equivalent version of (5.93) (although there are some typos) was first given in [37, (6.3)] without reference.

(ii) When $\mathbf{n} = (n, n, n)$ and $\alpha = 0$ a direct combinatorial computation [61] leads to

$$\mathscr{L}(\mathbf{n}; 0) = |\mathscr{D}(\mathbf{n})| = n!^3 \sum_{k=0}^{n} \binom{n}{k}^3.$$

Invoking (5.76) we obtain the following formula for *Franel numbers* [2, p. 43]:

$$(-1)^n \int_0^{\infty} (L_n^{(0)}(x))^3 e^{-x} dx = \sum_{k=0}^{n} \binom{n}{k}^3.$$

(iii) When $\delta = 0$ the EGF of Meixner–Pollaczek polynomials (5.37) reduces to

$$\sum_{n \geq 0} P_n(x; 0, \eta) \frac{t^n}{n!} = (1+t^2)^{-\eta/2} \exp(x \arctan t),$$

while (5.92) becomes

$$\mathscr{P}(\mathbf{n}; 0, \eta) = \begin{cases} \frac{n_1! n_2! n_3! (\eta)_s}{(s-n_1)!(s-n_2)!(s-n_3)!} & \text{if } n_1+n_2+n_3 = 2s; \\ 0 & \text{otherwise,} \end{cases}$$

where s is certain non-negative integer. It follows from (5.72) and

(5.73) that the above formula is equivalent to

$$P_n(x;0,\eta)P_m(x;0,\eta) =$$

$$\sum_{k\geq 0} k! \binom{m}{k}\binom{n}{k}(\eta+m+n-2k)_k P_{m+n-2k}(x;0,\eta). \qquad (5.94)$$

(iv) In view of Theorem 5.30, the monic OPS $\{P_n(x)\}$ with respect to the moments (5.56) is defined by the three-term recurrence relation for $n \geq 0$:

$$P_{n+1}(x) = (x-(\alpha\beta+nu_3+nu_4))P_n(x) - n(\beta+n-1)u_1u_2 P_{n-1}(x), \qquad (5.95)$$

with $P_0(x) = 1$ and $P_{-1}(x) = 0$. Using (5.56) and (5.95) a unified combinatorial proof of Theorem 5.39 was constructed in [44].

5.5 Combinatorics of some q-polynomials

5.5.1 Al-Salam–Chihara polynomials

Since our q-Hermite, q-Charlier and q-Laguerre polynomials are special or rescaled special Al-Salam–Chihara polynomials, we first recall that the Al-Salam–Chihara polynomials $Q_n(x) := Q_n(x;t_1,t_2|q)$ are defined by the recurrence relation [45, Chapter 3]:

$$\begin{cases} Q_0(x) = 1, \quad Q_{-1}(x) = 0, \\ Q_{n+1}(x) = (2x-(t_1+t_2)q^n)Q_n(x) - (1-q^n)(1-t_1t_2q^{n-1})Q_{n-1}(x), \\ \qquad n \geq 0. \end{cases}$$

$$(5.96)$$

Let $Q_n(x) = 2^n p_n(x)$ then the monic OPS $\{p_n(x)\}$ satisfy

$$xp_n(x) = p_{n+1}(x) + \frac{1}{2}(t_1+t_2)q^n p_n(x) + \frac{1}{4}(1-q^n)(1-t_1t_2q^{n-1})p_{n-1}(x). \qquad (5.97)$$

Recall the q-shifted factorials

$$(a;q)_0 := 1, \quad (a;q)_n := \prod_{k=0}^{n-1}(1-aq^{k-1}), \quad n \geq 1.$$

The Al-Salam–Chihara polynomials have the following explicit expressions:

$$Q_n(x;t_1,t_2|q) = \frac{(t_1t_2;q)_n}{t_1^n}\sum_{k=0}^{n}\frac{(q^{-n};q)_k(t_1u;q)_k(t_1u^{-1};q)_k}{(t_1t_2;q)_k(q;q)_k}q^k,$$

where $x = \frac{u+u^{-1}}{2}$ or $x = \cos\theta$ if $u = e^{i\theta}$. They have the following generating function

$$G(t,x) = \sum_{n=0}^{\infty} Q_n(x;t_1,t_2|q)\frac{t^n}{(q;q)_n} = \frac{(t_1t,t_2t;q)_\infty}{(te^{i\theta},te^{-i\theta};q)_\infty}.$$

They are orthogonal with respect to the linear functional $\mathscr{L}_{ac}$:

$$\mathscr{L}_{ac}(x^n) = \frac{(q;q)_\infty}{2\pi}\int_0^\pi \cos^n\theta \frac{(t_1t_2,e^{2i\theta},e^{-2i\theta};q)_\infty}{(t_1e^{i\theta},t_1e^{-i\theta},t_2e^{i\theta},t_2e^{-i\theta};q)_\infty}d\theta, \quad (5.98)$$

where $x = \cos\theta$. Equivalently, the Al-Salam–Chihara polynomials $Q_n(x;t_1,t_2|q)$ are orthogonal on $[-1,1]$ with respect to the probability measure

$$\frac{(q,t_1t_2;q)_\infty}{2\pi}\prod_{k=0}^\infty \frac{1-2(2x^2-1)q^k+q^{2k}}{[1-2xt_1q^k+t_1^2q^{2k}][1-2xt_2q^k+t_2^2q^{2k}]}\frac{dx}{\sqrt{1-x^2}}. \quad (5.99)$$

Note that

$$\mathscr{L}_{ac}(Q_n(x)^2) = (q;q)_n(t_1t_2;q)_n.$$

The following linearization formula is established in [42, Theorem 1].

Theorem 5.42 ([42]) *We have*

$$Q_{n_1}(x)Q_{n_2}(x) = \sum_{n_3\geq 0} C_{n_1,n_2}^{n_3}(t_1,t_2;q)Q_{n_3}(x), \quad (5.100)$$

where

$$C_{n_1,n_2}^{n_3}(\alpha,\beta;q)$$
$$= (-1)^{n_1+n_2+n_3}\frac{(q;q)_{n_1}(q;q)_{n_2}}{(t_1t_2;q)_{n_3}}$$
$$\times \sum_{m_2,m_3}\frac{(t_1t_2;q)_{n_1+m_3}t_1^{m_2}t_2^{n_3+n_2-n_1-m_2-2m_3}q^{\binom{m_2}{2}+\binom{n_3+n_2-n_1-m_2-2m_3}{2}}}{(q;q)_{n_3+n_2-n_1-m_2-2m_3}(q;q)_{m_2}(q;q)_{m_3+n_1-n_3}(q;q)_{m_3+n_1-n_2}(q;q)_{m_3}}.$$

5.5.2 *Moments of continuous q-Hermite, q-Charlier and q-Laguerre polynomials*

The continuous q-Hermite polynomials $H_n(x|q)$ are the special Al-Salam–Chihara polynomials $Q_n(x;0,0|q)$, so they are generated by

$$\begin{cases} H_0(x|q) = 1, \quad H_{-1}(x|q) = 0, \\ H_{n+1}(x|q) = 2xH_n(x|q) - (1-q^n)H_{n-1}(x|q), \quad n\geq 0. \end{cases} \quad (5.101)$$

They satisfy the orthogonality relation

$$\frac{(q;q)_\infty}{2\pi} \int_0^\pi H_n(\cos\theta|q)H_m(\cos\theta|q)(e^{2\imath\theta}, e^{-2\imath\theta}; q)_\infty d\theta = (q;q)_n \delta_{m,n}.$$

(5.102)

If we rescale the continuous q-Hermite polynomials by

$$\tilde{H}_n(x|q) = H_n(ax/2|q)/a^n, \qquad a = \sqrt{1-q},$$

then (5.101) reads

$$x\tilde{H}_n(x|q) = \tilde{H}_{n+1}(x|q) + [n]_q \tilde{H}_{n-1}(x|q).$$

The moment functional for $\tilde{H}_n(x|q)$ is

$$\mathscr{L}_h(f) := N(x;q) \int_{-2/\sqrt{1-q}}^{2/\sqrt{1-q}} f(x) \prod_{k=0}^\infty \{1 + (2 - x^2(1-q))q^k + q^{2k}\} dx,$$

(5.103)

where

$$N(x;q) = \frac{\sqrt{1-q}}{4\pi} \frac{(q;q)_\infty}{\sqrt{1 - (1-q)x^2/4}}.$$

The q-Charlier polynomials $C_n(x|q) := C_n(x,a|q)$ are defined recursively by

$$C_{n+1}(x|q) = (x - a - [n]_q)C_n(x|q) - a[n]_q C_{n-1}(x|q),$$

(5.104)

where $C_{-1}(x|q) = 0$ and $C_0(x|q) = 1$. Comparing with (5.96) we see that this is a rescaled version of the Al-Salam–Chihara polynomials:

$$C_n(x|q) = \left(\frac{a}{1-q}\right)^{n/2} Q_n\left(\frac{1}{2}\sqrt{\frac{1-q}{a}}(x - a - \frac{1}{1-q}); \frac{-1}{\sqrt{a(1-q)}}, 0 \,\middle|\, q\right).$$

(5.105)

The first values of these polynomials are

$$C_1(x|q) = x - a,$$
$$C_2(x|q) = x^2 - (2a+1)x + a^2,$$
$$C_3(x|q) = x^3 - (q + 3a + 2)x^2 + (aq + 3a^2 + 2a + q + 1)x - a^3.$$

For any non-negative integer n set

$$[n]_q := \frac{1-q^n}{1-q} = 1 + q + \cdots + q^{n-1}$$

(5.106)

and for $0 \le k \le n$,

$$\begin{bmatrix} n \\ k \end{bmatrix}_q = \frac{(q;q)_n}{(q;q)_k (q;q)_{n-k}}. \tag{5.107}$$

The explicit formula of $C_n(x|q)$ is

$$C_n(x|q) = \sum_{k=0}^{n} \begin{bmatrix} n \\ k \end{bmatrix}_q q^{k(k-n)} (-a)^{n-k} \prod_{i=0}^{k-1} \left(x - [i]_q + a(q^{-i} - 1) \right).$$

We define $u_1(x)$ and $v_1(x)$ by

$$\begin{aligned} u_1(x) &= \frac{1-q}{2a} x^2 - \frac{a(1-q)+1}{a} x + \frac{1+a^2(1-q)^2}{2a(1-q)}, \\ v_1(x) &= \frac{1}{2} \sqrt{\frac{1-q}{a}} \left(x - a - \frac{1}{1-q} \right). \end{aligned} \tag{5.108}$$

The moment functional for $C_n(x|q)$ is

$$\begin{aligned} \mathcal{L}_c(f) = {} & \frac{(q;q)_\infty}{2\pi} \frac{1}{2} \sqrt{\frac{1-q}{a}} \\ & \times \int_{A_-}^{A_+} \prod_{k=0}^{\infty} \frac{[1 - 2u_1(x)q^k + q^{2k}]f(x)}{1 + 2v_1(x)q^k/(\sqrt{a(1-q)}) + q^{2k}/a(1-q)} \frac{dx}{\sqrt{1 - v_1(x)^2}}, \end{aligned} \tag{5.109}$$

where

$$A_\pm = a + \frac{1}{1-q} \pm 2 \sqrt{\frac{a}{1-q}}.$$

The q-Laguerre polynomials $L_n(x|q) := L_n(x, y|q)$ are defined by the recurrence:

$$L_{n+1}(x|q) = (x - y[n+1]_q - [n]_q)L_n(x|q) - y[n]_q^2 L_{n-1}(x|q), \tag{5.110}$$

with the initial conditions $L_{-1}(x|q) = 0$ and $L_0(x|q) = 1$. Hence, they are the re-scaled Al-Salam–Chihara polynomials:

$$L_n(x|q) = \left(\frac{\sqrt{y}}{q-1} \right)^n Q_n \left(\frac{(q-1)x + y + 1}{2\sqrt{y}}; \frac{1}{\sqrt{y}}, \sqrt{y}q \, | \, q \right). \tag{5.111}$$

One then deduces the explicit formula:

$$L_n(x|q) = \sum_{k=0}^{n} (-1)^{n-k} \frac{n!_q}{k!_q} \begin{bmatrix} n \\ k \end{bmatrix}_q q^{k(k-n)} y^{n-k} \prod_{j=0}^{k-1} \left(x - (1 - yq^{-j})[j]_q \right). \tag{5.112}$$

The first values of these q-Laguerre polynomials are given by

$$L_1(x; q) = x - y,$$
$$L_2(x; q) = x^2 - (1 + 2y + qy)x + (1 + q)y^2,$$
$$L_3(x; q) = x^3 - (q^2y + 3y + q + 2 + 2qy)x^2$$
$$+ (q^3y^2 + yq^2 + q + 2qy + 3q^2y^2 + 1 + 4qy^2 + 2y + 3y^2)x$$
$$- (2q^2 + 2q + q^3 + 1)y^3.$$

Define $u_2(x)$ and $v_2(x)$ by

$$u_2(x) = \frac{(1-q)^2}{2y}x^2 - \frac{(1-q)(1+y)}{y}x + \frac{y^2+1}{2y}, \quad v_2(x) = \frac{q-1}{2\sqrt{y}}x + \frac{y+1}{2\sqrt{y}}. \tag{5.113}$$

Then the moment functional in this case reads

$$\mathscr{L}_\ell(f) = \frac{(q,q;q)_\infty}{2\pi} \frac{1-q}{2\sqrt{y}}$$

$$\times \int_{B_-}^{B_+} \prod_{k=0}^\infty \frac{[1 - 2u_2(x)q^k + q^{2k}]f(x)}{[1 - 2v_2(x)q^k/\sqrt{y} + q^{2k}/y][1 - 2v_2(x)q^{k+1}\sqrt{y} + q^{2k+2}y]}$$

$$\times \frac{dx}{\sqrt{1 - v_2(x)^2}}, \tag{5.114}$$

where

$$B_\pm = \frac{(1 \pm \sqrt{y})^2}{1 - q}. \tag{5.115}$$

5.5.3 Linearization coefficients of continuous q-Hermite, q-Charlier and q-Laguerre polynomials

A matching of $[1 \ldots n]$ is a partition such that each block has only one or two elements. Let Π_n be the set of partitions of $[1 \ldots n]$ and $\mathcal{M}_n$ be the set of matchings of $[1 \ldots n]$. In what follows, we will write a partition by ordering its blocks in increasing order of the minima of blocks and within each block the elements are in increasing order. For example, the partition $\pi = \{1, 9, 10\} - \{2, 3, 7\} - \{4\} - \{5, 6, 11\} - \{8\}$ has five blocks. We can also depict a partition of $[1 \ldots n]$ by a graph on the vertex set $[1 \ldots n]$ such that there is an edge e joining i and j if and only if i and j are consecutive elements in a same block. An example is given in Figure 5.6.

Given a partition π of $[1 \ldots n]$, two edges $e_1 = (i_1, j_1)$ and $e_2 = (i_2, j_2)$ of π are said to form:

(i) a crossing with e_1 as the *initial edge* if $i_1 < i_2 < j_1 < j_2$;

Figure 5.6 Graph of the partition $\pi = \{1,9,10\} - \{2,3,7\} - \{4\} - \{5,6,11\} - \{8\}$.

(ii) a nesting with e_2 as *interior edge* if $i_1 < i_2 < j_2 < j_1$;

(iii) an alignment with e_1 as *initial edge* if $i_1 < j_1 \leq i_2 < j_2$.

Let $\mathrm{cro}(\pi)$ (resp. $\mathrm{ne}(\pi)$) be the number of crossings (resp. nestings) of π.

Remark 5.43 The aforementioned crossings and nestings are called *restricted crossings* and *restricted nestings* in [41].

The study of the moments of the continuous q-Hermite polynomials $\tilde{H}_n(x|q)$ goes back to Touchard [55] and Riordan [49]. From their continued fraction expansion it is easy to derive the following result

$$\mathscr{L}_h(x^n) = \sum_{\pi \in \mathscr{M}_n} q^{\mathrm{cro}(\pi)}. \tag{5.116}$$

Using (5.116) and a q-version of (5.39) Ismail, Stanton, Viennot [35] proved the following formula for the linearization coefficients of continuous q-Hermite polynomials.

Theorem 5.44 ([35]) *We have*

$$\mathscr{L}_h\big(\tilde{H}_{n_1}(x|q)\cdots\tilde{H}_{n_k}(x|q)\big) = \sum_{\pi \in \mathscr{M}(n_1,n_2,\dots,n_k)} q^{\mathrm{cro}(\pi)}. \tag{5.117}$$

Remark 5.45 A formula about the apparently more general integral

$$\mathscr{L}_h\big(x^n\tilde{H}_{n_1}(x|q)\cdots\tilde{H}_{n_k}(x|q)\big)$$

is given in [12, Theorem 7.3]. As $\tilde{H}_1(x|q) = x$ the latter integral can be written as $\mathscr{L}_h\big((\hat{H}_1(x|q))^n\tilde{H}_{n_1}(x|q)\cdots\tilde{H}_{n_k}(x|q)\big)$ and its combinatorial interpretation in [12, Theorem 7.3] follows from the above theorem.

For $k = 3$ the above formula gives Rogers' linearization formula

$$\tilde{H}_{n_1}(x|q)\,\tilde{H}_{n_2}(x|q) = \sum_{l=0}^{\min(n_1,n_2)} \begin{bmatrix} n_1 \\ l \end{bmatrix}_q \begin{bmatrix} n_2 \\ l \end{bmatrix}_q l!_q\,\tilde{H}_{n_1+n_2-2l}(x|q). \tag{5.118}$$

For $k = 4$ it gives a combinatorial evaluation of the Askey–Wilson integral (5.3).

The moments of q-Charlier polynomials have the following combinatorial interpretations.

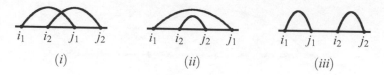

Figure 5.7 Crossing, nesting and alignments of two edges.

Theorem 5.46 ([41, 43])   *We have*

$$\mu_n(a) := \mathscr{L}_c(x^n) = \sum_{\pi \in \Pi_n} a^{|\pi|} q^{\mathrm{cro}(\pi)} = \sum_{\pi \in \Pi_n} a^{|\pi|} q^{\mathrm{ne}(\pi)},$$

where Π_n denotes the set of partitions of $[1\ldots n]$.

The first values of $\mu_n(a)$ are as follows:

$$\mu_1(a) = a, \quad \mu_2(a) = a + a^2, \quad \mu_3(a) = a + 3a + a^3,$$
$$\mu_4(a) = a + (6+q)a^2 + 6a^3 + a^4.$$

Definition 5.47   A *Charlier configuration* on $[1\ldots n]$ is a pair (B,σ) such that $B \subset [1\ldots n]$ and σ is a permutation on $[1\ldots n] \setminus B$ such that:

- each $i \in B$ is written as $[i]$;
- each cycle of σ is written as $(a\,\sigma(a)\,\sigma^2(a)\ldots\sigma^{l-1}(a))$ with $\sigma^l(a) = a$, and a is the maximum of the cycle;
- order the cycles and elements of B in decreasing order of their maxima.

A pair (i,j), $i > j$, is called a *Charlier-inversion* in (B,σ) if i is not a maximum of any cycle of π and i appears to the left of j in (B,σ). Let $w(B,\sigma)$ denote the number of Charlier-inversions in (B,σ).

Example 5.48   Let $n = 9$, $B = \{2,9\}$ and $\sigma = (6)(74)(8153)$. Then

$$(B,\sigma) = [9](8,153)(74)(6)[2],$$

which has the weight $(-1)^{9-3}x^3 a^2 q^{0+3+1+1} = a^2 q^5 x^3$, because three cycles have label x, two cycles have label a and there are five Charlier-inversions, namely, $(5,3),(5,4),(5,2),(3,2),(4,2)$.

Theorem 5.49 ([41])   *We have*

$$C_n(x,a;q) = \sum_{(B,\sigma)} (-1)^{n-\mathrm{cyc}\,\sigma} a^{|B|} x^{\mathrm{cyc}\,\sigma} q^{w(B,\sigma)},$$

where $B \subset [1\ldots n]$ and σ is a permutation on $[1\ldots n] \setminus B$.

The following linearization formula was first proved by Anshelevich [1] using a stochastic process.

Theorem 5.50 ([1])   *The linearization coefficients of q-Charlier polynomials are the generating functions for the inhomogeneous partitions:*

$$\mathscr{L}_c\left(C_{n_1}(x,a;q)\cdots C_{n_k}(x,a;q)\right) = \sum_{\pi \in \Pi(n_1, n_2, \dots, n_k)} q^{\mathrm{rc}(\pi)} a^{|\pi|}. \qquad (5.119)$$

For example, if $k = 3$ and $n_1 = n_2 = 2$ and $n_3 = 1$, then $\Pi(2,2,1)$ is

$$\{\{(1,3,5)(2,4)\}, \{(1,4,5)(2,3)\}, \{(2,3,5)(1,4)\}, \{(2,4,5)(1,3)\}\}.$$

It is easy to see that the corresponding generating function in (5.119) is

$$a^2 q^2 + a^2 + a^2 q + a^2 q = a^2 (1+q)^2.$$

If $k = 2$, equation (5.119) gives the orthogonality relation. When $k = 3$, the following explicit formula for the generating function in (5.119) is proved combinatorially in [41].

Theorem 5.51 ([41])   *We have*

$$\sum_{\pi \in \Pi(n_1, n_2, n_3)} q^{\mathrm{cro}(\pi)} a^{|\pi|}$$

$$= \sum_{l \geq 0} \frac{n_1!_q n_2!_q n_3!_q \, a^{l+n_3} q^{\binom{n_1+n_2-n_3-2l}{2}}}{l!_q (n_3 - n_1 + l)!_q (n_3 - n_2 + l)!_q (n_1 + n_2 - n_3 - 2l)!_q}. \qquad (5.120)$$

De Médicis, Stanton, and White [14] had considered another sequence of combinatorial q-Charlier polynomials which are a rescaled version of the Al-Salam–Carlitz polynomials. Their q-Charlier polynomials have a natural q-Stirling number associated with their moments with a simple explicit formula. However these polynomials have a complicated and non-positive linearization formula.

We define the *crossing number* of a permutation $\sigma \in \mathfrak{S}_n$ by

$$\mathrm{cro}(\sigma) = \sum_{i=1}^{n} \#\{j \mid j < i \leq \sigma(j) < \sigma(i)\} + \sum_{i=1}^{n} \#\{j \mid j > i > \sigma(j) > \sigma(i)\},$$

and the number of *weak excedances* of σ by

$$\mathrm{wex}(\sigma) = \#\{i \mid 1 \leq i \leq n \text{ and } i \leq \sigma(i)\}.$$

For the permutation σ, factorized as product of disjoint cycles in Figure 5.8, we have $\mathrm{wex}(\sigma) = 8$ and $\mathrm{cro}(\sigma) = 7$. The following formula for the moments of the q-Laguerre polynomials $L_n(x; q)$ defined in (5.111) is given in [42]; see [11] for an equivalent continued fraction expansion:

$$\mathscr{L}_\ell(x^n) = \sum_{\sigma \in \mathfrak{S}_n} y^{\mathrm{wex}(\sigma)} q^{\mathrm{cro}(\sigma)}. \qquad (5.121)$$

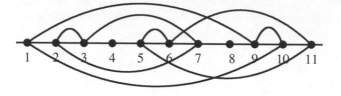

Figure 5.8 Diagram of $\sigma = (1,9,10)(2,3,7)(4)(5,6,11)(8)$..

A combinatorial interpretation of these q-Laguerre polynomials can be derived from the Simion and Stanton's combinatorial model for the $a = s = u = 1$ and $r = t = q$ special case of the quadrabasic Laguerre polynomials [51, p. 313]. The following linearization formula was proved in [42].

Theorem 5.52 ([42]) *The linearization coefficients of the q-Laguerre polynomials are*

$$\mathscr{L}_\ell(L_{n_1}(x;q)\cdots L_{n_k}(x;q)) = \sum_{\sigma \in \mathscr{D}(n_1,\dots,n_k)} y^{\mathrm{wex}(\sigma)} q^{\mathrm{cro}(\sigma)}. \qquad (5.122)$$

This is proved by induction on the sequence $(n_1,\dots,n_k)$ starting from $(n_1,\dots,n_k) = (1,\dots,1)$. The hard part is to show that the right-hand side of (5.122) is symmetric on $n_1,\dots,n_k$. Here, we only check the $n_1 = \cdots = n_k = 1$ case. Let

$$d_n(y,q) = \sum_{\sigma \in \mathscr{D}_n} y^{\mathrm{wex}(\sigma)} q^{\mathrm{cro}(\sigma)}.$$

Then we have to show that $\mathscr{L}_q((x-y)^n) = d_n(y,q)$. Note that

$$\mathscr{L}_\ell((x-y)^n) = \sum_{k=0}^n (-1)^{n-k} \binom{n}{k} y^{n-k} \mathscr{L}_\ell(x^k).$$

By binomial inversion, this is equivalent to

$$\mathscr{L}_\ell(x^n) = \sum_{k=0}^n \binom{n}{k} y^k d_{n-k}(y,q).$$

The last identity is obvious by classifying the permutations according to the fixed points. When $k = 3$ we derive the following explicit double sum formula from Theorem 5.42.

Theorem 5.53 ([42])   *We have*

$$\mathscr{L}_\ell(L_{n_1}(x;q)L_{n_2}(x;q)L_{n_3}(x;q))$$
$$-\sum_s \frac{n_1!_q\,n_2!_q\,n_3!_q\,s!_q\,y^s}{(n_1+n_2+n_3-2s)!_q(s-n_3)!_q(s-n_2)!_q(s-n_1)!_q}$$
$$\times \sum_k \begin{bmatrix} n_1+n_2+n_3-2s \\ k \end{bmatrix}_q y^k q^{\binom{k+1}{2}+\binom{n_1+n_2+n_3-2s-k}{2}}.$$

If $\mathbf{n} = (2,2,1)$, Theorem 5.53 provides

$$\mathscr{L}_\ell(L_2(x;q)L_2(x;q)L_1(x;q)) = (1+q)^3(1+qy)y^2.$$

A pictorial verification of the above computation is given in the appendix of [42].

5.5.4 A curious q-analogue of Hermite polynomials

For an element w in the *Weyl algebra* generated by D and U with relation $DU = UD+1$, the normally ordered form is $w = \sum_{i,j\geq 0} c_{i,j} U^i D^j$. For example, if U is the multiplication operator X on the polynomials in x defined by $Xp(x) = xp(x)$ for $p(x) \in \mathbb{K}[x]$, and D is the differential operator $\frac{d}{dx}$, then

$$DU - UD = \mathbf{1}, \tag{5.123}$$

where $\mathbf{1}$ is the identity operator. It is well known (see [9, p. 145]) that the normalized Hermite polynomials $h_n(x) := 2^{-n/2}H_n(x/\sqrt{2})$ are given by

$$h_n(x) = (-1)^n e^{x^2/2} \frac{d^n}{dx^n} e^{-x^2/2} = \left(X - \frac{d}{dx}\right)^n \cdot 1.$$

A curious family of q-Hermite polynomials was studied in [10] by considering a q-analogue of the above formula. It is convenient to use the rescaled Hermite polynomials $h_n(x,s) := s^{n/2}h_n(x/\sqrt{s})$. Hence, they satisfy the recurrence relation

$$h_{n+1}(x,s) = xh_n(x,s) - snh_{n-1}(x,s). \tag{5.124}$$

and have the explicit formula

$$h_n(x,s) = \sum_{k=0}^n \binom{n}{2k}(-s)^k(2k-1)!!x^{n-2k}.$$

Let $D = \frac{d}{dx}$ be the differential operator on $\mathbb{K}[x]$; then $Dh_n(x,s) = nh_{n-1}(x,s)$ and

$$h_n(x,s) = (x - sD)^n \cdot 1, \tag{5.125}$$

$$h_n(x + sD, s) \cdot 1 = x^n. \tag{5.126}$$

From the generating function $\sum_{n \geq 0} H_n(x,s) \frac{t^n}{n!} = e^{xt - st^2/2}$ and the identity

$$e^{st^2/2} \sum_{n \geq 0} h_n(x,s) \frac{t^n}{n!} = e^{xt}$$

we immediately derive

$$h_n(x,s) = \sum_{k=0}^{n} \frac{n!(-s)^k}{2^k k!(n-2k)!} x^{n-2k}, \tag{5.127}$$

$$x^n = \sum_{k=0}^{\lfloor n/2 \rfloor} \frac{n!s^k}{2^k k!(n-2k)!} H_{n-2k}(x,s). \tag{5.128}$$

Consider a q-Weyl algebra generated by two generators U and D satisfying the commutation relation

$$DU - qUD = \mathbf{1}. \tag{5.129}$$

Clearly, if U and D are respectively the *multiplication* and *q-differential* operators D_q on $\mathbb{K}[x]$, then the above identity holds true. Recall that the q-differential operator D_q on $\mathbb{R}[x]$ is defined by

$$D_q f(x) = (f(qx) - f(x))/(q-1)x.$$

Quite many combinatorial models related to (5.123) and (5.129) have been given in [6].

Given a matching $\alpha \in \mathcal{M}_n$ we define the statistic $c(\alpha)$ by

$$c(\alpha) = \sum_{a \in \mathrm{iso}(\alpha)} |\{\text{edges } (i,j) : i < a < j\}|,$$

where $\mathrm{iso}(\alpha)$ denotes the set of isolated vertices of α, and let $\mathcal{M}(n,k)$ denote the set of matchings of $[1 \ldots n]$ with k isolated vertices. For instance, the matching $\alpha = \{1,7\} - \{2,5\} - \{3\} - \{4,8\} - \{6\}$ of $[1..8]$ is depicted in Figure 5.9. We have $c(\alpha) = 2$ and the crossing number $\mathrm{cro}(\alpha)$ is equal to 1.

In 2005 Varvak [56] studied the combinatorial formula in the normal ordering of two non commutative operators U and D satisfying (5.129) and proved the following *q-Weyl binomial formula.*

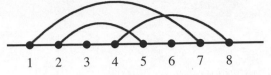

Figure 5.9 A matching of $[1..8]$ with two isolated points.

Theorem 5.54 ([56]) *We have*

$$(D+U)^n = \sum_{m,k \geq 0} \left\{ {n \atop m} \right\}_{k,q} U^{n-m-k} D^{m-k}.$$

where

$$\left\{ {n \atop m} \right\}_{k,q} = \left[{n-2k \atop m-k} \right]_q \cdot \sum_{\alpha \in \mathcal{M}(n,n-2k)} q^{c(\alpha)+\mathrm{cro}(\alpha)}.$$

On the other hand, the following q-analogue of Hermite polynomials was introduced in [10]:

$$h_n(x,s|q) := (x - sD_q)^n \cdot 1. \tag{5.130}$$

The first few terms of the sequence $\{h_n(x,s|q)\}$ are

$$\begin{aligned}
&1, \quad x, \quad -s+x^2, \quad -(2+q)sx+x^3, \\
&(1+q+q^2)s^2 - (3+2q+q^2)sx^2 + x^4, \\
(3+4q+4q^2+3q^3+q^4)s^2x &- (4+3q+2q^2+q^3)sx^3 + x^5.
\end{aligned}$$

It turns out that the above q-Hermite polynomials $h_n(x,s|q)$ are connected to Varvak's binomial formula as follows.

Theorem 5.55 ([10]) *We have*

$$h_n(x,-1|q) = \sum_{\sigma \in \mathcal{M}(n)} x^{\mathrm{iso}(\sigma)} q^{c(\sigma)+\mathrm{cro}(\sigma)}. \tag{5.131}$$

Although the polynomials $h_n(x,-1|q)$ are not an OPS, they have many interesting properties. They are actually related to the continuous q-Hermite polynomials $\tilde{H}_n(x|q)$. Let

$$h_n(x,-1|q) = \sum_{k \equiv n \pmod 2} c(n,k,q)x^k, \tag{5.132}$$

$$\tilde{H}_n(x|q) = \sum_{k \equiv n \pmod 2} b(n,k,q)x^k(-1)^{\frac{n-k}{2}}. \tag{5.133}$$

The following inverse relation was established in [10].

Theorem 5.56 ([10]) *The matrices*

$$(c(i,j,q))_{i,j=0}^{n-1} \quad and \quad (b(i,j,q)(-1)^{\frac{i-j}{2}})_{i,j=0}^{n-1}$$

are mutually inverse. In other words, we have

$$x^n = \sum_{k \geq 0} c(n,k,q)\tilde{H}_{n-2k}(x|q). \tag{5.134}$$

An explicit formula for $c(n,k,q)$ was found in [10, 38]. In the rest of this section we shall give an (easy) proof of this formula based on the above theorem. Let $U_n(x) = \sin((n+1)\theta)/\sin\theta$, $x = \cos\theta$, be the Chebyshev polynomials of the second kind. Note that

$$2xU_n(x) = U_{n+1}(x) + U_{n-1}(x) \quad (n \geq 0), \tag{5.135}$$

with $U_{-1}(x) = 0$ and $U_0(x) = 1$ and

$$U_n(x) = \sum_{k \geq 0} (-1)^k \binom{n-k}{k} (2x)^{n-2k}.$$

The following formulae are two special cases of Rogers' connection formula for continuous q-ultraspherical polynomials [32, (13.3.1)], but can be verified straightforwardly once discovered.

Lemma 5.57 *We have*

$$H_n(x|q) = \sum_{k=0}^{\lfloor n/2 \rfloor} \left(\begin{bmatrix} n \\ k \end{bmatrix}_q - \begin{bmatrix} n \\ k-1 \end{bmatrix}_q \right) U_{n-2k}(x), \tag{5.136}$$

$$U_n(x) = \sum_{k=0}^{\lfloor n/2 \rfloor} (-1)^k q^{\binom{k+1}{2}} \begin{bmatrix} n-k \\ k \end{bmatrix}_q H_{n-2k}(x|q). \tag{5.137}$$

Proof We show that the right-hand side of (5.136) satisfies the recurrence displayed in (5.101). This is obvious for $n = 0$. Assume that (5.136) is true until $n \geq 0$. Substituting (5.136) into $2xH_n(x|q) - (1-q^n)H_{n-1}(x|q)$ and using (5.135) we have

$$\sum_{k \geq 0} \left(\begin{bmatrix} n \\ k \end{bmatrix}_q - \begin{bmatrix} n \\ k-1 \end{bmatrix}_q \right) (U_{n+1-2k}(x) + U_{n-1-2k}(x))$$

$$- (1-q^n) \sum_{k \geq 0} \left(\begin{bmatrix} n-1 \\ k \end{bmatrix}_q - \begin{bmatrix} n-1 \\ k-1 \end{bmatrix}_q \right) U_{n-1-2k}(x).$$

The coefficient of $U_{n+1-2k}(x)$ in the last expression is equal to

$$\begin{bmatrix} n \\ k \end{bmatrix}_q - \begin{bmatrix} n \\ k-2 \end{bmatrix}_q - (1-q^n)\begin{bmatrix} n-1 \\ k-1 \end{bmatrix}_q + (1-q^n)\begin{bmatrix} n-1 \\ k-2 \end{bmatrix}_q$$

$$= \begin{bmatrix} n+1 \\ k \end{bmatrix}_q - \begin{bmatrix} n+1 \\ k-1 \end{bmatrix}_q.$$

The proof of (5.136) is thus completed. Similarly we can verify that the right-hand side of (5.137) satisfies the recurrence (5.135). $\qquad\square$

When $q = 1$, as $H_n(x|1) = (2x)^n$, the equation (5.136) reduces to

$$(2x)^n = \sum_{k=0}^{\lfloor n/2 \rfloor}\left(\binom{n}{k} - \binom{n}{k-1}\right)U_{n-2k}(x),$$

substituting (5.137) into the above identity we obtain

$$(2x)^n =$$

$$\sum_{l\geq 0} H_{n-2l}(x|q) \sum_{j=0}^{l}(-1)^j q^{\binom{j+1}{2}}\left(\binom{n}{l-j} - \binom{n}{l-j-1}\right)\begin{bmatrix} n-2l+j \\ j \end{bmatrix}_q.$$
(5.138)

Since $\tilde{H}_n(x|q) = a^{-n}H_n(ax/2|q)$ with $a = \sqrt{1-q}$, we derive immediately from (5.134) and (5.138) the following result, which was originally proved in [10, 38] with more work.

Theorem 5.58 ([10, 38]) *If $k \equiv n \pmod 2$ then*

$$c(n,k,q)$$
$$= \sum_{\alpha\in.\mathscr{M}(n,k)} q^{c(\alpha)+\mathrm{cro}(\alpha)} \tag{5.139}$$

$$= (1-q)^{-\frac{n-k}{2}}\sum_{j\geq 0}\left(\binom{n}{\frac{n-k-2j}{2}} - \binom{n}{\frac{n-k-2j-2}{2}}\right)(-1)^j q^{\binom{j+1}{2}}\begin{bmatrix} k+j \\ k \end{bmatrix}_q.$$
(5.140)

Remark 5.59   (a) When n is even, the $k = 0$ case of (5.139) reduces to the Touchard–Riordan formula:

$$\sum_{\alpha\in M(2n,0)} q^{\mathrm{cro}(\alpha)} = \frac{1}{(1-q)^n}\sum_{k=0}^{n}\left(\binom{2n}{n-k} - \binom{2n}{n-k-1}\right)q^{\frac{k(k+1)}{2}}.$$

(b) Formula (5.139) is equivalent to the evaluation of the generalized moments of the continuous q-Hermite polynomials $\mathscr{L}(x^n H_m(x|q))$. Kim and Stanton [40] have extended this to Askey–Wilson polynomials using a bootstrapping method.

5.5.5 Combinatorics of continued fractions and γ-positivity

Recall that the Eulerian polynomials $(A_n(t))_{n\geq0}$ are defined by

$$\sum_{n=0}^{\infty} A_n(t)\frac{x^n}{n!} = \frac{1-t}{1-te^{(1-t)x}},$$

and have the following continued fraction expansion [16, 57]:

$$\sum_{n=0}^{\infty} A_{n+1}(t)x^n = \cfrac{1}{1-(1+t)x-\cfrac{1\cdot2tx^2}{1-2(1+t)x-\cfrac{2\cdot3tx^2}{\ddots}}}, \qquad (5.141)$$

where $b_k = (k+1)(1+t)$, $\lambda_k = k(k+1)t$. It follows immediately from (5.29) that there are positive integers $\gamma_{n,k}$ such that

$$A_{n+1}(t) = \sum_{k=0}^{\lfloor n/2\rfloor} \gamma_{n,k}t^k(1+t)^{n-2k}. \qquad (5.142)$$

This connection hints that continued fractions can shed light on the γ-positivity of Eulerian polynomials. We first review some motivations of this notion.

A finite sequence of positive integers $\{a_0,\dots,a_n\}$ is *unimodal* if there exists an index i' such that $a_i \leq a_{i+1}$ if $i < i'$ and $a_i > a_{i+1}$ otherwise; it is *log-concave* if $a_i^2 \geq a_{i-1}a_{i+1}$ for all $i \in [n-1]$. A polynomial $p(x) = a_0+a_1x+\cdots+a_nx^n$ with non-negative coefficients is unimodal (resp. log-concave) if and only if the sequence of its coefficients is unimodal (resp. log-concave). It is well known that a polynomial with non-negative coefficients and with only *real roots* is log-concave and that log-concavity implies unimodality. The set of polynomials whose coefficients are symmetric with center of symmetry $d/2$ is a vector space with a basis given by $\{t^k(1+t)^{d-2k}\}_{k=0}^{\lfloor d/2\rfloor}$. If a polynomial $p(t)$ has non-negative coefficients when expanded in this basis, then the coefficients in the standard basis $\{t^k\}_{k=0}^{\infty}$ form a unimodal sequence (being a non-negative sum of symmetric and unimodal sequences with the same center of symmetry). Hence, the expansion (5.142) implies both the symmetry and the unimodality of the Eulerian numbers. In two recent papers [52, 53] the combinatorics of continued fractions is exploited to derive the γ-positivity of some q-Eulerian polynomials.

For any permutation $\sigma \in \mathfrak{S}_n$, written as the word $\sigma = \sigma(1)\dots\sigma(n)$, the entry $i \in [1\dots n]$ is called a *descent* of σ if $i < n$ and $\sigma(i) > \sigma(i+1)$; the entry $i \in [1 \ldots n]$ is called an *excedance* of σ if $i < \sigma(i)$. Denote the number of descents, excedances in σ by des σ, exc σ respectively. It is well known [22] that the statistics des, exc and drop have the same distribution on $\mathfrak{S}_n$ and their common enumerative polynomial is the Eulerian polynomial:

$$A_n(t) = \sum_{\sigma \in \mathfrak{S}_n} t^{\mathrm{des}\,\sigma} = \sum_{\sigma \in \mathfrak{S}_n} t^{\mathrm{exc}\,\sigma}. \tag{5.143}$$

Define the set $\mathrm{DD}_{n,k} := \{\sigma \in \mathfrak{S}_n : \mathrm{des}\,\sigma = k \text{ and } \mathrm{ddes}(\sigma 0) = 0\}$, where $\mathrm{ddes}(\sigma 0)$ is the number of *double descents* in the word $\sigma(1)\cdots\sigma(n)0$, i.e., indices i, $1 < i \leq n$, such that $\sigma(i-1) > \sigma(i) > \sigma(i+1)$ with $\sigma(n+1) = 0$. For example,

$$\mathrm{DD}_{4,1} = \{1324, 1423, 2314, 2413, 3412, 2134, 3124, 4123\}.$$

Foata and Schützenberger [22] proved the following expansion formula

$$A_n(t) = \sum_{k=0}^{\lfloor (n-1)/2 \rfloor} |\mathrm{DD}_{n,k}| t^k (1+t)^{n-1-2k}. \tag{5.144}$$

Recall that for $\sigma \in \mathfrak{S}_n$ the inversion number inv σ is the number of pairs (i,j) such that $i < j$ and $\sigma(i) > \sigma(j)$, the statistic (31-2) σ is the number of pairs (i,j) such that $2 \leq i < j \leq n$ and $\sigma(i-1) > \sigma(j) > \sigma(i)$, similarly, the statistic (2-13) σ is the number of pairs (i,j) such that $1 \leq i < j \leq n-1$ and $\sigma(j+1) > \sigma(i) > \sigma(j)$. In [53] the following q-analogue of (5.144) is established using the combinatorics of the J-continued fraction expansions of their ordinary generating functions.

Theorem 5.60 ([53]) *We have*

$$\sum_{\sigma \in \mathfrak{S}_n} t^{\mathrm{exc}\,\sigma} q^{\mathrm{inv}\,\sigma - \mathrm{exc}\,\sigma} = \sum_{0 \leq k \leq (n-1)/2} \gamma_{n,k}(q) t^k (1+t)^{n-1-2k}, \tag{5.145}$$

where the coefficient $\gamma_{n,k}(q)$ *has the following combinatorial interpretation*

$$\gamma_{n,k}(q) = \sum_{\sigma \in \mathrm{DD}_{n,k}} q^{2\,(2\text{-}13)\,\sigma + (31\text{-}2)\,\sigma}$$

and is divisible by $q^k (1+q)^k$ *for* $0 \leq k \leq (n-1)/2$.

Remark 5.61 Lin and Fu [47] have recently shown that the γ-positivity (5.145) has a counterpart in the subclass of 321-avoiding permutations σ in $\mathfrak{S}_n$, namely, there does not exist indices $i < j < k$ such that $\sigma_i > \sigma_j > \sigma_k$.

5.6 Some open problems

We conclude this chapter with some open problems or future directions.

(i) Besides the continuous q-Hermite, q-Charlier and q-Laguerre polynomials it would be interesting to know whether there are other rescaled Al-Salam–Chihara polynomials which have interesting linearization coefficients.

(ii) The q-Laguerre polynomials and their moments have known combinatorial meanings. It would be desirable to find a bijective proof of the linearization coefficients formula for q-Laguerre polynomials (5.122) in the same vein as the bijective proofs for the linearization coefficients of q-Hermite [35] and q-Charlier polynomials [41].

(iii) In 1954 Carlitz (see [20]) introduced the q-Eulerian polynomials $A_n(z, q)$ via

$$\sum_{k=0}^{\infty} ([k+1]_q)^n z^k = \frac{A_n(z, q)}{(z; q)_{n+1}}.$$

We remark that Foata's formula [20, (5.3)]

$$(1-q)^{-n} \sum_{k=0}^{n} \binom{n}{k} (-q)^k \frac{(z; q)_k}{(zq; q)_k} = \frac{A_n(z, q)}{(zq; q)_n} \tag{5.146}$$

is equivalent to saying that the fractions $A_n(z, q)/(zq; q)_n$ are the moments of the rescaled little q-Jacobi polynomials $p_n(x; a, b|q)$; see [45, p. 442]. This can been seen as follows: the moments of the latter polynomials are equal to

$$\mu_n = \frac{(aq; q)_\infty}{(abq^2; q)_\infty} \sum_{k \geq 0} \frac{(bq; q)_k}{(q; q)_k} (aq^{n+1})^k = \frac{(aq; q)_n}{(abq^2; q)_n}.$$

By (5.10) the moments of $(1 - 1/q)^{-n} p_n(1/q + (1 - 1/q)x; a, 1|q)$ are given by

$$\hat{\mu}_n = (1-q)^{-n} \sum_{k=0}^{n} \binom{n}{k} (-q)^k \frac{(aq; q)_k}{(aq^2; q)_k},$$

which is the left-hand side of (5.146) upon substituting $a = z/q$.

Also, it is shown in [8] that Carlitz's q-Bernoulli numbers are moments of a rescaled version of q-Hahn polynomials. One consequence of this connection with moments is the remarkable formulae of the corresponding Hankel determinants. It would be interesting to see how the other remarkable sequences are related to the moments of

Askey–Wilson polynomials. Note that Corteel, Kim and Stanton's recent survey [12] has also related open problems on the moments of Askey–Wilson polynomials.

(iv) A remarkable extension of Askey–Wilson integral (5.3) is the following formula due to Nassrallah–Rahman [48]:

$$\frac{(q;q)_\infty}{2\pi} \int_0^\pi \frac{h(\cos 2\theta; 1)h(\cos \theta; t_1 t_2 t_3 t_4 t_5)}{h(\cos \theta; t_1, t_2, t_3, t_4, t_5)} d\theta \qquad (5.147)$$

$$= \frac{(t_1 t_2 t_3 t_4, t_1 t_2 t_3 t_5, t_1 t_2 t_4 t_5, t_1 t_3 t_4 t_5, t_2 t_3 t_4 t_5; q)_\infty}{\prod_{1 \le r < s \le 5}(t_r t_s; q)_\infty},$$

where $\max\{|t_1|, |t_2|, |t_3|, |t_4|, |t_5|\} < 1$. Is there a combinatorial proof of (5.147) in the spirit of Ismail–Stanton–Viennot's proof of (5.3)?

Acknowledgements. This survey is based on my lectures at the OPSF Summer School, which was held at the University of Maryland, College Park, MD, July 11–15, 2016. I would like to thank Mourad Ismail and Howard Cohl for their kind invitation and for showing patience during the preparation of these notes. I am also indebted to Howard Cohl for numerically checking the linearization formula for Laguerre polynomials which illustrated some errors initially.

References

[1] M. Anshelevich, Linearization coefficients for orthogonal polynomials using stochastic processes. *Ann. Probab.* **33** (2005), no. 1, 114–136.

[2] R. Askey, *Orthogonal Polynomials and Special Functions*. SIAM, Philadelphia, PA, 1975. vii+110 pp.

[3] R. Askey, M.E.H. Ismail, Permutation problems and special functions. *Canad. J. Math.* **28** (1976), no. 4, 853–874.

[4] R. Askey, J.A. Wilson, Some basic hypergeometric orthogonal polynomials that generalize Jacobi polynomials. *Mem. Amer. Math. Soc.*, **54** (1985), no. 319, iv+55 pp.

[5] R. Azor, J. Gillis, J.D. Victor, Combinatorial applications of Hermite polynomials. *SIAM J. Math. Anal.* **13** (1982), no. 5, 879–890.

[6] P. Blasiak, Ph. Flajolet, Combinatorial models of creation–annihilation. *Sém. Lothar. Combin.* **65** (2010/12), Art. B65c, 78 pp.

[7] L. Carlitz, A set of polynomials in three variables. *Houston J. Math.* **4** (1978) no. 1, 11–33.

[8] F. Chapoton, J. Zeng, Nombres de q-Bernoulli–Carlitz et fractions continues, to appear in *J. Théor. Nombres de Bordeaux*, arXiv:1507.04123.

[9] T. Chihara, *An Introduction to Orthogonal Polynomials*. Gordon and Breach, New York, London, Paris, 1978.

[10] J. Cigler, J. Zeng, A curious q-analogue of Hermite polynomials. *J. Combin. Theory Ser. A* **118** (2011), no. 1, 9–26.

[11] S. Corteel, Crossings and alignments of permutations. *Adv. in Appl. Math.* **38** (2007), no. 2, 149–163.

[12] S. Corteel, J.S. Kim, D. Stanton, Moments of orthogonal polynomials and combinatorics. In *Recent Trends in Combinatorics*, 545–578, IMA Vol. Math. Appl., **159**, Springer, 2016.

[13] R.J. Clarke, E. Steingrímsson, J. Zeng, New Euler–Mahonian statistics on permutations and words. *Adv. in Appl. Math.* **18** (1997), no. 3, 237–270.

[14] A. de Médicis, D. Stanton, D. White, The combinatorics of q-Charlier polynomials. *J. Combin. Theory Ser. A* **69** (1995), no. 1, 87–114.

[15] S. Even, J. Gillis, Derangements and Laguerre polynomials. *Math. Proc. Cambridge Philos. Soc.* **79** (1976), no. 1, 135–143.

[16] Ph. Flajolet, Combinatorial aspects of continued fractions. *Discrete Math.*, **32** (1980), 125–161.

[17] Ph. Flajolet, R. Sedgewick, *Analytic Combinatorics*. Cambridge University Press, Cambridge, 2009.

[18] D. Foata, A combinatorial proof of the Mehler formula. *J. Combin. Theory Ser. A* **24** (1978), no. 3, 367–376.

[19] D. Foata, *Combinatoire des identités sur les polynômes orthogonaux*. In *Proc. Internat. Congress of Math.* [Warsaw. August 16–24, 1983], vol. 2, p. 1541–1553. PWN – Polish Scientific Publications, Warsaw and North-Holland, Amsterdam, 1984.

[20] D. Foata, Eulerian polynomials: from Euler's time to the present. In *The Legacy of Alladi Ramakrishnan in the Mathematical Sciences*. Edited by Krishnaswami Alladi, John R. Klauder and Calyampudi R. Rao. Springer, New York, 2010, pp. 253–273.

[21] D. Foata and A.M. Garsia, A combinatorial approach to the Mehler formulas for Hermite polynomials. In *Relations between Combinatorics and Other Parts of Mathematics*, (Proc. Sympos. Pure Math., Ohio State Univ., Columbus, Ohio, 1978), Amer. Math. Soc., Providence, RI, 1979, pp. 163–179.

[22] D. Foata, M.-P. Schützenberger, *Théorie Géométrique des Polynômes Eulériens*. Lecture Notes in Mathematics, Vol. 138 Springer-Verlag, Berlin–New York 1970.

[23] D. Foata, V. Strehl, Combinatorics of Laguerre polynomials. In *Enumeration and Design* (Waterloo, Ont., 1982), 123–140, Academic Press, Toronto, ON, 1984.

[24] D. Foata and D. Zeilberger, Laguerre polynomials, weighted derangements and positivity. *SIAM J. Disc. Math.* **1** (1988), 425–433.

[25] D. Foata and D. Zeilberger, Denert's permutation statistic is indeed Euler–Mahonian. *Stud. Appl. Math.* **83** (1990), no. 1, 31–59.

[26] J. Françon, G. Viennot, Permutations selon leurs pics, creux, doubles montées et double descentes, nombres d'Euler et nombres de Genocchi. *Discrete Math.* **28** (1979), no. 1, 21–35.

[27] I. Gessel, G. Viennot, Binomial determinants, paths, and hook length formulae. *Adv. in Math.* **58** (1985), no. 3, 300–321.

[28] I.M. Gessel, Generalized rook polynomials and orthogonal polynomials. In *q-series and Partitions* (Minneapolis, MN, 1988), 159–176, IMA Vol. Math. Appl., **18**, Springer, New York, 1989.

[29] J. Gillis, G. Weiss, Products of Laguerre polynomials. *Math. Comput.* **14** (1960), 60–63.

[30] I.P. Goulden, D.M. Jackson, *Combinatorial enumeration*. With a foreword by Gian-Carlo Rota. Reprint of the 1983 original. Dover Publications, Inc., Mineola, NY, 2004.

[31] V.J.W. Guo, M. Ishikawa, H. Tagawa, Hiroyuki, J. Zeng, A quadratic formula for basic hypergeometric series related to Askey–Wilson polynomials. *Proc. Amer. Math. Soc.* **143** (2015), no. 5, 2003–2015.

[32] M.E.H. Ismail, *Classical and Quantum Orthogonal Polynomials in One Variable*, paperback edition, Cambridge University Press, Cambridge, 2009.

[33] M.E.H. Ismail, A. Kasraoui, J. Zeng, Separation of variables and combinatorics of linearization coefficients of orthogonal polynomials. *J. Combin. Theory Ser. A* **120** (2013), no. 3, 561–599.

[34] M.E.H. Ismail and D. Stanton, More orthogonal polynomials as moments. In *Mathematical Essays in Honor of Gian-Carlo Rota* (Cambridge, MA, 1996), Progr. Math., vol. 161, Birkhäuser Boston, Boston, MA, 1998, pp. 377–396.

[35] M.E.H. Ismail, D. Stanton, G. Viennot, The combinatorics of *q*-Hermite polynomials and the Askey–Wilson integral. *European J. Combin.* **8**, 1987, 379–392.

[36] M.E.H. Ismail, J. Zeng, Addition theorems via continued fractions. *Trans. Amer. Math. Soc.* **362** (2010), no. 2, 957–983.

[37] M.E.H. Ismail, J. Zeng, A combinatorial approach to the 2*D*-Hermite and 2*D*-Laguerre polynomials. *Adv. in Appl. Math.* **64** (2015), 70–88.

[38] M. Josuat-Vergès, Rook placements in Young diagrams and permutation enumeration. *Adv. in Appl. Math.* **47** (2011), no. 1, 1–22.

[39] A. Joyal, Une théorie combinatoire des séries formelles. *Adv. in Math.* **42** (1981), no. 1, 1–82.

[40] J.S. Kim, D. Stanton, Bootstrapping and Askey–Wilson polynomials. *J. Math. Anal. Appl.* **421** (1) (2015), 501–520.

[41] D. Kim, D. Stanton, J. Zeng, The combinatorics of the Al-Salam–Chahara *q*-Charlier polynomials. *Sém. Lothar. Combin.* **54** (2005/07), Art. B54i, 15 pp.

[42] A. Kasraoui, D. Stanton, J. Zeng, The combinatorics of Al-Salam–Chihara *q*-Laguerre polynomials. *Adv. in Appl. Math.* **47** (2011), no. 2, 216–239.

[43] A Kasraoui, J. Zeng, Distribution of crossings, nestings and alignments of two edges in matchings and partitions. *Electron. J. Combin.* **13** (1) (2006), R33.

[44] D.-S. Kim, J. Zeng, A combinatorial formula for the linearization coefficients of general Sheffer polynomials. *European J. Combin.* **22** (2001), no. 3, 313–332.

[45] R. Koekoek, P.A. Lesky, R.F. Swarttouw, *Hypergeometric Orthogonal*

Polynomials and their q-Analogues. With a foreword by Tom H. Koornwinder. Springer Monographs in Mathematics. Springer-Verlag, Berlin, 2010. xx+578 pp.

[46] G. Ksavrelof, J. Zeng, Two involutions for signed excedance numbers. *Sém. Lothar. Combin.* **49** (2002/04), Art. B49e, 8 pp.

[47] Z. Lin, S. Fu, On 1212-Avoiding Restricted Growth Functions. *Electron. J. Combin.* **24**(1) (2017), #P1.53.

[48] B. Nassrallah, M. Rahman, Projection formulas, a reproducing kernel and a generating function for q-Wilson polynomials. *SIAM J. Math. Anal.* **16** (1) (1985), 186–197.

[49] J. Riordan, The distribution of crossings of chords joining pairs of $2n$ points on a circle. *Math. Comput.*, **29** (129) (1975), pp. 215–222.

[50] S. Roman, *The Umbral Calculus*, Pure and Applied Mathematics, **111**. Academic Press, Inc., New York, 1984

[51] R. Simion and D. Stanton, Octabasic Laguerre polynomials and permutation statistics. *J. Comp. Appl. Math.* **68** (1996), 297–329.

[52] H.S. Shin, J. Zeng, The symmetric and unimodal expansion of Eulerian polynomials via continued fractions. *European J. Combin.* **33** (2012), no. 2, 111–127.

[53] H.S. Shin, J. Zeng, Symmetric unimodal expansions of excedances in colored permutations. *European J. Combin.* **52** (2016), part A, 174–196.

[54] V. Strehl, Lacunary Laguerre Series from a Combinatorial Perspective. *Sém. Lothar. Combin.*, **B76c** (2017), 39 pp.

[55] J. Touchard, Sur un problème de configurations et sur les fractions continues. *Canad. J. Math.*, **4** (1952), 2–25

[56] A. Varvak, Rook numbers and the normal ordering problem. *J. Combin. Theory Ser. A* **112** (2005), no. 2, 292–307.

[57] G. Viennot, Une théorie combinatoire des polynômes orthogonaux généraux. Lecture Notes, 1983, Université du Québec à Montréal.

[58] H.S. Wall, *Analytic Theory of Continued Fractions*. D. Van Nostrand Company, Inc., 1948; reprinted (1973) by Chelsea Publishing Company.

[59] H.S. Wilf, *Generatingfunctionology*. Second edition. Academic Press, Inc., Boston, MA, 1994.

[60] A. Wünsche, Laguerre 2D-functions and their application in quantum optics. *J. Phys. A* **31** (1998), no. 40, 8267–8287.

[61] J. Zeng, Calcul saalschützien des partitions et des dérangements colorés. *SIAM J. Discrete Math.* **3** (1990), no. 1, 149–156.

[62] J. Zeng, Linéarisation de produits de polynômes de Meixner, Krawtchouk, et Charlier. *SIAM J. Math. Anal.* **21** (1990), no. 5, 1349–1368.

[63] J. Zeng, The β-extension of the multivariable Lagrange inversion formula. *Stud. Appl. Math.* **84** (1991), no. 2, 167–182.

[64] J. Zeng, Weighted derangements and the linearization coefficients of orthogonal Sheffer polynomials. *Proc. London Math. Soc. (3)* **65** (1992), no. 1, 1–22.

[65] J. Zeng, Enumérations de permutations et J-fractions continues. *European J. Combin.* **14** (1993), no. 4, 373–382.

Printed in the United States
By Bookmasters